Porous Models for Wave-seabed Interactions

Dong-Sheng Jeng

Porous Models for Wave-seabed Interactions

上海交通大學出版社
SHANGHAI JIAO TONG UNIVERSITY PRESS

Springer

Dong-Sheng Jeng
Shanghai Jiao Tong University
Shanghai, People's Republic of China

ISBN 978-3-642-33592-1 ISBN 978-3-642-33593-8 (eBook)
DOI 10.1007/978-3-642-33593-8
Springer Heidelberg New York Dordrecht London

Jointly published with Shanghai Jiao Tong University Press
ISBN: 978-7-313-09006-5 Shanghai Jiao Tong University Press

Library of Congress Control Number: 2012951680

Printed on acid-free paper

Springer is part of Springer Science+Business Media (www.springer.com)

To Tina, Cathryn and Joy.

Preface

The phenomenon of the wave-seabed interaction has attracted extreme attention among coastal and geotechnical engineers in recent years. Lots of research activities in the area have been carried out by numerous groups in the world. Understanding of mechanisms and processes of the wave-seabed interaction problem is particularly important for marine geotechnical engineers involved in the design of foundation around marine infrastructures. The aim of this book is to provide readers a comprehensive theoretical background for the wave-induced soil response in marine sediments covering various aspects.

This book consists of ten chapters. The first two chapters give background of the research topic, recent advances and possible future research agenda in the area. These chapters provide junior researchers and someone who starts moving into this discipline an overall picture of the research topic and a starting point for postgraduate students. Chapters 3 and 4 present detailed mathematical formulations for the wave-induced soil response, including pore pressure, effective stresses and soil displacements, and seabed instability such as shear failure and liquefaction. These two chapters provide practical engineers a simple and effective analytical tool for the wave-induced seabed instability around marine infrastructures. Chapter 5 presents analytical and numerical models for the seabed response with variable soil characteristics and cross-anisotropic soil behavior. In Chaps. 6 and 7, dynamic soil behavior and Coulomb-damping effects are considered and the applicable ranges of dynamic models are clarified. Then, random wave-induced seabed response in marine sediments with two commonly used wave spectra is explored in Chap. 8. Chapter 9 presents the analytical solution of the wave-induced pore pressure accumulation (pore pressure build-up) and a simplified formula for engineering practice is suggested. Chapter 10 clarifies the process of the wave-induced post-liquefaction (progressive liquefaction).

The first draft of this book was the lecture note I used for the short course delivered at Institute of Mechanics, Chinese Academy of Sciences, Beijing China in 2004. The course was designed for postgraduate students and junior researchers. Part of contents presented in this book have been delivered for several undergraduate and postgraduate courses at different universities in Australia and UK, including Griffith University, University of Sydney and University of Dundee.

Numerous co-workers have made significant contributions to part of materials included in this book. These include: Professor B.R. Seymour (The University of British Coulumb, Canada), Professor J.R.C. Hsu (The University of Western Australia), Professors C.P. Tsai and Y.S. Lin (National Chung-Hsing University, Taiwan), Professor T.L. Lee (Leader University, Taiwan) and Professor K. Zen (Kyushu University, Japan) and others.

The contents covered in this book include major research outcomes of numerous research projects sponsored by Australian Research Council Grants (Australia), EPSRC (UK), National Nature Science Foundation of China. Several postgraduate students, Yu Zhang, Chengcong Liao and Ning Yang, at Center for Marine Geotechnical Engineering, Shanghai Jiao Tong University, China, helped me to finalize the whole manuscripts at the final stage. I am grateful to Professor Mark Randolph (Centre for Offshore Foundation Systems, The University of Western Australia) for his kindly support and help during the most difficult time (1997–1999) in my academic career. Finally, I appreciate the continuous supports from my parents, wife and daughters.

Shanghai, People's Republic of China Dong-Sheng Jeng

Contents

Chapter 1
Introduction

Abstract In this chapter, the background and importance of marine geotechnics with an emphasis on the phenomenon of wave-seabed interactions and its application are outlined. Several hot topics in this area are also suggested for future research development.

Keywords Marine geotechnics · Porous flow · Wave-seabed interactions

1.1 Introduction

When a coastal structure is installed in marine environments, the presence of the structure will alter the flow patterns in its immediate neighborhood. The flow condition around the structure does not only affect the wave force acting on the structure, but also induce sea floor instability. The former one has been the main concern in the design of coastal structures, which has been intensively studied by coastal and structural engineers. However, the latter involves the foundations of the structure, which has attracted attention from coastal geotechnical engineers in recent years.

In the past few decades, considerable efforts have been devoted to the phenomenon of the wave–soil–structure interactions. The major reason for the growing interest is that many coastal structures (such as vertical walls, caissons, offshore mono-piles and pipelines, etc.) have been damaged by the wave-induced seabed response, rather than from construction deficiencies [9, 13, 21]. It is common to observe that concrete armor blocks at the toes of many marine structures have been found to subside into the seabed. Wave-induced liquefaction and shear failure have been identified as culprit for this problem [19]. Another reason is that the poroelastic theories for wave-soil interaction have been applied to field measurements, such as the determination of the shear modulus of soil [24] and the directional spectra of ocean surface waves [15], as well as acoustic wave propagating through porous media [25].

When water waves propagate in the ocean, they generate significant dynamic pressures on the sea floor. This pressure field induces pore water pressure and effective stresses within the seabed. With excess pore-pressure and diminishing vertical effective stress, part of the seabed may become unstable or even liquefied. Once

D.-S. Jeng, *Porous Models for Wave-seabed Interactions*,
DOI 10.1007/978-3-642-33593-8_1,

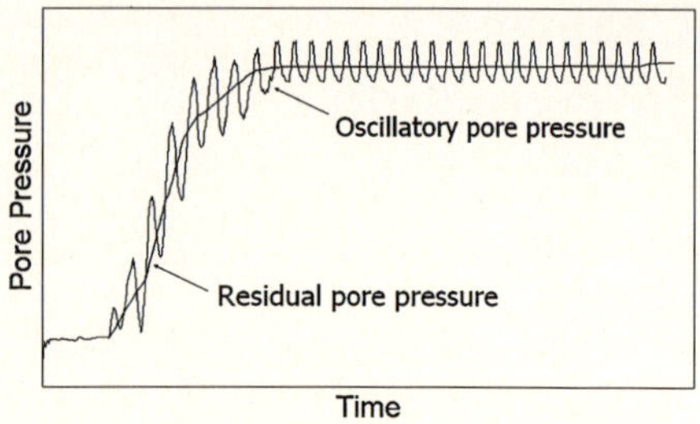

Fig. 1.1 Conceptual sketch of two different mechanisms of pore pressure (not in scale)

liquefaction occurs, the soil particles are likely to be carried away as a fluid by any prevailing bottom current or mass transport owing to the action of ocean waves.

Generally speaking, two mechanisms of the wave-induced soil response have been observed in the laboratory and field measurements, depending on the manner that the pore pressure is generated, as illustrated in Fig. 1.1 [14, 26, 27]. One is caused by the progressive nature of the excess pore pressure, which appears at the initial stage of cyclic loading. The other is generated by the oscillatory pore pressure, which is accompanied by the amplitude damping and phase lag in the pore pressure. This type of soil response appears periodically during a storm sequence. In most marine sediments, the wave-induced soil response is oscillatory in nature, except for some special cases of non-cohesive sediments with loose to medium density [18].

The occurrence of seabed instability is a widespread phenomenon in ocean environments [20]. There is evidence of ocean floor instability in a wide variety of offshore regions, from shallow water, near-shore zones, continental slopes and beyond to deep ocean floors. Seabed instability has been responsible for the damage and destruction of offshore structures [2, 3, 9].

Many variables affect the wave-induced soil response in a porous seabed. One of them is the soil permeability, which is a measure of how rapidly fluid is transmitted through voids between grains. Marine sediments below the water-seabed interface undergo consolidation owing to both the overburden soil pressure and the water pressure above it. This will result in a decrease in void ratio and porosity, accompanied by an increase in specific gravity of the soil. An example of the soil permeability varying with buried depth (z) was reported by [8] for marine sediments in the Gulf of Mexico. Similar evidence for soil consolidation versus depth has also been reported in the literature [4, 5, 17].

Shear modulus is another important parameter in determining the wave-induced soil response. It is a proportionality coefficient in shear stress-shear strain relationship. For the consolidation problem, the medium whose shear modulus increases linearly with depth, called *Gibson Soil*, has been studied [7, 10]. In fact, the rigidity of soil in a natural seabed generally increases with depth as a consequence of the increasing effective overburden pressure. Some evidence for the shear modulus of soils varying strongly with depth has been reported [1, 12, 22].

In reality, most marine sediments display certain degrees of anisotropy, having different elastic properties in the vertical and horizontal directions. This is caused by the manner of their deposition, particular grain shapes and stress history. However, many materials show more limited forms of anisotropy. For example, a material having the same properties in any horizontal direction but different in the vertical direction is referred to as cross-anisotropy [16]. When a material is deposited vertically and then subjected to uniform horizontal stress, it is expected to exhibit a vertical axis of symmetry and then be transversely isotropic [11]. It is noted that the anisotropical soil behavior discussed here is the mechanical behavior due to change of stress. There can also be anisotropy in permeability and variability in porosity.

A natural seabed normally consists of multiple layers in which different soil properties exist in each layer. For instance, the sediment in the Ekofisk oil field in North Sea has an upper layer about 75 m, consisting of a mixture of sand and clay. Below this, a clay layer was observed [6]. Moreover, it is common to place concrete armor blocks or gravel on the seabed at the toes of marine structures, in order to protect the seabed. In recent years, some marine geotechnical engineers have even suggested pre-mixing or replacing part of the seabed sediment by coarser materials to minimize the effect of wave-induced liquefaction [23]. Thus, it is obvious that the soil column has to be treated as a multi-layered medium in both natural and artificial seabed. However, the effects of a cover layer on the wave-induced soil response and seabed instability have rarely been investigated in engineering applications.

To date, non-homogeneous soil characteristics (either variable soil characteristics or multi-layered) and anisotropic soil behavior have been considered in existing poro-elastic models for the wave-induced seabed response. The relevant studies will be reviewed in latter sections.

Although the phenomenon of wave-induced seabed instability has received great attention among coastal and geotechnical engineers since 1980's, preliminary experiments and theories for such a problem have only been available for two-dimensional progressive waves. Recently, significant progress has been made towards the development of both analytical and numerical approaches for some simple modes of instability in the vicinity of marine structures. However, to date, a systemic review on the development of the sea floor dynamics in the vicinity of coastal structure, providing a research guide for coastal geotechnical engineers, has not been available.

1.2 Hot Research Topics

In the area of marine geotechnical engineering, despiting the recent developments, in the author's opinion, more intensive and advanced researches are desired in the following two topics:

- Most existing models for wave-seabed interactions have been based on the constitutive models used in onshore geotechnical engineering. However, a nature seabed is under a much complicated environmental loadings such as wave, current and seismic loadings. Furthermore, the seabed in normally under seawater. The question arising is that whether the existing constitutive relations can be applied to offshore environments or not. To date, an appropriate soil constitutive model for marine geotechnical engineering is still not available in the public literature.

- Most previous studies for wave-seabed interactions have been limited to uncoupled approached or so-called integrated model, rather than a fully coupling models. In fact, the seabed properties are changed by environmental loadings as well as they will affect the environmental loading. Most existing studies have considered the former mechanism, but ignore the latter one. Although the influence of seabed characteristics on the environmental loading may not be significant, however, without n appropriate approach, its effects are unknown and may underestimate the seabed instability.

In addition to the above fundamental issues in the areas of wave-seabed interactions, the applications of the porous models to another practical engineering issue, wave-seabed-structure interaction, is also a vital problem in the file dof marine geotechnics.

1.3 Outline of the Book

The present book is an attempt to give a comprehensive account of wave-seabed interaction around marine structure. It also takes into consideration all state-of-the art knowledge. We shall start off with the literature review (Chap. 2). These include a detailed review and summary of existing work. Next, we shall concentrate on the mathematical models of wave-induced soil response in an isotropic and homogeneous seabed (Chaps. 3 and 4), which will be followed bytheoretical models for a seabed of variable soil characteristics and soil behavior (Chap. 5). In Chaps. 6 and 7, dynamic soil behavior and Coulomb-damping effects are considered and the applicable ranges of dynamic models are clarified. Then, random wave-induced seabed response in marine sediments with two commonly used wave spectra is explored in Chap. 8. In Chap. 9, the mechanism of pore pressure accumulation (pore pressure build-up) and its analytical solution are discussed. Finally, the process of the wave-induced post-liquefaction (progressive liquefaction) is clarified in Chap. 10.

References

1. Badiey, M., Zen, K., Yamazaki, H., Suzuki, H.: Field and laboratory measurement of shear modulus profile in seabed. Rep. Port Harb. Res. Inst. Jpn. **29**(4), 3–26 (1990)
2. Barends, F.B.J.: Interaction between ocean waves and sea-bed. In: Proceedings of the International Conference on Geotechnical Engineering for Coastal Development-Theory and Practice on Soft Ground (Geot-Coastal'91), Yokohama, Japan, vol. 2, pp. 1091–1108 (1991)
3. Bea, R.G., Wright, S.G., Sircar, O.P., Niedoroda, A.W.: Wave-induced slides in south pass block 70, Mississippi Delta. J. Geotech. Geoenviron. Eng. **109**(1), 617–644 (1975)
4. Bennett, R.H., Fischer, K.M., Li, H., Lamber, D.H., Hulbert, M.H., Yamamoto, T., Badiey, M.: In-situ porosity and permeability of selected carbonate sediments: Great Bahama Bank. Part 2: Microfabric. Mar. Geotechnol. **9**(1), 29–47 (1990)
5. Bennett, R.H., Li, H., Lamber, D.H., Fischer, K.M., Walter, D.J., Hickox, C.E., Hulbert, M.H., Yamamoto, T., Badiey, M.: In-situ porosity and permeability of selected carbonate sediments: Great Bahama Bank. Part 1: Measurements. Mar. Geotechnol. **9**(1), 1–28 (1990)

6. Bjerrum, J.: Geotechnical problem involved in foundations of structures in the North Sea. Géotechnique **23**(3), 319–358 (1973)
7. Booker, J.R., Small, J.C.: An investigation of the stability of numerical solutions of Biot's equations of consolidation. Int. J. Solids Struct. **11**(7), 907–917 (1975)
8. Bryant, W.R., Hottman, W., Trabant, P.: Permeability of unconsolidated and consolidated sediments, Gulf of Mexico. Mar. Geotechnol. **1**(1), 1–14 (1975)
9. Christian, J.T., Taylor, P.K., Yen, J.K.C., Erali, D.R.: Large diameter underwater pipeline for nuclear power plant designed against soil liquefaction. In: Proceeding of Offshore Technology Conference, pp. 597–606 (1974)
10. Gibson, R.E.: Some results concerning displacements and stresses in a non-homogeneous elastic half-space. Géotechnique **17**, 58–67 (1967)
11. Graham, J., Houlsby, G.T.: Anisotropic elasticity of a natural clay. Géotechnique **33**(2), 165–180 (1983)
12. Hardin, B.O., Drnevich, V.P.: Shear modulus and damping in soils: Design equations and curves. J. Soil Mech. Found. Div. **98**(7), 667–692 (1972)
13. Lundgren, H., Lindhardt, J.H.C., Romold, C.J.: Stability of breakwaters on porous foundation. In: Proceeding of 12th International Conference on Soil Mechanics and Foundation Engineering, vol. 1, pp. 451–454 (1989)
14. Nago, H., Maeno, S., Matsumoto, T., Hachiman, Y.: Liquefaction and densification of loosely deposited sand bed under water pressure variation. In: Proceeding of the 3rd International Offshore and Polar Engineering Conference, Singapore, vol. I, pp. 578–584 (1993)
15. Nye, T., Yamamoto, T.: Field test of buried ocean-wave directional spectrometer system. J. Waterw. Port Coast. Ocean Eng. **120**(5), 451–466 (1994)
16. Pickering, D.J.: Anisotropic elastic parameters for soil. Géotechnique **20**(3), 271–276 (1970)
17. Samarasinghe, A.M., Huang, Y.H., Drnevich, V.P.: Permeability and consolidation of normal consolidated soils. J. Geotech. Eng. Div. **108**(6), 835–849 (1982)
18. Seed, H.B., Rahman, M.S.: Wave-induced pore pressure in relation to ocean floor stability of cohesionless soils. Mar. Geotechnol. **3**(2), 123–150 (1978)
19. Silvester, R., Hsu, J.R.C.: Sines revisited. J. Waterw. Port Coast. Ocean Eng. **115**(3), 327–344 (1989)
20. Silvester, R., Hsu, J.R.C.: Coastal Stabilization. PTR Prentice-Hall, New Jersey (1993)
21. Smith, A.W., Gordon, A.D.: Large breakwater toe failures. J. Waterways Harbor Coast. Eng. Div. **109**(2), 253–255 (1983)
22. Suzuki, H., Ando, K., Kitahara, M., Badiey, M., Yamamoto, T.: Shear modulus profile measurement of the shallow-water seabed in Japan. In: Proceedings of the International Conference on Geotechnical Engineering for Coastal Development-Theory and Practice on Soft Ground (Geot-Coastal'91), vol. 1, pp. 99–104 (1991)
23. Umehara, Y., Zen, K., Yoshizawa, H.: Design concept of treated ground by premixing method. In: Proceedings of the International Conference on Geotechnical Engineering for Coastal Development-Theory and Practice on Soft Ground (Geot-Coastal'91), vol. 1, pp. 519–524 (1991)
24. Yamamoto, T., Trevorrow, M.: Experimental verifications of bottom shear modulus profiler (BSMP) method. In: Proceedings of the International Conference on Geotechnical Engineering for Coastal Development-Theory and Practice on Soft Ground (Geot-Coastal'91), vol. 1, pp. 123–128 (1991)
25. Yamamoto, T., Turgut, A.: Acoustic wave propagation through porous media with arbitrary pore size distributions. J. Acoust. Soc. Am. **83**(5), 1744–1751 (1988)
26. Zen, K., Yamazaki, H.: Mechanism of wave-induced liquefaction and densification in seabed. Soil Found. **30**(4), 90–104 (1990)
27. Zen, K., Yamazaki, H.: Oscillatory pore pressure and liquefaction in seabed induced by ocean waves. Soil Found. **30**(4), 147 161 (1990)

Chapter 2
Recent Advances

Abstract In this chapter, a comprehensive literature review for the wave-seabed interactions in marine sediments, including wave-seabed-structure interactions, was provided. In this review, the major outcomes of the existing research were summarized. The objective of this chapter is to provide readers an overall understanding of the topic.

Keywords Literature review · Pore pressure · Seabed instability · Wave-seabed interactions

2.1 Introduction

The present book is an attempt to give a comprehensive account of wave-seabed interaction around marine structure. It also takes into consideration all state-of-the art knowledge. We shall start off with the literature review (the present chapter). These include a detailed review and summary of existing work.

2.2 Waves Propagating over a Porous Seabed: Theoretical Models (Transient Mechanism)

Numerous models for the wave-induced seabed response have been developed with various assumptions since 1940's. The assumptions and the leading governing equations for each model will be summarized here:

- *Un-coupled model (or drained model):*

In the model, both pore fluid and soil are considered as incompressible medium, and the accelerations due to fluid and soil motion are ignored. This leads to an uncoupled model, in which the *Laplace's Equation* is the governing equation.

Another similar approach is to include the compressibility of pore fluid, which leads to *Diffusion Equation*.

D.-S. Jeng, *Porous Models for Wave-seabed Interactions*,
DOI 10.1007/978-3-642-33593-8_2,

- *Consolidation model (or quasi-static model):*

  In this model, both pore fluid and soil are considered to be compressible, but the accelerations due to fluid and soil motion are ignored. This assumption leads to Biot's consolidation equation.

  Since the acceleration has been ignored, this model is also called as quasi-static model. This model has been commonly used since 1978 [130].

- *u–p approximation:*

  In this model, the acceleration due to pore fluid is ignored in the general formulation. This model was first proposed by Zienkiewicz *et al.* [141] with a one-dimensional analysis, and been extended to two-dimensional analysis for wave-seabed interaction [46, 48].

- *Dynamic model:*

  The full set of governing equations established by Biot [5, 6] is employed in the analysis, in which both accelerations due to pore fluid and soil motion are included. Since this model is rather complicated, only a few investigations have been available in the literature [8, 38, 47, 121].

- *Poro-elastoplastic model*:

  Besides linear poro-elastic models, some advanced models such as the poro-elastoplastic model for the wave-induced seabed response have been developed recently [101–105]. The poro-elastoplastic models will provide better predictions of the potential of the wave-induced seabed [135] instability, which is normally a large deformation.

In the aforementioned models, both analytical and numerical models have been employed to obtain the wave-induced soil response. In the following sections, most previous theoretical investigations for the wave-induced seabed response will be reviewed under the heading of each model.

2.2.1 Un-coupled Models (or Drained Models)

Un-coupled models have been used as the first approximation in the area of wave-seabed interaction. In the model, either pore fluid or soil has been considered as incompressible medium. The accelerations due to both pore fluid and soil motion are ignored in this approach. Since the governing equation is either *Laplace's Equation* or the *Diffusion Equation*, for which the analytical solutions have been well developed, most previous investigations with these assumptions are analytical approximations.

Based on the assumptions of a rigid, permeable sandy seabed and incompressible pore fluid, Laplace's equation is the governing equation for the wave-induced pore pressure. Using a linear wave theory, Putnam [89] presented a simple solution for an isotropic porous seabed of finite thickness and concluded that a significant loss of wave energy occurred in the presence of a porous sandy seabed due to viscous percolation of fluid. The percolation was activated by the pressure variation at the interface between sea water and seabed. The solution indicated that pressure distribution within the seabed depended only on the wave characteristics and geometry of the sand layer, and not on the properties of the seabed. However, a possible error of the wave height was observed in Putnam's paper, which overestimated the dissipation function by a factor of four [97].

Based on the same assumptions as Putnam [89], the wave-induced pore pressure for a porous seabed of finite thickness with anisotropic permeability was examined by Sleath [108]. He also conducted laboratory experiments to verify the theory, resulting in the discovery of a phase lag (less than 10 degrees) in the pore pressure. However, these experimental results were inconsistent with his theoretical results. This inconsistence is due to the assumptions of his theoretical approach.

Considering the viscous effect of the boundary layer and energy balance, Liu [63] modeled the flow in a permeable bed and determined the damping rate for an infinite seabed. Continuity of pressure and velocity are required as boundary conditions at the interface up to order $O(\sqrt{\nu})$, where ν is the viscosity of the pore fluid. His results indicated that there was no relationship between pore pressure and permeability, whilst fluid velocity depended on the porosity and permeability. However, Liu's solution [63] only considered the pressure condition whilst neglecting shear stress [28]. Thus, it may not be a complete analysis of viscous flow.

With the same framework of Liu [63], Liu [64] further developed a solution for the damping of the wave-induced pressure in a two-layered porous seabed. Compared with the solution for an infinite seabed [63], the pore pressure was found to depend on both the permeability and thickness of the upper layer only to a small degree. However, he only considered the case of a two-layered seabed with uniform permeability in each layer. Later, based on the generalized Darcy's equation [11], the boundary layers between the seabed surface and the impervious stratum are included in the model [65]. It was concluded that the spatial damping rate depended strongly on the permeability and the water depth, when the physical wave number remained approximately the same over its rigid bottom.

Because the aforementioned continuum approach disregards the seepage flow and pore pressure development in soils, they provide no information on the effective stress, which is essential when elastic waves in soils are considered. However, it has been reported that the internal friction form is independent of strain rate [110]. The experimental data suggested that the soil internal friction is not of viscous friction form, but of Coulomb friction type.

From a different aspect, Massel [72] took into account the non-linear damping and the inertia terms in the momentum equation for a rigid porous seabed. His results indicated that the effect of permeability on the pressure distribution in the seabed was negligible and that they were essentially the same as that from Laplace's equation.

In describing the wave-induced soil response, Mallard [71] used a set of elastic displacement equations based on the assumption of stress equilibrium within the soil, neglecting the effect of soil inertia on the response. Dawson [12] considered the soil inertia term in the model of Mallard [71], and concluded that the effect of the soil inertia term could not generally be ignored without committing a serious error for the case of an incompressible soil.

Another type of un-coupled model has been proposed by Nakamura *et al.* [79] and Moshagen and Torum [76], based on assumptions of compressible pore fluid and a non-deformable porous seabed. This results in the heat conduction equation or diffusion equation for pore pressure. Among these, Nakamura *et al.* [79] compared theoretical results of pore pressure with laboratory experiments in fine and coarse sandy beds. The experimental results for the latter showed no phase lag and agreed reasonably well with the solution from the Laplace's equation. The data for fine sand exhibited a large pressure attenuation and phase lag, which agreed reasonably well with the diffusion theory. However, an unexplainable pressure discontinuity exists near the seabed surface in their experimental data. As Yamamoto *et al.* [130] pointed out, the waves generated in the experiments of Nakamura *et al.* [79] were too steep. Therefore, the state of stresses in sandy beds under wave crests and troughs might have reached the limit of equilibrium or the state of liquefaction, thus causing a large pressure drop. Furthermore, a critical error was found in their calculations. The compressibility of the water used in the calculation was 980 times that of the real water. Yamamoto *et al.* [130] showed that the false agreement reported might be explained by the existence of a small amount of air in the sand used.

Moshagen and Torum [76] considered the wave-induced flow in a porous medium under the assumption of compressible pore fluid and an incompressible soil. They found that the inclusion of pore fluid compressibility in the analysis of the wave-induced pore pressures in a porous soil significantly altered the vertical seepage forces acting on the soil. However, the assumption made by Moshagen and Torum [76] regarding the relative compressibility of the pore fluid and the soil skeleton appeared somewhat unrealistic [88]. Thus, serious doubts arise about the validity of Moshagen and Torum's conclusion.

All the aforementioned theories assumed that the seabed is a rigid porous medium. Because these approaches do not permit the coupling of pore-fluid motion and soil motion, the governing equation for the pore pressure is Laplace's equation for incompressible fluid, or the diffusion equation for compressible pore fluid for hydraulically isotropic (i.e., isotropical permeability) seabeds. However, such solutions for pore pressure are limited to a particular case of soil and wave conditions, i.e. Laplace's equation for very permeable beds such as coarse sand, or a diffusion equation for poorly permeable beds such as clay. Furthermore, these approaches provide no information for the effective stresses and soil displacements in the seabed.

2.2.2 *Biot's Consolidation Model (Quasi-Static Model)*

The second model for the wave-induced seabed response is based on the assumption of compressible pore fluid and soil, but ignoring the accelerations due to pore fluid and soil motion. This model has been widely used to investigate the wave-induced soil response since 1970's. The methodologies in solving the governing equations can be summarized into three divisions: direct analytical solution, boundary-layer approximation and numerical modeling.

2.2.2.1 Direct Analytical Solution

Most previous investigations with Biot's consolidation equations have been directly solved and obtain the wave-induced pore pressure, soil displacements and effective stresses. This approach was first developed by Yamamoto *et al.* [130], Yamamoto [126] and Madsen [67], who considered compressible pore fluid in a compressible porous medium. A three-dimensional general consolidation equation [4] and storage equation [124] were adopted in these studies, in which only progressive waves were examined. Among these, Madsen [67] considered a hydraulically anisotropic and unsaturated porous bed, whilst Yamamoto *et al.* [130] studied an isotropic medium. Both considered only an infinite thickness. Moreover, Yamamoto [126] investigated soil response in a homogeneous soil of finite thickness under isotropic and partially saturated conditions. However, Yamamoto's solution [126] was cast in a semi-analytical manner that did not have a closed form.

Yamamoto *et al.* [130] concluded that when the stiffness of a porous medium is much smaller than that of the pore fluid (for example, saturated soft soils), the soil response is independent of the permeability and has no phase lag. On the other hand, when the stiffness of a porous medium is much larger than that of the pore fluid (for example, partially saturated dense sands), pore pressure attenuates rapidly. In the latter case, the phase lag increases linearly with the distance from the seabed surface.

Madsen [67] investigated a hydraulically anisotropic and partially saturated seabed. He found that this had an appreciable effect on the nature of the wave-induced effective stresses in coarse sand. For all soils, the effect of partial saturation on soil response may be significant.

Yamamoto [127] developed a semi-analytical solution for a non-homogeneous layered porous seabed, together with a comprehensive verification using data obtained from Mississippi Delta. Yamamoto [127] pointed out that a layer of concrete blocks had significant effect on the wave-induced soil response. However, the assumption of treating the concrete blocks as soils seems unrealistic because the properties of concrete blocks are quite different from soil.

Okusa [83] used the compatibility equation under elastic conditions and reduced the governing equation of Yamamoto *et al.* [130] to a fourth-order differential linear equation. It is noted that Okusa [83] was based on plane stress conditions, while Yamamoto *et al.* [130] was based on plane strain conditions. Okusa [83] found that the wave-induced pore pressure and effective stresses consisted of two parts. The first depended only on the wave characteristics and the second was related to both the sediment and the wave characteristics. He reported that the wave-induced soil response depended only on the wave conditions, not on the soil characteristics for a fully saturated and isotropic sandy seabed of infinite thickness. However, this conclusion is invalid for an isotropic seabed of finite thickness, even under a saturated condition [18, 39].

Rahman *et al.* [93] summarized the previous work with the direct analytical framework in a semi-analytical analysis. In their model, a general layered seabed is considered, which is particularly important for the design of a cover layer for seabed protection.

The aforementioned investigations have been limited to an isotropic homogeneous seabed, which may be an idealized case. The major difficulty to analysis the wave-induced soil response in a seabed with variable permeability has been the governing equation includes variable coefficients. By employing the VS function [123], Seymour *et al.* [107] derived an analytical solution for such a condition. In their study, only fine sand is considered. In their model, the first-order derivation of permeability respect to vertical distance was excluded, which has been reported to play an important role in the evaluation of wave-induced soil response in coarser material [61]. Later, Jeng and Seymour [49, 50] further developed analytical solutions for general soils in seabed of both infinite and finite thickness. They concluded that the relative difference of the wave-induced pore pressure between variable and uniform permeability might up to 23 % of the amplitude of the wave pressure at the seabed surface.

Recently, another analytical solution for the wave-induced seabed response with variable permeability was proposed by Kitano and Mase [52] and Kitano *et al.* [53]. However, their one-dimensional model is limited to exponential distribution decay of the permeability, although it provides a simpler formulation than that of [49, 50].

The inclusion of cross-anisotropic soil behavior in the wave-seabed interaction can also be handled analytically. Jeng [30, 32] may have been the first to derive the analytical solution for the wave-seabed interaction in a cross-anisotropic seabed. His numerical results show that the conventional solution with the assumption of isotropic soil behavior may overestimate the pore pressure, but underestimate the effective stresses. Consideration of cross-anisotropic soil behavior is particular important in determining the wave-induced soil displacements. An identical approximation is also proposed by Yuhi and Ishida [134].

A simplified analytical solution for the soil response in a cross-anisotropic seabed was proposed by Yuhi and Ishida [136]. In the model, two parameters related to boundary layer thickness and the stiffness ratio were introduced. However, their model is only applicable in the region near the seabed surface (for example, $|z|/L < 0.02$, L is the wavelength of ocean waves), outside this region, the relative difference between the simplified solution [136] and exact solution[30] is not negligible, as reported in Jeng [37].

2.2.2.2 Boundary-Layer Approximation

Since the direct analytical solution for the wave-induced seabed response is involved with complicated mathematical presentations, especially for a seabed of finite thickness, it is even impossible to have a closed-form solution for a layered seabed. An alternative approximation, the boundary-layer approximation, was proposed by Mei and Foda [73]. The principle of the boundary-layer approximation is to divide the whole soil domain into two regions: inner and outer regions. In the inner region (near the seabed surface, defined by the boundary layer thickness), the full solution is required. On the other hand, a simplified solution is sufficient in the outer region. This solution agrees well with that of Yamamoto [126] for fine sand. However, it may lose accuracy for all soils in unsaturated conditions and for coarse sand under saturated conditions [126]. This shortcoming may be attributed to the solution being only suitable for a seabed with low permeability, for which a scaling was carried out [28]. However, their solutions are more convenient for engineering application due to their much simpler form, compared to those of Hsu and Jeng [26].

Although the boundary-layer approximation proposed by Mei and Foda [73] is a simple yet fairly accurate analysis, it was restricted to low frequency waves only. Huang and Chwang [27] investigated Biot's equation for the acoustic problem and obtained three uncoupled Helmholtz equations to represent each of the three kinds of waves. Their approach is applicable for the complete range of wave frequencies.

Later, Huang and Song [28] applied this approach to investigate the problem of linear water waves in a channel of constant depth propagating over a horizontal poro-plastic bed of infinite thickness. In the general solution presented, five non-dimensional physical parameters were defined. One of them represented the relative stiffness of solid and fluid and another expressed penetrability, whilst the other three revealed Mach numbers for two longitudinal waves and one transverse wave of the porous medium of low soil permeability. However, their solution was restricted to a porous seabed of infinite thickness. Later, Song and Huang [109] further applied the boundary-layer approximation to examine the mechanism of laminar poro-elastic media flow under wave loading.

Recently, Kitano and Mase [51] derived another set of analytical solution to investigate the influence of cross-anisotropic soil behavior on the wave-induced soil response through the boundary-layer approximation. Their model agreed well with the exact solution proposed by Jeng [32] for fine sand.

Besides the direct analytical solution and boundary layer approximation, Sumer and Cheng [112] proposed a random-walk model for the wave-induced pore pressure in marine sediments. The principle of the model is based on the shear stress in the marine sediment to determine the build-up pore pressure. However, the determination of the shear stress relied on other analytical solutions such as that of Hsu and Jeng [26].

2.2.2.3 Numerical Methods

Numerical methods, including a finite difference method, a finite element method and a boundary element method, is another type of approximation which renders numerical results at discrete points for the wave-induced soil response and seabed instability.

Finite Difference Method Madga [66] developed a one-dimensional finite difference model for the wave-induced pore pressure in a highly saturated sandy bed. He concluded that the time phase in pressure generation is dominated by the degree of saturation, compressibility of the soil skeleton and soil permeability.

Zen and Yamazaki [138, 139] simplified a two-dimensional boundary value problem to one dimension, based on the assumption of the seabed thickness being very small compared with the wavelength. A numerical model (finite difference method) was also established, which was only applicable to a single layer of porous seabed.

Finite Element Method Gatmiri [18] developed a simplified finite element model for the wave-induced effective stresses and pore pressure in an isotropic and saturated permeable seabed. Two important conclusions were drawn from his paper. First, there exists a critical bed thickness about 0.2 times the wavelength, in which the horizontal movement of the soil skeleton is a maximum and where the unstable state occurs. Second, the soil response is affected by soil characteristics even in a hydraulically isotropic and saturated seabed of finite thickness. This result complemented the solution for a seabed of infinite thickness reported by Okusa [83]. However, the general trend of pore pressure distribution versus the seabed thickness in [18] was found to be inconsistent [39].

Gatmiri [19] further extended his numerical model to consider the soil response in a cross-anisotropic saturated seabed. The numerical results showed that the effects of cross-anisotropic soil parameters are significant and the soil response was affected by the combined parameters in different ways. Compared with the cross-anisotropy, the effect of hydraulic anisotropic permeability on the variation of effective stresses may be insignificant.

A possible error in the results of Gatmiri's model [18, 19] may have stemmed from the boundary condition used. The lateral boundaries at $x/L = 0, 1$ in his paper, "$v = 0$, $p = 0$ and u free", were used in his model. However, it has been proved that there is a phase lag in soil response in a fully saturated seabed of finite thickness [39]. This implies that the lateral boundary conditions, $v = 0$, $p = 0$, are invalid in a porous seabed of finite thickness. Thus, the numerical results of [18, 19] seem doubtful. In fact, this obstacle can be overcome by using the principle of repeatability [142], as suggested by the author [62].

Thomas [115, 116] developed a one-dimensional finite element method for a two layered unsaturated seabed. His result agreed well with the analytical solutions of Yamamoto [126] and Okusa [83]. It also suggested that the stiffer sediment in the top layer dominated the response of the bottom layer in a two-layered seabed.

With a similar framework, a series of one-dimensional finite element models was developed for the wave-induced seabed response in a non-homogeneous seabed [42, 43, 60, 61]. The results of these numerical models agree well with previous two-dimensional experimental data and analytical solutions. In the models, the permeability and shear modulus are considered to vary with burial depth. [43] have also examined the influence of non-linear wave components on the soil response. In addition, the combined effect of cross-anisotropic soil behavior and non-homogeneous soil characteristics on the wave-induced soil response was examined by Jeng and Lin [44].

The advantage of the above one-dimensional finite element models is the reduced computational time. However, such one-dimensional models cannot apply to the case with a structure. Thus, a two-dimensional finite element model, was proposed [62] by employing the principle of repeatability [142].

Boundary Element Method Raman-Nair and Sabin [96] proposed a boundary element technique for computing the wave-induced effective stresses and pore pressures within the slope by using Biot's theory of poro-elasticity. The boundary element method was verified with analytical solutions for flat beds. The results indicated that there was no significant difference in the extent of the failure zones for fine and coarse sand, although in some cases the failure zone in fine sand was slightly deeper.

2.2.3 *u–p* Approximation

Based on Biot's poro-elastic theory [5, 6], Zienkiewicz *et al.* [141] proposed a one-dimensional so-called *u–p* approximation for waves propagating over a porous media. Their paper provided a simplified approximation for such a problem. Based on the numerical examples, Zienkiewicz [141] concluded that neither the *u–p* approximation nor the dynamic approximation is required for the case of ocean waves over a seabed. However, in their examples, the compressive wave velocity under ocean wave loading was fixed at 1000 m/s, which is applicable during an earthquake, but may not be a reasonable value for most ocean environments. Thus, the conclusion drawn from the examples was in doubt.

Sakai *et al.* [100] extended Mei and Foda's boundary-layer approximation [73] to examine the effect of inertia and gravity force of the ocean waves on the seabed response, and verified it with results of their numerical model (finite element method). They concluded that the inertia term (acceleration) could be neglected in normal wave conditions except for breaking, whilst the gravity term can practically be ignored.

Later, Sakai *et al.* [99] modified the boundary-layer approximation to take into account the effect of wave-induced bottom shear stress which cannot be neglected in the surf zone. In his paper, it was simply assumed that the amplitude of the wave-induced bottom shear stresses was proportional to the wave pressure, without any phase lag. However, it has been reported that the wave-induced bottom shear stress has a phase lag of 45° to the wave-induced pressure at the seabed surface [23].

Recently, the one-dimensional *u–p* approximation [141] was further extended to two-dimensional waves over a porous seabed [46, 48]. The seabed was considered to be either infinite or finite thickness. It was found that the relative differences between *u–p* approximation and quasi-static solution may reach 17 % of the amplitude of wave pressure.

2.2.4 Dynamic Models

The major difference between dynamic analysis and *u–p* approximation has been the inclusion of acceleration due to pore fluid. With a similar framework to [141], Jeng and his co-workers [8, 38, 47] investigated the effects of dynamic soil behavior on the wave-induced soil response through a two-dimensional analysis. Although their dynamic solution can provide a better prediction of the wave-induced soil response, their solutions were lengthy and difficult to be applied in engineering practice.

Based on the boundary-layer approximation, Huang and his co-workers [9, 25, 28] further developed a series of analytical solutions for waves propagating over a soft poro-elastic bed. Both linear and non-linear wave loading were considered. Their approximation provided a simpler formulation, compared with the full closed-form solution.

Alternatively, based on the governing equations derived by Mei and Foda [73], Yuhi and Ishida [135] directly solved the boundary value problem, rather than using a boundary-layer approximation. The seabed was considered to be of infinite thickness. The characteristics of three waves within the soil column, including two compressive waves and one shear wave, was discussed in detail. This work has been later extended to the case of finite thickness [40].

Besides the linear poro-elastic models, Yamamoto and his co-workers [129, 131] carried out a series of studies on the non-linear mechanism and the Coulomb type friction failure for the interaction between waves and porous seabed. With similar framework, the non-linear interaction between waves and soil was examined [128]. Yamamoto and Takahashi [131] applied their solutions to the case of acoustic waves propagating through porous media. Recently, these solutions have also been applied to back calculate the shear modulus of marine sediments [132]. Marine geotechnical engineers have used this procedure, named as Bottom Shear Modulus Profile (BSMP), in field measurements.

Lee *et al.* [56] extended Yamamoto's model with Coulomb-damping friction [129] to a case of finite thickness. They demonstrated the significant of Coulomb-damping friction on the wave-induced soil response in fine sand, not in graveled bed. An identical work has also been reported by Lin [58].

Detailed of comparative evaluation of various analytical solutions for the wave-induced seabed response is given in [8, 38].

2.2.5 Poro-Elastoplastic Models

All aforementioned poro-elastic models have been limited to small deformations, which is an idealized condition. However, large deformations will be a more important issue for engineering problems, especially under the action of a storm. In that case, poro-elastoplastic models are required to provide a better estimation of the soil response. Since a poro-elastoplastic model is more complicated than a poro-elastic model, especially in the problem of wave-seabed interaction, to date, only a few investigations have been published.

Sekiguchi *et al.* [105] may have been the first to derive an analytical solution for waves propagating over a poro-elastoplastic seabed with Laplace's transformation. Their model clearly demonstrated the difference between poro-elastic and poro-elastoplastic models. Also, overall agreement between their model and centrifugal tests was reported. However, their model was based on the assumption of the seabed thickness being much less than the wavelength, which is questionable [31].

With the non-associative bounding surface plastic constitutive model, Yang and Poonoshasb [133] developed a numerical code to investigate the standing wave-induced soil response in a seabed. Their model demonstrated the significant influence of permeability on the wave-induced seabed response. However, no comparison between elastic and their plastic models was made in their paper. Thus, the importance of plastic model was not observed.

Recently, Li *et al.* [57] established a poro-elastoplastic model for wave propagation in saturated media, using the Drucker-Prager criterion to describe non-linear constitutive behavior of pressure-dependent elastoplasticity for the solid skeleton. Based on their model, they further investigated the stationary discontinuities and flutter instability of wave propagation in a saturated seabed, and concluded that the flutter instability may occur prior to stationary discontinuity only if the deviative derivative effective stress normal to the surface of discontinuity is compressive.

Besides the poro-elastoplastic models, discontinuous deformational analysis (DDA) is another alternative approach for larger deformation. However, the conventional DDA models have been based on total stress analysis, which may be suitable for rock mechanics, rather than wave-soil interaction. Thus, the conventional DDA model needs to be modified by including the pore pressure and become effective stress analysis, as proposed by Jeng *et al.* [45]. The relative difference between discontinuous deformation and conventional continuous approaches was reported to be significant under large wave loading [45].

2.3 Waves Propagating over a Porous Seabed: Theoretical Model (Residual Mechanism)

Based on the analytical solution proposed by Sumer and Cheng [112], Cheng *et al.* [10] developed a finite difference model to investigate the wave-induced build-up pore pressure in marine sediments. They concluded that the pore pressure build-up is more sensitive to shear stress near the seabed surface in a deep soil column. It is noted that the governing equation for the pore pressure used in the model is one-dimensional. However, the shear stress required in their pore pressure model was obtained from two-dimensional Biot's consolidation. This inconsistency may lead to some doubts on regarding their model.

2.4 Waves Propagating over a Porous Seabed: Physical Modeling

Theoretical investigations usually involve some assumptions to simplify the realistic problem to become solvable. However, physical models, including laboratory experiments and field measurements, have commonly been conducted by researchers in the early stage of a series of studies on this topic. From these, a limited number of data can be chosen to verify the theoretical models.

2.4.1 Field Measurements

Using conventional techniques, properties of seabed deposits have been estimated from laboratory tests of bored samples. However, these are usually accompanied by many difficulties such as: (1) disturbances from remolding and swelling caused by sampling; (2) difficulty in controlling the degree of saturation of the sand samples and (3) performing compression tests at low confining pressure in obtaining mechanical properties of the surface sediment on the seabed. Therefore, in-situ measurements are recommended to estimate the properties of seabed deposits.

Bennett [2] measured pore pressure and hydrostatic pressure of silty clay in the Mississippi Delta. Piezometric measurements in submarine sediments at a selected site revealed the presence of high excess pore-water pressures. Its presence was not only due to short-term rapid fluctuations in excess pore pressure during active storm conditions, but also due to long-term changes in excess pore pressure following the passage of Hurricane Eloise through the Gulf of Mexico. Relatively rapid fluctuations were observed in hydrostatic and dynamic pore-pressures during storm activity, with the maximum pore pressure variations measured approximately one-half of that observed in hydrostatic conditions. This suggested an energy loss or damping effect of the wave through the sediments during the storm.

Bennett and Faris [3] implanted a shallow-water piezometer in the Mississippi Delta silty clay to measure pore water and hydrostatic pressures for a period of approximately eight months (March to November 1977). The field measurements indicated that water surface activity by tides and short period waves were observed to produce significant fluctuations in pore pressure.

Okusa and Uchida [85] measured the pore-water pressure in disturbed silty sand sediment at 1.5 m below the sea floor in about 12 m deep water. Concurrently, they monitored wave pressure in the water at about 10 m above the sea floor, over a period 1976–1978 in Shimizu Harbor, central Japan. They reported that the damping for a long-period wave was smaller than for a short-period one. The time lag was relatively clear for the former and not for the latter. The curves of wave pressure were slightly different from the fluctuations of the differential pore-water pressure.

Later, based on the field measurements of pore pressure in submarine sediments under various marine conditions, including mixtures of silt, sand and gravel, Okusa *et al.* [84] concluded that the rate of damping increased with shorter period waves. The pore pressures measured in the sediments decreased faster than that predicted by the theories available (for example, [67, 130]). This might have come from the existence of pore gas in the sediment or other damping mechanisms.

Okusa [82] has also reported several observations of the wave-induced pore pressure, which were measured in sand and gravel sediment at a depth of about 4 m from the seabed surface in the water about 1 m deep. The results revealed that the wave-induced sea floor pressure was transferred into the sediment with damping, which was slightly greater than that predicted by Madsen [67], coupled with a time lag. Furthermore, permeability and deformation properties of submarine sediments would appear to play an important role in the damping transfer of wave pressure at the sea floor to the pore pressure with a time lag.

Maeno and Hasegawa [69] proposed a new method for predicting physical and mechanical properties of the deposits near the surface of a seabed by means of in-situ measurements of pore pressure in Nabae beach, Japan. They pointed out that low frequency components of the wave-induced pressure fluctuations propagate into the seabed more easily than the high frequency components. Furthermore, the time lag between the pore pressure in the seabed and the wave pressure at the seabed surface increased as the frequency decreased.

Zen and Yamazaki [140] also conducted field observations of wave-induced pore pressure and effective stress at Hazaki, Japan. They concluded that the effective stress in the seabed varied periodically in accordance with the propagation of ocean waves. Furthermore, wave-induced liquefaction was closely related to the upward seepage flow induced in the seabed during the passage of wave troughs.

2.4.2 Laboratory Experiments

Laboratory experiments have been extensively used for estimating the soil behavior under wave action. In general, three different experimental approaches have been used. They are: wave tank experiments, compressive tests, and centrifugal wave modeling. Wave tank experiments have been commonly used by coastal engineers [13, 68, 108, 119], which provide the spatial distribution of the wave-induced pore pressure within a seabed, but lack accuracy in determining soil parameters. Compressive tests have been widely used by geotechnical engineers [138, 139], which can provide a better estimation of pore pressure, but are unable to provide the spatial distribution of pore pressure. Both wave tank experiments and compressive tests have the scale problem under one gravity acceleration environment. Centrifugal wave modeling is a new advanced experimental approach, in which the experiments are conducted under N times gravitational acceleration. Such an experiment does not only provide the identical stress distribution to that in the field, but also provides the spatial distribution of soil response. However, the development of wave experiments in centrifuges is still not mature, and is a challenging experimental and instrumentation task.

The previous experimental approaches are summarized in the following sections.

2.4.2.1 Wave Tank Experiments

There have been numerous investigations for the wave-induced pore pressure based on water tank experiments. Among these, Sleath [108] conducted an experiment to verify his theory. He pointed out that the wave-induced pore pressure was out of the phase with the wave profile by about 10 degrees.

Later, Tsui and Helfrich [119] conducted experiments for loose and dense sands and confirmed the existence of a phase lag. They discovered that the maximum phase lag might reach one-third of the wave period.

Maeno and Hasegawa [68] proposed an empirical equation for the wave-induced pore pressure in some sandy beds, using second-order Stokes wave theory. In this, the pore pressure from the seabed surface to the bottom of the sand layer was expressed as a function of the wave steepness and two experimental parameters, which depended on the permeability of the bed. This empirical equation involved both wave characteristics and drainage conditions of the sandy bed.

Demars [13] measured the wave-induced pore pressure and stresses in a sandy bed. From their laboratory data, they concluded that elastic theories provided a reasonable method for estimating the total vertical and horizontal stresses in a sandy bed in which dilation of the grain matrix is small. Furthermore, their theoretical solution of the total stress for a seabed of infinite depth provided a lower bound for stresses in a sandy bed of finite depth.

Tzang [120] conducted a series of wave tank experiments to study the fluidization of silty soil under wave loading. His experimental results demonstrated the relationship between the build-up pore pressure and fluidization in a soft seabed.

2.4.2.2 Compressive Tests

The compressive test has been commonly used by geotechnical engineers, due to a better estimation of soil characteristics. Zen and Yamazaki [138, 139] also conducted a series of experiments to investigate the wave-induced pore pressure. They reported that the oscillatory pore pressure could be represented by two non-dimensional parameters, namely, coefficients of drainage and propagation. It was also observed that the number of waves did not affect the oscillatory pore pressure significantly. However, the amplitude of input pressure in their model tests is too large, compared with a realistic ocean environment. This is a shortcoming of compressive tests, and has been questioned by coastal engineers.

2.4.2.3 Centrifugal Wave Experiment

Centrifugal modeling is a new technology in geotechnical engineering. It has been widely applied to simulate various geotechnical problems and solute transport phenomena. Sekiguchi and his co-workers [87, 106] may have been the first to attempt conducting wave experiments in centrifuge. They developed the fundamental framework of the centrifugal wave experiments, and their experimental data has been commonly used for the verification of theoretical results. Later, the technology of wave generation in a centrifuge has been further improved and under better controlled experimental environments [101–104]. The phenomenon of wave-induced liquefaction can be physically observed under multi-gravity acceleration. Recently, centrifugal modeling has further applied to investigate the wave-induced progressive liquefaction [103]. However, sample preparation is a difficult task for centrifugal modeling, while a well controlled wave generation system is another challenge.

2.5 Waves Propagating over a Porous Seabed: Wave Damping and Seepage Flux

2.5.1 Wave Damping in a Porous Seabed

When ocean wave propagates on a porous seabed, not only the wave pressure penetrates into the seabed, but also the soil column modifies the wave characteristics. This phenomenon has attracted attention from ocean engineers since 1958 [15, 17, 20, 24, 78]. However, these early researches were based on the analytical solution with assumption of rigid permeable seabed. Under such a condition, the wave dispersion equation needs to be modified as

$$\omega^2 - gk \tanh kd = -i\omega \frac{k_z}{g}\big(gk - \omega^2 \tanh kd\big) \tanh kd. \tag{2.1}$$

It is noted that the modified wave dispersion equation consists of soil permeability and wave characteristics. Thus, the wavelength will be modified by the soil permeability. However, this modification in the wavelength is minor, because the soil model is an un-couple approach. Viscous effects in the wave damping problem, based on the similar assumptions, have been summarized in [16].

Recently, the effects of soil characteristics on the wave characteristics such as wavelength, wave profile and wave pressure have been discussed [34–36, 55]. Among these, both quasi-static soil behavior [34, 36] and dynamic soil behavior [35, 55] have been considered. The new linear wave dispersion equation was derived as

$$\tanh kd = \frac{\omega^2 - gkF}{gk - \omega^2 F} \tag{2.2}$$

where F is the function of soil characteristics. The above equation indicates that the wave characteristics will be modified by soil parameters. Detailed information for wave damping in a porous seabed will be given in Chap. 7.

The phenomenon of wave damping in Coulomb friction seabed was also studied by Yamamoto and others [128, 129, 131]. However, only the damping ratio (k_c/k_r)

was investigated in their studies. Later, the modification of wave characteristics due to soil was examined by Lee *et al.* [56]. They concluded that the response of soil characteristics on the wave field is important in graveled seabeds, which agrees with the conclusions drawn from the solution without Coulomb-damping friction. Detailed comparison of various models for wave damping is given in Lin and Jeng [59] and summarized in Chap. 7.

2.5.2 Wave-Driven Seepage Flux in Sediments

Sediments in bays, estuaries, and in the seabed near river inlets are often contaminated. Many inorganic contaminants (notably heavy metals) do not decompose. Under certain conditions, these accumulated substances can be released back into the receiving body of water through mass transfer processes at the seabed. The mass transfer rate is largely controlled by the seepage flux exchange between the sediment and the seawater [86]. Increased wave action and higher sediment hydraulic conductivity generally cause larger transfer rates. Clearly, quantification of the mass transfer rate is a key factor in water quality modeling.

Mu *et al.* [77] may be the first to investigate the wave driven seepage flux at the interface of water and sediments, based on quasi-static models. The volume of seepage exchange between seawater and seabed driven by the water waves, that is, the volume exchange at $z = 0$. This can be calculated by

$$V_z = \int_{-T/4}^{T/4} \int_{-T/4}^{T/4} \frac{k_z}{\gamma_w} \frac{\partial p}{\partial z} \, dx \, dt. \tag{2.3}$$

It is noted that in (2.3) is the "volume displacement" per unit width at interface between water and sediment.

The wave-driven seepage flux in marine sediment was also studied by Jeng *et al.* [41] with a dynamic model. The wave damping was also included. The numerical results concluded that the seepage flux across the seabed boundary is larger in shallow water and for shorter waves. Also, the magnitude of the seepage flux is also affected by the soil type and degree of saturation. In general, it is larger in coarse sands than in fine sands. The framework has been later extended to the case of cross-anisotropic seabed [40].

2.6 Wave-Induced Seabed Instability

Evaluation of seabed instability is an important part of the foundation design for many marine facilities, because some structures might have failed owing to seabed instability and concomitant subsidence. In general, the sea floor instability has been classified into three mechanisms: shear failure, liquefaction and scour [90, 91, 113, 114]. Since the mechanism of scour around coastal structures has been recently reviewed by Sumer and Fredsøe [113], we only discuss the mechanism of shear failure and liquefaction.

2.6.1 Shear Failure

The shear stresses at a point within the marine sediment, induced by gravity forces and storms, may be significant enough to overcome its shear resistance, resulting in seabed instability. This type of seabed instability, referred to as shear failure, may produce a horizontal movement of sediment.

Henkel [21] may have been the first to identify the important role of water waves on submarine landslides. He proposed a total stress analysis for the Birdfoot Delta of Mississippi, based on the principle of limiting equilibrium and the assumption of a circular failure surface under the action of a standing wave. He pointed out that the pressure changes on the bottom were large enough to cause shear failure in soft sediments. Based on this concept, [125] used a finite element method to compute the wave-induced stresses and displacements appropriate to an almost horizontal slope type of failure. However, in the limiting equilibrium analysis, no information concerning the deformation or the extent of failure was available.

Mitchell [75] demonstrated that wave action was instrumental in initiating mass movement of submarine sediments in model studies, but indicated little correlation between the results of model tests and analytical predictions. From this study, they concluded that the driving force in submarine land slides was primarily gravitational, and that a progressive wave train initiated failure by reducing the soil strength.

As pointed by Mitchell [74], cyclic subsurface pressures due to wave action caused remolding and loss of strength in fine grained submarine sediments. The depth of remolding generally increased with subsurface pressure. Furthermore, in sloping offshore sediments, the remolding due to wave action leads to slope instability. The depth of potential slope failure increased with increase subsurface pressure, in general accordance with the relation proposed by Henkel [21].

Bea [1] described failure of a major offshore platform caused by a sea floor slide and hurricane wave forces. As they reported, it was developed during the passage of Hurricane Camille (August, 1969) through South Pass Block 70, Mississippi River Delta. Large soil movements extended to depths in excess of 30 m and lateral soil translations exceeded 1000 m. Three offshore platforms failed as a result of soil movements and hurricane forces. They compared the results of the previous models, such as limit equilibrium, finite element and layered continuum analytical models. Ultimate strength and plastic design methods employing load factors on specified loading conditions were suggested to supplement or replace other elastic design methods and allowable stresses in design of a slide-resistant platform.

Lee and Edwards [54] proposed a regional method to assess offshore slope stability using properties of short core sample analyzed in the laboratory with wave parameters. Gravity corewere taken with conventional oceanographic sampling equipment in which pipes penetrated the sea floor as a result of momentum gained during a free-fall phase. However, this method was not applicable to all offshore situations, but was shown to be useful in certain simple depositional environments [54].

Rahman and his co-workers [90, 95] using simplified analyses, formulated the necessary conditions for wave-induced instability. The results showed that a seabed with cohesive sediments might suffer from shear failure, leading to slope instability. However, a seabed with non-cohesive sediments was unlikely to become unstable due to shear failure [90].

From a different aspect, Sulisz [111] considered the effect of permeability on the stability of rubble-mound bases. Their approach consisted of theoretical modeling of the interaction of water waves with a caisson-type breakwater and its rubble foundation, and then employing predicted flow field in a stability model to determine the required stone size. Their results showed that the permeability of a rubble-base had substantial effect on its stability. In general, the stability increases with the increase of the rubble-base permeability at intermediate water depths and increase with the decrease of the rubble-base permeability in shallow water. However, the permeability of the armor layer has only little effect on the rubble-base stability.

Despite aforementioned investigations, Hoeg [22] presented a review of key geotechnical issues in offshore engineering. He pointed out trends and needs for improvements in theory and practice related to offshore site investigations, seafloor instability and pipeline safety, single- and multi-based gravity structures, pile foundations and platform anchorage, jack-up platforms and arctic offshore constructions.

Based upon the analytical solution proposed by Hsu and Jeng [26] and Mohr-Coulomb's failure criterion, Jeng [33] investigated the wave-induced shear failure in front of a breakwater. The influence of wave and soil characteristics on th wave-induced maximum shear failure depth was examined through a parametric study. The results are summarized in Chap. 4.

2.6.2 Liquefaction

When excess pore pressure is equal to the downward effective soil weight, liquefaction may occur. This causes a quick sand effect in which external loads cannot be supported. As this occurs, the soil particles in a suspended state may be readily transported as fluid. The liquefaction is an extreme form of seabed instability, which may cause a vertical movement of sediment.

Liquefaction is the state where the material has lost all its structural strength. It behaves as a heavy liquid with no rigidity, and can therefore flow. It has been submitted that when pore water pressure increases until the effective stress becomes zero, then liquefaction can occur in non-cohesive soils.

In summary, two different criteria for liquefied state have been used in the past: the first criterion is based on the concept of effective stress, as given by Okusa [83] for two-dimensional cases, in which a liquefied state is reached when the vertical effective normal stress is greater than the submerged weight of the soil deposits. This criterion was extended to a three-dimensional case by Tsai [117]. The second criterion is based on the concept of excess pore pressure, as suggested by Zen and Yamazaki [138, 139] for two-dimensional case. This was modified by Jeng [33] to a three-dimensional case.

The possibility of wave-induced liquefaction occurring in saturated seabed sediments was first recognized and analyzed by Bjerrum [7], in connection with the foundation design for deep-water structures in the North Sea.

Nataraja *et al.* [81] suggested a simplified procedure for ocean wave-induced liquefaction analysis. Based on the concepts proposed, they concluded that the existing data on cyclic shear strength of liquefiable soils under seismic loading could be used to estimate the cyclic shear strength under wave loading conditions. Later, Nataraja and Gill [80] summarized the features of the simplified procedure for ocean wave-induced liquefaction analysis and examined the procedure by using data from four projects.

Employing the theoretical solution for a porous seabed of infinite thickness [130], Ishihara and Yamazaki [29] proposed a methodology for evaluating the magnitude of cyclic stress and wave-induced liquefaction on the basis of design storm parameters specified at an offshore location in deep water. This methodology for liquefaction analysis was applied to some typical cases in which several storm conditions were considered. They pointed out that: "For a medium-dense deposit of sand with 70 percent relative density, liquefaction could extend down to a depth of 17.7 m at a location of 14 m water depth."

Finn *et al.* [14] reviewed previous methods for estimating the stability of an ocean floor under wave loading and the determination of the wave-induced pore-water pressures. They developed two computer programs for computing transient and residual soil response and estimated liquefaction potential in each. The numerical results indicated that the distribution of the transient pore-water pressures, induced by the wave loading and the associated effective stress field, might be investigated by Biot's theory of poro-plastic soil.

Umchara *et al.* [122] proposed a method to evaluate liquefaction resistance for partially drained conditions. Their results indicated that soil strength increases with partial drainage and can be well represented by the relative density and a coefficient of drainage effect. The authors also suggested that a thin silty layer located at the sub-surface may play an important role in causing liquefaction of the deposits.

Based on the solution for an infinite seabed, Okusa [83] suggested two simple criteria to gauge the liquefaction depth. First, the "liquefied state" was reached when the effective vertical normal stress was greater than the submerged weight of the soil deposit. This criterion is not valid for cohesive soils because the lateral earth pressure was not accounted for. The second criterion was given by that the geostatic pressure being less than that of the wave-induced effective mean normal stress. These criteria were applied to examine the liquefaction potential for marine deposits in the North Sea and Mississippi Delta.

A case study of a damaged breakwater was demonstrated by Zen *et al.* [137]. It was suggested that the wave-induced liquefaction of the foundation with its subsequent slip circle failure was one of the possible causes of the collapse. The case study showed that the liquefaction might be induced by ocean waves only under the condition that the pore pressure dissipation or redistribution was restricted by the existence of layers with low permeability.

Maeno *et al.* [70], employing the empirical formula of Maeno and Hasegawa [69], discussed the effects of wave steepness on the wave-induced liquefaction depth. They pointed out that the liquefaction depth predicted by the theoretical solution of Yamamoto *et al.* [130] was considerably less than this empirical formula for small waves. In the case of large waves (for wave steepness greater than 0.2) the former is consistent with the latter. They also concluded that wave height and wave period play an important role on the wave-induced liquefaction.

Tsotsos *et al.* [118] developed a numerical model for the evaluation of pore pressure generation and liquefaction potential in the sea floor due to cyclic wave action. The analysis included the development of both transient and residual pore-water pressures, and the simultaneous partial pressure dissipation. The paper revealed that soil permeability had a very significant influence on pore-water pressure generation and liquefaction because high permeability prevented the development of excess pore-water pressure.

Sakai *et al.* [98], using Mei and Foda's boundary-layer approximation [73] examined the wave-induced momentary liquefaction. They concluded that it increases as the soil stiffness ($G\beta$) increases. The maximum liquefaction depth reached around half of the wave height in surf zone conditions.

According to Rahman [90, 91], a fully saturated seabed with non-cohesive sediments could experience liquefaction (associated with progressive build-up of pore pressure) in the region of vulnerable water depth, the liquefaction potential increased as the wave period increased. However, a partly saturated seabed may experience transient liquefaction (associated with oscillatory pore pressure). The liquefaction potential increased with a decrease in degree of saturation and with an increase of wave period. In his paper, the critical values of the relative water depth (d/L) in which the wave-induced seabed instability occurred were also presented in graphs. In fact, the conclusions of his papers were drawn from Okusa's criteria for liquefaction [83].

Jeng [33] examined these two criteria for several different cases, together with the field data from Zen and Yamazaki [140]. He concluded that the criterion suggested by Okusa [83] and Tsai [117] may only be suitable for a seabed with large thickness. He also found that no liquefaction occurs in a saturated seabed, except under the following conditions: (i) very shallow water, (ii) large wave, (iii) a seabed with very low permeability.

Despite the aforementioned theoretical approaches, an alternative approach, probabilistic liquefaction risk analysis, including fuzzy variables, was proposed by Rahman and Jabery [92, 94]. In their study, the combined effect of fuzzy variables is introduced into the basic model by evaluating an average fuzzy factor, and then transforming it into equivalent random variables. Thereafter, the general framework of probabilistic risk analysis is used to formulate the probability of liquefaction. The inclusion of fuzzy variables into the probabilistic liquefaction analysis represents an enhancement of estimation of liquefaction potential in practical circumstances.

References

1. Bea, R.G., Wright, S.G., Sircar, O.P., Niedoroda, A.W.: Wave-induced slides in south pass block 70, Mississippi Delta. J. Geotech. Eng. **109**(1), 617–644 (1975)
2. Bennett, R.H.: Pore-water pressure measurements: Mississippi Delta submarine sediments. Mar. Geotechnol. **2**, 177–189 (1978)
3. Bennett, R.H., Faris, J.R.: Ambient and dynamic pore pressures in fine-grained submarine sediments: Mississippi. Appl. Ocean Res. **1**(3), 115–123 (1979)
4. Biot, M.A.: General theory of three-dimensional consolidation. J. Appl. Phys. **26**(2), 155–164 (1941)
5. Biot, M.A.: Theory of propagation of elastic waves in a fluid-saturated porous solid, Part I: Low frequency range. J. Acoust. Soc. Am. **28**, 168–177 (1956)
6. Biot, M.A.: Theory of propagation of elastic waves in a fluid-saturated porous solid, Part II: High frequency range. J. Acoust. Soc. Am. **28**, 178–191 (1956)
7. Bjerrum, J.: Geotechnical problem involved in foundations of structures in the North Sea. Géotechnique **23**(3), 319–358 (1973)
8. Cha, D.H., Jeng, D.S., Rahman, M.S., Sekiguchi, H., Zen, K., Yamazaki, H.: Effects of dynamic soil behaviour on the wave-induced seabed response. Int. J. Ocean Eng. Technol. **16**(21–33), 5 (2002)
9. Chen, T.W., Huang, L.H., Song, C.H.: Dynamic response of poroelastic bed to nonlinear water waves. J. Eng. Mech. **123**(10), 1041–1049 (1997)
10. Cheng, L., Sumer, B.M., Fredsöe, J.: Solution of pore pressure build up due to progressive waves. Int. J. Numer. Anal. Methods Geomech. **25**, 885–907 (2001)
11. Dagan, G.: The generalization of Darcy law for nonuniform flows. Water Resour. Res. **15**(1), 1–7 (1979)
12. Dawson, T.H.: Wave propagation over a deformable sea floor. Ocean Eng. **5**, 227–234 (1978)
13. Demars, K.R., Vanover, E.A.: Measurement of wave-induced pressures and stresses in a sand bed. Mar. Geotechnol. **6**(1), 29–59 (1985)
14. Finn, W.D.L., Siddharthan, R., Martin, G.R.: Response of seafloor to ocean waves. J. Geotech. Eng. **109**(4), 556–572 (1983)
15. Flaten, G., Rygg, O.B.: Dispersive shallow water waves over a porous sea bed. Coast. Eng. **15**, 347–369 (1991)
16. Foda, M.A.: Sea Floor Dynamics, vol. 1, pp. 77–123. World Scientific, Singapore (1995)
17. Gade, H.: Effects of a non-rigid impermeable bottom on plane surface waves in shallow water. J. Mar. Res. **16**, 61–82 (1958)
18. Gatmiri, B.: A simplified finite element analysis of wave-induced effective stress and pore pressures in permeable sea beds. Géotechnique **40**(1), 15–30 (1990)
19. Gatmiri, B.: Response of cross-anisotropic seabed to ocean waves. J. Geotech. Eng. **118**(9), 1295–1314 (1992)
20. Gu, Z., Wang, H.: Gravity waves over porous bottoms. Coast. Eng. **15**, 497–524 (1991)
21. Henkel, D.H.: The role of waves in causing submarine landslides. Géotechnique **20**(1), 75–80 (1970)
22. Hoeg, K.: Geotechnical issues in offshore engineering, In: ASTM Special Technical Publication, vol. 923, pp. 7–50 (1986)
23. Horikawa, K.: Nearshore Dynamics and Coastal Processes. University of Tokyo Press, Japan (1988)
24. Hsiao, S.V., Shemdin, O.H.: Interaction of ocean waves soil bottom. J. Phys. Oceanogr. **10**, 605–610 (1980)
25. Hsieh, P.C., Huang, L.H., Wang, T.W.: Dynamic response of soft poroelastic bed to linear water waves—a boundary layer approximation. Int. J. Numer. Anal. Methods Geomech. **25**, 651–674 (2001)
26. Hsu, J.R.C., Jeng, D.S.: Wave-induced soil response in an unsaturated anisotropic seabed of finite thickness. Int. J. Numer. Anal. Methods Geomech. **18**(11), 785–807 (1994)

27. Huang, L.H., Chwang, A.T.: Trapping and absorption of sound waves. II: A sphere covered with a porous layer. Wave Motion **12**, 401–414 (1990)
28. Huang, L.H., Song, C.H.: Dynamic response of poro-plastic bed to water waves. J. Hydraul. Eng. **119**(9), 1003–1020 (1993)
29. Ishihara, K., Yamazaki, A.: Analysis of wave-induced liquefaction in seabed deposit of sand. Soil Found. **24**(3), 85–100 (1984)
30. Jeng, D.S.: Wave-induced liquefaction potential in a cross-anisotropic seabed. J. Chin. Inst. Eng. **19**(1), 59–70 (1996)
31. Jeng, D.S.: Discussion of "Response of poro-elastic beds to standing wave" by Sekiguchi et al. Soil Found. **37**(2), 139 (1997)
32. Jeng, D.S.: Soil response in cross-anisotropic seabed due to standing waves. J. Geotech. Geoenviron. Eng. **123**(1), 9–19 (1997)
33. Jeng, D.S.: Wave-induced seabed instability in front of a breakwater. Ocean Eng. **24**(10), 887–917 (1997)
34. Jeng, D.S.: On calculating the length of a short-crested wave over a porous seabed. Appl. Ocean Res. **22**(2), 63–73 (2000)
35. Jeng, D.S.: A new wave dispersion equation: Effects of soil characteristics. J. Offshore Mech. Arct. Eng. **125**(4), 177–181 (2001)
36. Jeng, D.S.: Wave dispersion equation in a porous seabed. Ocean Eng. **28**(12), 1585–1599 (2001)
37. Jeng, D.S.: Discussion to "Simplified solution of wave-induced seabed response in anisotropic seabed" by Yuhi and Ishida. J. Waterw. Port Coast. Ocean Eng. **129**(3), 151–153 (2003)
38. Jeng, D.S., Cha, D.H.: Effects of dynamic soil behavior and wave non-linearity on the wave-induced pore pressure and effective stresses in porous seabed. Ocean Eng. **30**(16), 2065–2089 (2003)
39. Jeng, D.S., Hsu, J.R.C.: Wave-induced soil response in a nearly saturated seabed of finite thickness. Géotechnique **46**(3), 427–440 (1996)
40. Jeng, D.S., Lee, T.L.: Dynamic response of porous seabed to ocean waves. Comput. Geotech. **28**(2), 99–128 (2001)
41. Jeng, D.S., Li, L., Barry, D.A.: Wave-induced seepage into seabed. Int. J. Numer. Anal. Methods Geomech. **25**, 771–787 (2001)
42. Jeng, D.S., Lin, Y.S.: Finite element modelling for water waves-soil interaction. Soil Dyn. Earthq. Eng. **15**(5), 283–300 (1996)
43. Jeng, D.S., Lin, Y.S.: Non-linear wave-induced response of porous seabed: A finite element analysis. Int. J. Numer. Anal. Methods Geomech. **21**(1), 15–42 (1997)
44. Jeng, D.S., Lin, Y.S.: Poroelastic analysis for wave-seabed interaction problem. Comput. Geotech. **26**(1), 43–64 (2000)
45. Jeng, D.S., Oh, Y.N., Chen, S.: DDA-ESM for wave-induced pore pressure and effective stresses in a porous seabed. In: The Second International Conference on Advances in Structural Engineering and Mechanics (ASEM02). (CD ROM) Korea (2002)
46. Jeng, D.S., Rahman, M.S.: Effective stresses in a porous seabed of finite thickness: Inertia effects. Can. Geotech. J. **37**(4), 1388–1397 (2000)
47. Jeng, D.S., Rahman, M.S.: Wave-induced oscillatory soil response: Difference between quasi-static and dynamic solutions. In: Computer Methods and Advances in Geomechanics, vol. 2, pp. 1103–1106. A.A. Balkema, The Netherlands (2001)
48. Jeng, D.S., Rahman, M.S., Lee, T.L.: Effects of inertia forces on wave-induced seabed response. Int. J. Offshore Polar Eng. **9**(4), 307–313 (1999)
49. Jeng, D.S., Seymour, B.R.: Response in seabed of finite depth with variable permeability. J. Geotech. Geoenviron. Eng. **123**(10), 902–911 (1997)
50. Jeng, D.S., Seymour, B.R.: Wave-induced pore pressure and effective stresses in a porous seabed with variable permeability. J. Offshore Mech. Arct. Eng. **119**(4), 226–233 (1997)
51. Kitano, T., Mase, H.: Boundary-layer theory for anisotropic seabed response to sea waves. J. Waterw. Port Coast. Ocean Eng. **125**(4), 187–194 (1999)

52. Kitano, T., Mase, H.: Wave-induced porewater pressure in a seabed with inhomogeneous permeability. Ocean Eng. **28**, 279–296 (2001)
53. Kitano, T., Mase, H., Nakano, S.: Effects of inhomogeneity of permeability on wave-induced porewater pressure in seabed. In: Canadian Coastal Conference, pp. 611–623 (1999)
54. Lee, H.J., Edwards, B.D.: Regional method to assess offshore slope stability. J. Geotech. Eng. **112**(5), 489–509 (1986)
55. Lee, T.L., Tsai, C.P., Jeng, D.S.: Analytical solution of wave-induced seabed response: Effects of inertia forces. Ocean Eng. **29**(12), 1577–1601 (2002)
56. Lee, T.L., Tsai, C.P., Jeng, D.S.: Ocean waves propagating over a coulomb-damped poroelastic seabed of finite thickness: An analytical solution. Comput. Geotech. **29**(2), 119–149 (2002)
57. Li, X., Zhang, J., Zhang, H.: Instability of wave propagation in saturated poroelastoplastic media. Int. J. Numer. Anal. Methods Geomech. **26**, 563–578 (2002)
58. Lin, M.: The analysis of silt behaviour induced by water waves. Sci. China Ser. E **31**(1), 86–96 (2001)
59. Lin, M., Jeng, D.S.: Comparison of existing poroelastic models for wave damping in a porous seabed. Ocean Eng. **30**(11), 1335–1352 (2003)
60. Lin, Y.S., Jeng, D.S.: Response of poro-elastic seabed to a 3-d wave system: A finite element analysis. Coast. Eng. Jpn. **39**(2), 165–183 (1996)
61. Lin, Y.S., Jeng, D.S.: The effect of variable permeability on the wave-induced seabed response. Ocean Eng. **24**(7), 623–643 (1997)
62. Lin, Y.S., Jeng, D.S.: Effects of variable shear modulus on wave-induced seabed response. J. Chin. Inst. Chem. Eng. **24**(1), 109–115 (2000)
63. Liu, P.L.F.: Damping of water waves over porous bed. J. Hydraul. Div. **99**(12), 2263–2271 (1973)
64. Liu, P.L.F.: On gravity waves propagated over a layered permeable bed. Coast. Eng. **1**, 135–148 (1977)
65. Liu, P.L.F., Dalrymple, R.A.: The damping of gravity water-waves due to percolation. Coast. Eng. **8**(1), 33–49 (1984)
66. Madga, W.: On one-dimensional model of pore pressure generation in a highly saturated sandbed due to cyclic loading acting on a sand surface. I: Theoretical description and numerical approach. Tech. rep., Internal report No. 5, SFB-205, TP A13, Kusteningenieurwesen, University Hanover (1990)
67. Madsen, O.S.: Wave-induced pore pressures and effective stresses in a porous bed. Géotechnique **28**(4), 377–393 (1978)
68. Maeno, Y.H., Hasegawa, T.: Evaluation of wave-induced pore pressure in sand layer by wave steepness. Coast. Eng. Jpn. **28**, 31–44 (1985)
69. Maeno, Y.H., Hasegawa, T.: In-situ measurements of wave-induced pore pressure for predicting properties of seabed deposits. Coast. Eng. Jpn. **30**(1), 99–115 (1987)
70. Maeno, Y.H., Sakai, T., Mase, H.: Influences of wave steepness on wave-induced liquefaction in sand layer. In: Proceedings of 6th Symposium on Coastal and Ocean Management, Coastal Zone'89, pp. 3945–3957 (1989)
71. Mallard, W.W., Dalrymple, R.A.: Water waves propagating over a deformable bottom. In: Proceedings 9th Annual Offshore Technology Conference, pp. 141–145 (1977)
72. Massel, S.R.: Gravity waves propagated over permeable bottom. J. Waterways Harbors Coast. Eng. **102**(2), 111–121 (1976)
73. Mei, C.C., Foda, M.A.: Wave-induced response in a fluid-filled poro-elastic solid with a free surface-a boundary layer theory. Geophys. J. R. Astron. Soc. **66**, 597–631 (1981)
74. Mitchell, J.K., Hull, J.A.: Stability and bearing capacity of bottom sediments. In: Proceedings 14th International Conferences on Coastal Engineering, ASCE, vol. 2, pp. 1252–1273 (1974)
75. Mitchell, J.K., Tsui, K.K., Sangrey, D.A.: Failure of submarine slope under wave action. In: Proceedings 13th International Conferences on Coastal Engineering, ASCE, vol. 2, pp. 1515–1539 (1972)

76. Moshagen, H., Torum, A.: Wave induced pressures in permeable seabeds. J. Waterways Harbors Coast. Eng. Div. **101**(1), 49–57 (1975)
77. Mu, Y., Cheng, A.H.D., Badiey, M., Bennett, R.: Water wave driven seepage in sediment and parameter inversion based on pore pressure data. Int. J. Numer. Anal. Methods Geomech. **23**, 1655–1674 (1999)
78. Murray, J.D.: Viscous damping of gravity waves over a permeable bed. J. Geophys. Res. **70**(10), 2325–2331 (1965)
79. Nakamura, H., Onishi, R., Minamide, H.: On the seepage in the seabed due to waves. In: Proceedings of 20th Coastal Engineering Conference, J.S.C.E., pp. 421–428 (1973)
80. Nataraja, M.S., Gill, H.S.: Ocean wave-induced liquefaction analysis. J. Geotech. Eng. **109**(4), 573–590 (1983)
81. Nataraja, M.S., Singh, H., Maloney, D.: Ocean wave-induced liquefaction analysis: A simplified procedure. In: Proceedings of International Symposium on Soils under Cyclic and Transient Loadings, pp. 509–516 (1980)
82. Okusa, S.: Measurements of wave-induced pore pressure in submarine sediments under various marine conditions. Mar. Geotechnol. **6**(2), 119–144 (1985)
83. Okusa, S.: Wave-induced stress in unsaturated submarine sediments. Géotechnique **35**(4), 517–532 (1985)
84. Okusa, S., Nakamura, T., Fukue, M.: Measurements of Wave-Induced Pore Pressure and Coefficients of Permeability of Submarine Sediments During Reversing Flow, pp. 113–122. Graham and Trotman, London (1983)
85. Okusa, S., Uchida, A.: Pore-water pressure change in submarine sediments due to waves. Mar. Geotechnol. **4**(2), 145–161 (1980)
86. Oldham, C.E., Lavery, P.S.: Porewater nutrient fluxes in a shallow fetch-limited estuary. Mar. Ecol. Prog. Ser. **183**, 39–47 (1999)
87. Phillips, R., Sekiguchi, H.: Generation of water wave trains in drum centrifuge. In: Proceedings of International Symposium on Technology in Ocean Engineering (Techno-Ocean'92), vol. 1, pp. 29–34 (1992)
88. Prevost, J.H., Eide, O., Anderson, K.H.: Discussion on "Wave induced pressures in permeable seabeds" by Moshagen and Torum. J. Waterways Harbors Coast. Eng. Div. **101**(1975), 464–465 (1975)
89. Putnam, J.A.: Loss of wave energy due to percolation in a permeable sea bottom. Trans. Am. Geophys. Union **30**(3), 349–356 (1949)
90. Rahman, M.S.: Wave-induced instability of seabed: Mechanism and conditions. Mar. Geotechnol. **10**, 277–299 (1991)
91. Rahman, M.S.: Instability and movement of oceanfloor sediments: A review. Int. J. Offshore Polar Eng. **7**(3), 220–225 (1997)
92. Rahman, M.S., El-Zahaby, K.: Probabilistic liquefaction risk analysis including fuzzy variables. Soil Dyn. Earthq. Eng. **16**, 63–79 (1997)
93. Rahman, M.S., El-Zahaby, K., Booker, J.: A semi-analytical method for the wave-induced seabed response. Int. J. Numer. Anal. Methods Geomech. **18**, 213–236 (1994)
94. Rahman, M.S., Jabery, W.Y.: A simplified drained analysis for wave-induced liquefaction in ocean floor sands. Soil Found. **26**(1), 57–68 (1986)
95. Rahman, M.S., Jabery, W.Y.: Submarine landslide: Element of analysis. Mar. Geotechnol. **10**, 97–124 (1991)
96. Raman-Nair, W., Sabin, G.C.W.: Wave-induced failure of poro-plastic seabed slopes: A boundary element study. Proc. Inst. Civ. Eng. 2. Res. Theory **91**, 771–794 (1991)
97. Reid, R.O., Kajiura, K.: On the damping of gravity waves over a permeable sea bed. Trans. Am. Geophys. Union **38**, 662–666 (1957)
98. Sakai, T., Hatanaka, K., Mase, H.: Wave-induced effective stress in seabed and its momentary liquefaction. J. Waterw. Port Coast. Ocean Eng. **118**(2), 202–206 (1992)
99. Sakai, T., Hattori, A., Hatanaka, K.: Wave-induced transient pore-water pressure and seabed instability in the surf zone. In: Proceeding of International Conference on Geotechnical En-

gineering for Coastal Development-Theory and Practice on Soft Ground, Yokohama, Japan, vol. 1, pp. 627–632 (1991)
100. Sakai, T., Mase, H., Matsumto, A.: Effects of inertia and gravity on seabed response to ocean waves. In: Modelling Soil-Water-Structure Interaction, pp. 61–66. A.A. Balkema, The Netherlands (1988)
101. Sassa, S.: Fundamental studies of wave-induced liquefaction of sand beds. Ph.D. thesis, Kyoto University, Japan (2000)
102. Sassa, S., Sekiguchi, H.: Wave-induced liquefaction of beds of sand in a centrifuge. Géotechnique **49**(5), 621–638 (1999)
103. Sassa, S., Sekiguchi, H.: Analysis of wave-induced liquefaction of sand beds. Géotechnique **51**(2), 115–126 (2001)
104. Sassa, S., Sekiguchi, H., Miyamamot, J.: Analysis of progressive liquefaction as moving-boundary problem. Géotechnique **51**(10), 847–857 (2001)
105. Sekiguchi, H., Kita, K., Okamoto, O.: Response of poro-elastoplastic beds to standing waves. Soil Found. **35**(3), 31–42 (1995)
106. Sekiguchi, H., Phillips, R.: Generation of water waves in a drum centrifuge. In: Proceedings of the International Conference Centrifuge, pp. 343–350 (1991)
107. Seymour, B.R., Jeng, D.S., Hsu, J.R.C.: Transient soil response in a porous seabed with variable permeability. Ocean Eng. **23**(1), 27–46 (1996)
108. Sleath, J.F.A.: Wave-induced pressures in beds of sand. J. Hydraul. Div. **96**(2), 367–378 (1970)
109. Song, C.H., Huang, L.H.: Laminar poroelastic media flow. J. Eng. Mech. **126**(4), 358–366 (2000)
110. Stokoe, K.H., Isenhower, W.M., Hsu, J.R.: Dynamic properties of offshore silty samples. In: Proceedings 12th Annual Offshore Technology Conference, pp. 289–302 (1989)
111. Sulisz, W.: Effect of permeability on stability of rubble bases. J. Waterw. Port Coast. Ocean Eng. **121**(3), 162–166 (1995)
112. Sumer, B.M., Cheng, N.S.: A random-walk model for pore pressure accumulation in marine soils. In: The 9th International Offshore and Polar Engineering Conference (ISOPE99), vol. 1, pp. 521–528 (1999)
113. Sumer, B.M., Fredsøe, J.: Wave Scour Around Structures, vol. 4, pp. 191–248. World Scientific, Singapore (1997)
114. Sumer, B.M., Fredsoe, J.: The Mechanism of Scour in the Marine Environment. World Scientific, New Jersey (2002)
115. Thomas, S.D.: A finite element model for the analysis of wave induced stresses, displacements and pore pressure in an unsaturated seabed. I: Theory. Comput. Geotech. **8**(1), 1–38 (1989)
116. Thomas, S.D.: A finite element model for the analysis of wave induced stresses, displacements and pore pressure in an unsaturated seabed. II: Model verification. Comput. Geotech. **17**(1), 107–132 (1995)
117. Tsai, C.P.: Wave-induced liquefaction potential in a porous seabed in front of a breakwater. Ocean Eng. **22**(1), 1–18 (1995)
118. Tsotsos, S., Georgiadis, M., Damaskindou, A.: Numerical analysis of liquefaction potential of partially drained seafloor. Coast. Eng. **13**(2), 117–128 (1989)
119. Tsui, Y.T., Helfrich, S.C.: Wave-induced pore pressures in submerged sand layer. J. Geotech. Eng. **109**(4), 603–618 (1983)
120. Tzang, S.Y.: Water wave-induced soil fluidization in a cohesionless fine-grained seabed. Ph.D. thesis, University of California at Berkeley (1992)
121. Ulker, M.B.C., Rahman, M.S., Jeng, D.S.: Wave-induced response of seabed: Various formulations and their applicability. Appl. Ocean Res. **31**(1), 12–24 (2009)
122. Umehara, Y., Zen, K., Hamada, K.: Evaluation of soil liquefaction potentials in partially drained conditions. Soil Found. **25**(2), 57–72 (1985)
123. Varley, E., Seymour, B.R.: A method for obtaining exact solutions to partial differential equations with variable coefficients. Stud. Appl. Math. **78**, 183–225 (1988)

124. Verruijt, A.: Elastic storage of aquifers. In: De Wiest, R.J.M. (ed.) Flow Through Porous Media, pp. 331–376. Academic Press, New York (1969)
125. Wright, S.G., Dunham, R.S.: Bottom stability under wave induced loading. In: Proceedings of the 4th Annual Offshore Technology Conference, pp. 853–862 (1972)
126. Yamamoto, T.: Wave induced instability seabed. In: Proceedings A.S.C.E. Special Conference, Coastal Sediments'77, pp. 898–913 (1977)
127. Yamamoto, T.: Wave-induced pore pressures and effective stresses in inhomogeneous seabed foundations. Ocean Eng. **8**, 1–16 (1981)
128. Yamamoto, T.: Non-linear mechanics of ocean wave interactions with sediment beds. Appl. Ocean Res. **4**(2), 99–106 (1982)
129. Yamamoto, T.: On the response of a Coulomb-damped poro-plastic bed to water waves. Mar. Geotechnol. **5**(2), 93–130 (1983)
130. Yamamoto, T., Koning, H., Sellmeijer, H., Hijum, E.V.: On the response of a poro-elastic bed to water waves. J. Fluid Mech. **87**(1), 193–206 (1978)
131. Yamamoto, T., Takahashi, S.: Wave damping by soil motion. J. Waterw. Port Coast. Ocean Eng. **111**(1), 62–77 (1985)
132. Yamamoto, T., Trevorrow, M.: Experimental verifications of bottom shear modulus profiler (BSMP) method. In: Proceedings of the International Conference on Geotechnical Engineering for Coastal Development—Theory and Practice on Soft Ground (Geot-Coastal 91), vol. 1, pp. 123–128 (1991)
133. Yang, Q.S., Poorooshasb, H.B.: Seabed response to wave loading. In: Proceedings of the Seventh, International Offshore and Polar Engineering Conference, vol. 1, pp. 689–695 (1997). 1997
134. Yuhi, M., Ishida, H.: Theoretical analysis of the response of a cross-anisotropic seabed to ocean surface waves. In: Proceeding of Japanese Society of Civil Engineering (JSCE), vol. 572, pp. 49–61 (1997)
135. Yuhi, M., Ishida, H.: Analytical solution for wave-induced seabed response in a soil-water two-phase mixture. Coast. Eng. Res. J. **40**(4), 367–381 (1998)
136. Yuhi, M., Ishida, H.: Simplified solutions for wave-induced response of anisotropic seabed. J. Waterw. Port Coast. Ocean Eng. **128**(1), 46–50 (2002)
137. Zen, K., Umehara, Y., Finn, W.D.L.: A case study of the wave-induced liquefaction of sand layers under damaged breakwater. In: Proceeding 3rd Canadian Conference on Marine Geotechnical Engineering, pp. 505–520 (1985)
138. Zen, K., Yamazaki, H.: Mechanism of wave-induced liquefaction and densification in seabed. Soil Found. **30**(4), 90–104 (1990)
139. Zen, K., Yamazaki, H.: Oscillatory pore pressure and liquefaction in seabed induced by ocean waves. Soil Found. **30**(4), 147–161 (1990)
140. Zen, K., Yamazaki, H.: Field observation and analysis of wave-induced liquefaction in seabed. Soil Found. **31**(4), 161–179 (1991)
141. Zienkiewicz, O.C., Chang, C.T., Bettess, P.: Drained, undrained, consolidating and dynamic behaviour assumptions in soils. Géotechnique **30**(4), 385–395 (1980)
142. Zienkiewicz, O.C., Scott, F.C.: On the principle of repeatability and its application in analysis of turbine and pump impellers. Int. J. Numer. Methods Eng. **9**, 445–452 (1972)

Chapter 3
Wave-Induced Soil Response in an Isotropic Seabed

Abstract In this chapter, a series of analytical solutions for the short-crested wave-induced soil response, including pore pressure, effective stresses and soil displacements, were presented. The porous seabed was treated as infinite thickness, finite thickness and multi-layered medium. The reduced two-dimensional forms of the proposed mathematical solutions were verified with previous experimental data, other analytical solutions and numerical models. A new combined wave obliquity and hydraulically anisotropic parameter was obtained from the closed-form analytical solution. The effects of wave obliquity, wave and seabed characteristics on the soil response were discussed in detail. The proposed analytical solution provide practical engineers an effective tool as the first approximation for the evaluation of the wave-induced soil response in the field.

Keywords Short-crested waves · Pore pressure · Effective stresses · Biot's consolidation theory

3.1 Introduction

Water waves propagating in the ocean generate dynamic pressure fluctuations on the sea floor. These fluctuations further induce variations of pore pressure and effective stresses in the seabed. When pore pressure becomes excessive, accompanied by the reduction in effective stresses, it might cause seabed instability. Thus, an evaluation of wave-induced soil response is particularly important for marine geotechnical and coastal engineers involved in the design of foundations for offshore or nearshore structures.

When a progressive wave arrives obliquely at a marine structure, it is reflected from the structure, resulting in a short-crested wave system, as shown in Fig. 3.1a. This wave system has a free surface which fluctuates periodically not only in the direction of propagation (the x-direction), but also normal to the structure (the y-direction). This can result in double wave heights at crest intersections, large wave forces on the structure, complex patterns of the water-particle motions and mass-transport, thus enhancing the seabed instability [6, 13, 37, 43]. Details of the wave particle orbital motions have been reported in the literature [34]. The water particle orbital motions vary across the normal to the reflecting wall and differ for angles

D.-S. Jeng, *Porous Models for Wave-seabed Interactions*,
DOI 10.1007/978-3-642-33593-8_3,

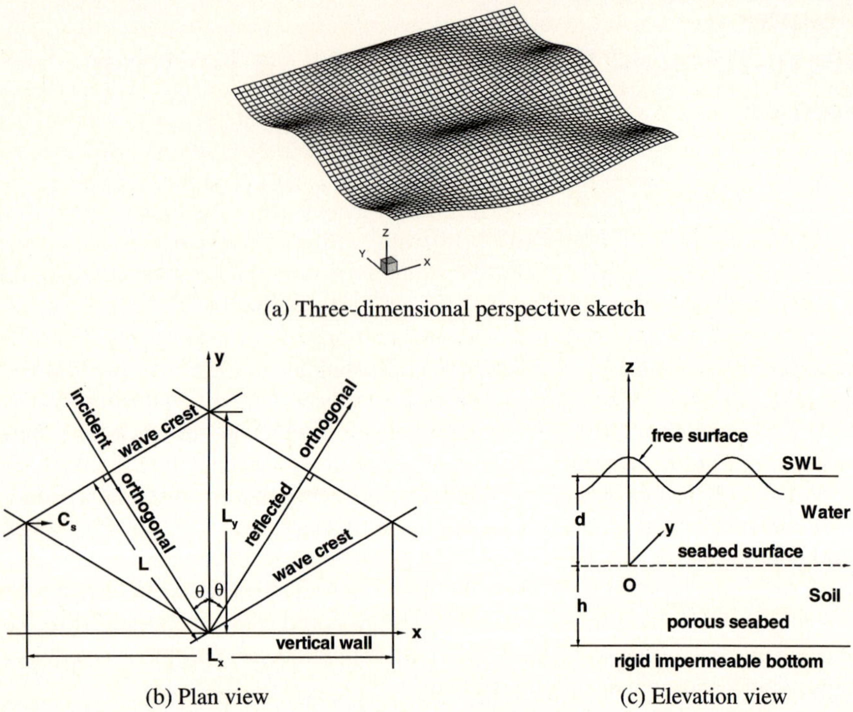

Fig. 3.1 Definition sketch of a short-crested wave system produced by full oblique reflection from a vertical wall, showing Cartesian co-ordinates used for wave-seabed interaction problem. (**a**) Three-dimensional perspective sketch, (**b**) plan and (**c**) elevation

of obliquity. At some small angles the interaction of reflected and incident waves is similar to the standing wave, in which case the maximum orbits occur in the nodal zone where orbits are normal to the wall. For larger angles between crest and face, the orbits parallel to it assume optimum velocities. The alignments of maximum scour thus vary normal to the wall [35].

Although the importance of the short-crested wave system has been recognized by coastal engineers in recent years, to date most previous investigations have dealt only with the kinematic and dynamic properties of this wave system [6, 13, 43], not with the seabed response on which the structure is based. Since numerous reports in the literature have suggested that many recent failure of breakwaters are probably due to foundation instability [21, 36, 38], the analysis of seabed response in front of marine structures deserves critical examinations.

In this chapter, a series of analytical solutions for the wave-induced soil response in front of a vertical wall is developed. Firstly, a brief description of the linear short-crested wave theory is given. The boundary value problem of the wave-soil interaction is then described. General solutions for soil response in a seabed of infinite, finite thickness, and a layered medium are formulated in Sect. 3.4, including the two-dimensional solutions of progressive and standing waves. Verification of the present theory against previous theoretical and experimental work is presented in Sect. 3.5, followed by the numerical results and detailed discussion of soil response.

3.2 A Short-Crested Wave System

Physically, a short-crested wave may be produced by two progressive wave trains angled to each other. This results in a three-dimensional wave with a new direction of propagation. Within a wind fetch, waves of different properties are generated. They grow along the fetch with components of different periods angled to the wind vector so producing a short-crested wave system. As the matured swell waves approach a marine structure obliquely, they are reflected from it, so causing a short-crested wave system in front of the structure. It is for these reasons that this three-dimensional wave system is more important than their two-dimensional counterparts that have been studied intensively in the past.

It has been shown that a short-crested wave system has a steep free-surface profile, complex water-particle motions [13, 30], and larger wave forces on a vertical wall than standing waves [6, 42]. Recently, numerous theories for analyzing short-crested wave systems have been presented [6, 7, 13, 24, 30, 31, 43]. However, both analytical solutions [6, 13, 24, 30] and numerical models [7, 31, 43] have been concerned with the dynamics and kinematics of the wave itself, rather than its effect on the seabed.

Attention has also been given to the scouring potential of short-crested waves [34–36]. Application of this wave system to bypass sediment across coastal inlets has been suggested [33]. All these studies have been of a practical nature and contain neither theories on liquefaction nor other important characteristics of the bed material.

Herein, a soil matrix subjected to a three-dimensional short-crested wave system is considered, as depicted in Fig. 3.1(b). A simple short-crested wave is produced by incident and reflected waves of equal height and period. The obliquity θ is measured between a wave orthogonal and the normal to the wall, or between the incident or reflected wave crest and the alignment of the wall. The combined wave crest propagates in the positive x-direction parallel to the wall, whilst the y-direction is measured normal to it (L_y is termed as the crest length and hence the term *short-crested*). The z-axis is positive upward from the *seabed surface*, as shown in Fig. 3.1(c).

Referring to a short-crested wave theory to the first-order [13], the velocity potential (ϕ) is given as

$$\phi = \frac{gH_s}{2\omega}\frac{\cosh kz}{\cosh kd}\sin(mkx - \omega t)\cos nky, \tag{3.1}$$

and the free surface elevation η is

$$\eta = \frac{H_s}{2}\cos(mkx - \omega t)\cos nky, \tag{3.2}$$

where H_s is the wave height of the short-crested wave, k is the wave number of the incident and reflected wave ($k = 2\pi/L$, in which L is the wavelength of the incident and the reflected waves), t is time and d is the water depth above the seabed surface.

The wave frequency ω ($= 2\pi/T$, where T is the wave period) satisfies the dispersion equation

$$\omega^2 = gk \tanh kd. \tag{3.3}$$

Two wave obliquity parameters m and n, relating to the components of wave number k in the x- and y-directions, respectively, are given by [13]

$$k_x = 2\pi/L_x = mk = k \sin\theta \quad \text{and} \quad k_y = 2\pi/L_y = nk = k \cos\theta, \tag{3.4}$$

where L_x and L_y are the wavelength and crest length of the short-crested waves, as shown in Fig. 3.1(b). From (3.4), the relationship

$$m^2 + n^2 = 1, \tag{3.5}$$

is applicable in a short-crested wave system.

Introducing the velocity potential ϕ and the surface elevation η into the Bernoulli equation, wave pressure P in the three-dimensional domain of this wave system is given by

$$P(x, y, z, t) = \frac{\gamma_w H_s}{2} \frac{\cosh kz}{\cosh kd} \cos(mkx - \omega t) \cos nky, \tag{3.6}$$

in which γ_w is the unit weight of water.

3.3 Boundary Value Problem

3.3.1 Governing Equations

To derive an analytical solution of the wave-seabed interaction problem described above, as in Fig. 3.1, some basic assumptions for the wave, structure and soil properties are necessary. They are:

1. The breakwater is a rigid structure of infinite length and is deeply embedded into the seabed.
2. The horizontal porous seabed is elastically isotropic, unsaturated, and hydraulically anisotropic.
3. The soil skeleton and pore fluid are compressible.
4. The inertial effect of sediments is neglected so that the soil skeleton generally obeys Hooke's law.
5. The flow in the porous bed obeys Darcy's law.
6. The wave pressure on the seabed surface is considered as the only external loading force in the wave-seabed interaction. It implies that the effect of boundary layer is ignored in this study.
7. Only linear wave theory is employed.

Since water waves normally produce a very slow propagating velocity through a sandy seabed, compared with that of shear waves, inertial forces are negligibly small compared with elastic forces in the equations of motion. Mynett and Mei [26] proved this assumption through a dimensional analysis. Thus, assumption (4) is acceptable for the problem of porous flow in a sandy seabed. A detailed information for the applicable range of different approaches can be found in [16, 45]. More detailed study for the effects of inertial forces on the wave-induced soil response will be discussed in Chap. 6.

Regarding assumption (6), it has been well documented that wave boundary layer only concentrates on a thin layer above the seabed surface. A detailed discussion on the effects of bottom shear stresses on the wave-induced soil response can be found in [53].

For assumption (7), the linear wave theory is commonly used as a first approximation in engineering practice. However, it has been agreed that it has its limits of application. The influence of non-linear wave components on the wave-induced soil response has been dealt with by Jeng and Lin [19] using a finite element method.

The consolidation equation [3] and storage equation [46] are generally accepted as those governing the flow of compressible pore fluid in a compressible porous medium. For a three-dimensional problem, and treating the porous bed as hydraulically anisotropic, with permeabilities K_x, K_y and K_z in the x-, y- and z-directions, respectively, the governing equation can be expressed as

$$\frac{K_x}{K_z}\frac{\partial^2 p}{\partial x^2}+\frac{K_y}{K_z}\frac{\partial^2 p}{\partial y^2}+\frac{\partial^2 p}{\partial z^2}-\frac{\gamma_w n'\beta}{K_z}\frac{\partial p}{\partial t}=\frac{\gamma_w}{K_z}\frac{\partial \varepsilon}{\partial t}, \tag{3.7}$$

where p is the wave-induced pore pressure; γ_w is the unit weight of the pore-water; n' is the soil porosity; and ε is the volume strain defined by

$$\varepsilon=\frac{\partial u}{\partial x}+\frac{\partial v}{\partial y}+\frac{\partial w}{\partial z}, \tag{3.8}$$

where u, v and w are the soil displacements in the x-, y-, and z-directions, respectively.

The compressibility of the pore fluid, β, is related to the apparent bulk modulus of the pore fluid K' and the degree of saturation S_r [46], such that

$$\beta=\frac{1}{K'}=\frac{1}{K_w}+\frac{1-S_r}{P_{wo}}, \tag{3.9}$$

where K_w is the true bulk modulus of elasticity of water (which may be taken as 1.95×10^9 N/m^2, [51]), P_{wo} is the absolute water pressure. If the soil skeleton is completely air-free, i.e. fully saturated, then $\beta=1/K_w$, since $S_r=1$.

The equations for overall equilibrium in a poro-elastic medium, relating to the effective stresses and pore pressure, are given by

$$\frac{\partial \sigma'_x}{\partial x}+\frac{\partial \tau_{xy}}{\partial y}+\frac{\partial \tau_{xz}}{\partial z}=\frac{\partial p}{\partial x}, \tag{3.10}$$

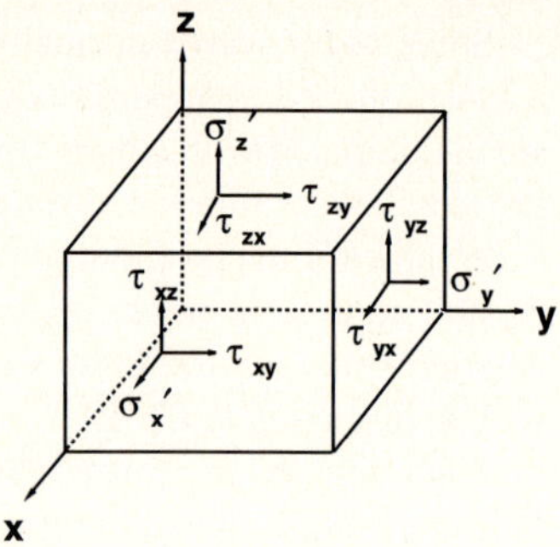

Fig. 3.2 Definition sketch of a stress block on a soil element in the domain of a soil matrix

$$\frac{\partial \tau_{xy}}{\partial x} + \frac{\partial \sigma'_y}{\partial y} + \frac{\partial \tau_{yz}}{\partial z} = \frac{\partial p}{\partial y}, \tag{3.11}$$

$$\frac{\partial \tau_{xz}}{\partial x} + \frac{\partial \tau_{yz}}{\partial y} + \frac{\partial \sigma'_z}{\partial z} = \frac{\partial p}{\partial z}, \tag{3.12}$$

where Cauchy stress tensor on the adjacent faces of a stress element consists of three effective normal stresses and six shear stress components. The shear stresses are expressed in double subscripts τ_{rs}, denoting the stress in the s-direction on a plane perpendicular to the r-axis, as seen in Fig. 3.2.

Based on the generalized Hooke's law, the relationships between elastic incremental effective stresses and soil displacements are given by

$$\sigma'_x = 2G\left[\frac{\partial u}{\partial x} + \frac{\mu}{1-2\mu}\varepsilon\right], \qquad \sigma'_y = 2G\left[\frac{\partial v}{\partial y} + \frac{\mu}{1-2\mu}\varepsilon\right], \tag{3.13}$$

$$\sigma'_z = 2G\left[\frac{\partial w}{\partial z} + \frac{\mu}{1-2\mu}\varepsilon\right], \qquad \tau_{xz} = G\left[\frac{\partial u}{\partial z} + \frac{\partial w}{\partial x}\right] = \tau_{zx}, \tag{3.14}$$

$$\tau_{yz} = G\left[\frac{\partial v}{\partial z} + \frac{\partial w}{\partial y}\right] = \tau_{zy}, \qquad \tau_{xy} = G\left[\frac{\partial u}{\partial y} + \frac{\partial v}{\partial x}\right] = \tau_{yx}, \tag{3.15}$$

where the shear modulus G is related to Young's modulus E by the Poisson's ratio μ in the form of $E/2(1+\mu)$.

Substituting (3.13)–(3.15) into (3.10)–(3.12), the equations of force equilibrium become

$$G\nabla^2 u + \frac{G}{(1-2\mu)}\frac{\partial \varepsilon}{\partial x} = \frac{\partial p}{\partial x}, \tag{3.16}$$

$$G\nabla^2 v + \frac{G}{(1-2\mu)}\frac{\partial \varepsilon}{\partial y} = \frac{\partial p}{\partial y}, \tag{3.17}$$

and

$$G\nabla^2 w + \frac{G}{(1-2\mu)}\frac{\partial \varepsilon}{\partial z} = \frac{\partial p}{\partial z}, \tag{3.18}$$

in the x-, y- and z-directions, respectively.

It is important to note that the equations of force equilibrium, (3.16)–(3.18), and the stress-strain relationships, (3.13)–(3.15), are only valid for an isotropic seabed. For a cross-anisotropic seabed, these equations must be modified, as considered in Chap. 5.

3.3.2 Boundary Conditions

For a homogeneous soil matrix, mathematical expressions for the wave-induced soil response can be derived, with appropriate boundary conditions. In general, two of these are required for a seabed of infinite and finite thickness, i.e., at a rigid impermeable bottom (**BBC**: Bottom Boundary Condition) and at the seabed surface (**SBC**: Surface Boundary Condition). An additional matching condition (**MBC**: Matching Boundary Condition) between any two adjacent layers is required for a layered seabed (Fig. 3.3).

(a) SBC: Boundary Conditions at Seabed Surface ($z = 0$) It is commonly accepted that vertical effective normal stresses and shear stresses vanish at the seabed surface, i.e.

$$\sigma_z' = \tau_{xz} = \tau_{yz} = 0, \quad \text{at } z = 0, \tag{3.19}$$

and the pore pressure is equal to the wave pressure at the seabed surface,

$$\begin{aligned} p &= \frac{\gamma_w H_s}{2\cosh kd} \cos nky \cos(mkx - \omega t) \\ &= p_o \cos nky \cos(mkx - \omega t) \quad \text{at } z = 0, \end{aligned} \tag{3.20}$$

where "$\cos nky \cos(mkx - \omega t)$" denotes the spatial and temporal variations in wave pressure within the three-dimensional short-crested wave field described above. p_o is the amplitude of the dynamic wave pressure based on the short-crested wave theory to a first-order.

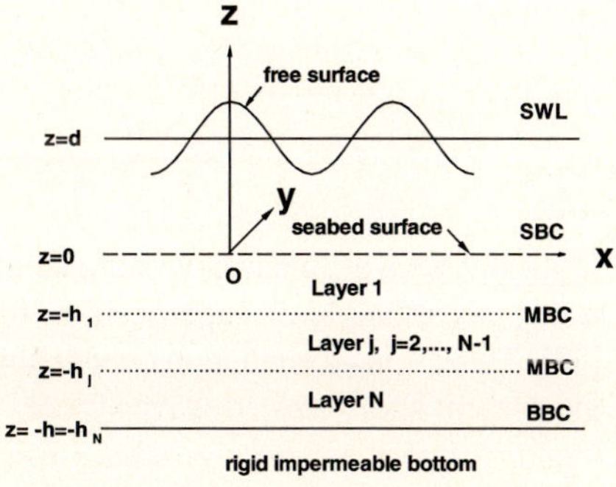

Fig. 3.3 A porous seabed with multiple sub-layers, showing different soil properties in each layer

(b) BBC: Boundary Condition at the Bottom ($z = -h$ or $z \to -\infty$) For the soil resting on an impermeable rigid bottom, zero displacements and no vertical flow occur at the horizontal bottom. For a seabed of infinite thickness,

$$u = v = w = p = 0 \quad \text{as } z \to -\infty, \tag{3.21}$$

For a seabed of finite thickness and in a layered medium,

$$u = v = w = 0, \quad \text{and} \quad \frac{\partial p}{\partial z} = 0 \quad \text{as } z = -h, \tag{3.22}$$

(c) MBC: Matching Boundary Conditions ($z = -h_j$) At the interface between any two consecutive layers in the seabed, the soil displacements, normal effective stresses and pore pressure should be concordant. Thus:

$$\begin{aligned} &p_j = p_{j+1}, \quad u_j = u_{j+1}, \quad v_j = v_{j+1}, \quad w_j = w_{j+1}, \quad \sigma'_{z,j} = \sigma'_{z,j+1}, \\ &\tau_{xz,j} = \tau_{xz,j+1}, \quad \tau_{yz} = \tau_{yz,j+1}, \quad K_{z,j}\frac{\partial p_j}{\partial z} = K_{z,j+1}\frac{\partial p_{j+1}}{\partial z}, \\ &\text{at } z = -h_j, \ j = 1, 2, 3, \ldots, N-1, \end{aligned} \tag{3.23}$$

in which subscript "j" denotes the jth sub-layer in a N-layered seabed.

3.4 General Solutions

The boundary value problem, describing the wave-seabed interaction, can be solved, based on the governing equations (3.7), (3.16)–(3.18) and boundary conditions (3.19)–(3.23). The analytical solution for the wave-induced pore pressure and soil displacements can first be obtained, from which, the effective stresses can then be found from (3.13)–(3.15). Closed-form analytical solutions for soil response in a porous seabed of infinite and finite thickness are provided, while a semi-analytical solution for a layered seabed is developed by solving the coefficient matrix numerically.

3.4.1 Basic Theoretical Framework

In the analysis of linear governing equations, it is expedient to perform the analysis using the principle of superposition for functions expressed in complex variables [22]. Thus, it is advantageous in dealing with complicated wave conditions, such as the present short-crested wave system. Since this wave system is produced by two identical linear progressive waves interacting at an angle to each other, the wave pressure at the seabed surface given by (3.6) can be expressed as the sum of the reflected $(mkx + nky - \omega t)$ and incident $(mkx - nky - \omega t)$ components, i.e.,

$$P = \frac{p_o}{2}\mathbf{Re}\left\{e^{i(mkx+nky-\omega t)} + e^{i(mkx-nky-\omega t)}\right\}, \tag{3.24}$$

where **Re** stands for the real part of the function and $i\ (=\sqrt{-1})$ denotes the complex variable. Only the real parts of the complex solution are utilized. For the sake of simplicity, this is applied to the real part of all complex solutions in the following analysis, unless specified otherwise.

Based on the principle of superposition, according to (3.24), the wave-induced pore pressure and soil displacements can be expressed as [12, 22]

$$\begin{Bmatrix} p \\ u \\ v \\ w \end{Bmatrix} = \begin{Bmatrix} P_1 \\ U_1 \\ V_1 \\ W_1 \end{Bmatrix} e^{i(mkx+nky-\omega t)} + \begin{Bmatrix} P_2 \\ U_2 \\ V_2 \\ W_2 \end{Bmatrix} e^{i(mkx-nky-\omega t)}. \tag{3.25}$$

The first parts (with subscript 1) on the right-hand side of (3.25) represent the contribution from the reflected component, while the second portions (with subscript 2) are for the incident component. The solutions of P, U, V and W for these two parts (i.e., P_1, U_1, V_1, W_1 and P_2, U_2, V_2, W_2) will be found separately and then superposed to yield the complete solution.

Substituting the first part (reflected waves) of (3.25) into (3.7) and (3.16)–(3.18), the final form of the governing equation for the soil displacement due to the reflected wave in the x-direction yields

$$\left(D^2 - k^2\right)^2\left(D^2 - \delta^2\right)U_1 = 0, \tag{3.26}$$

where $D\ (= d/dz)$ is the operator of differentiation. The coefficient δ is given by

$$\delta^2 = k^2 KMN - \frac{i\omega\gamma_w}{K_z}\left(n'\beta + \frac{1-2\mu}{2G(1-\mu)}\right), \tag{3.27}$$

in which $KMN = (K_x m^2 + K_y n^2)/K_z$.

The general solution of the sixth-order partial differential equation, (3.26), can be expressed in the following form:

$$U_1 = m\left\{(a_1 + a_2 z)\, e^{kz} + (a_3 + a_4 z)\, e^{-kz} + a_5\, e^{\delta z} + a_6\, e^{-\delta z}\right\}. \tag{3.28}$$

The six coefficients required for (3.28) can be determined by satisfying the boundary conditions specified in (3.19)–(3.23). Based on this framework, the analytical procedures and solutions for the wave-induced soil response in a porous seabed of infinite, finite thickness and layered seabed are presented systematically in the following subsections.

3.4.2 Soil Response in a Seabed of Infinite Thickness

Considering a porous seabed of infinite thickness, in which the bottom boundary condition at the impermeable rigid stratum, $z \to -\infty$, gives

$$a_3 = a_4 = a_6 = 0, \tag{3.29}$$

from which (3.28) can be simplified to

$$U_1 = m\{(a_1 + a_2 z)\, e^{kz} + a_5\, e^{\delta z}\}. \tag{3.30}$$

The soil displacement components in the y- and z-directions can also be obtained by a similar procedure. After some algebraic processes, this yields

$$V_1 = (b_1 + n a_2 z)\, e^{kz} + n a_5\, e^{\delta z}, \tag{3.31}$$

and

$$W_1 = -i\left\{\left(m^2 a_1 + n b_1 - \frac{1+2\lambda}{k} a_2 + a_2 z\right) e^{kz} + \frac{\delta}{k} a_5\, e^{\delta z}\right\}, \tag{3.32}$$

where

$$\lambda = \frac{(1-2\mu)\{k^2(1-KMN) + \frac{i\omega\gamma_w n'\beta}{K_z}\}}{k^2(1-KMN) + \frac{i\omega\gamma_w}{K_z}(n'\beta + \frac{1-2\mu}{G})}. \tag{3.33}$$

The remaining four unknowns, a_1, a_2, a_5 and b_1, in (3.30)–(3.32) are determined by the boundary conditions at the seabed surface ($z = 0$). Employing the same procedure as for the first part with subscript "1" above, solutions for the incident wave components (with subscript "2") can also be found.

Upon combining these solutions, it renders the complete expressions for the soil displacements as

$$u = \frac{imp_o}{2G}\left[(C_0^\infty + C_1^\infty z)\, e^{kz} + C_2^\infty\, e^{\delta z}\right] \cos nky\; e^{i(mkx-\omega t)}, \tag{3.34}$$

$$v = \frac{-np_o}{2G}\left[(C_0^\infty + C_1^\infty z)\, e^{kz} + C_2^\infty\, e^{\delta z}\right] \sin nky\; e^{i(mkx-\omega t)}, \tag{3.35}$$

$$w = \frac{p_o}{2G}\left[\left(C_0^\infty - \frac{1+2\lambda}{k} C_1^\infty + C_1^\infty z\right) e^{kz} + \frac{\delta}{k} C_2^\infty\, e^{\delta z}\right] \cos nky\; e^{i(mkx-\omega t)}, \tag{3.36}$$

where the superscripts "∞" denote the coefficients in the solution for a porous seabed of infinite thickness. The parameters δ and λ are coupled with soil properties and wave characteristics, as given by (3.27) and (3.33), respectively.

The wave-induced pore pressure is given by

$$p = \frac{p_o}{1-2\mu}\left[(1-2\mu-\lambda)C_1^\infty e^{kz} + \frac{\delta^2-k^2}{k}(1-\mu)C_2^\infty e^{\delta z}\right]\cos nky\, e^{i(mkx-\omega t)}, \tag{3.37}$$

where the effect of wave obliquity in terms of δ and λ can also be observed.

The effective normal stresses can be determined from (3.13)–(3.15), rendering

$$\sigma'_x = -p_o\left\{\left[m^2(C_0^\infty + C_1^\infty z) + \frac{2\mu\lambda}{1-2\mu}C_1^\infty\right]e^{kz} + \left(k^2m^2 - \frac{\mu(\delta^2-k^2)}{k(1-2\mu)}\right)C_2^\infty e^{\delta z}\right\}\cos nky\, e^{i(mkx-\omega t)}, \tag{3.38}$$

$$\sigma'_y = -p_o\left\{\left[n^2(C_0^\infty + C_1^\infty z) + \frac{2\mu\lambda}{1-2\mu}C_1^\infty\right]e^{kz} + \left(k^2n^2 - \frac{\mu(\delta^2-k^2)}{k(1-2\mu)}\right)C_2^\infty e^{\delta z}\right\}\cos nky\, e^{i(mkx-\omega t)}, \tag{3.39}$$

$$\sigma'_z = p_o\left\{\left[kC_0^\infty + C_1^\infty kz - \frac{2\lambda(1-\mu)}{1-2\mu}C_1^\infty\right]e^{kz} + \frac{1}{k(1-2\mu)}(\delta^2(1-\mu) - k^2\mu)C_2^\infty e^{\delta z}\right\}\cos nky\, e^{i(mkx-\omega t)}, \tag{3.40}$$

and the shear stresses are given by

$$\tau_{xz} = imp_o\left[\left(kC_0^\infty + (kz-\lambda)C_1^\infty\right)e^{kz} + \delta C_2^\infty e^{\delta z}\right]\cos nky\, e^{i(mkx-\omega t)}, \tag{3.41}$$

$$\tau_{yz} = -np_o\left[\left(kC_0^\infty + (kz-\lambda)C_1^\infty\right)e^{kz} + \delta C_2^\infty e^{\delta z}\right]\sin nky\, e^{i(mkx-\omega t)}, \tag{3.42}$$

$$\tau_{xy} = -imnkp_o\left[\left(C_0^\infty + C_1^\infty z\right)e^{kz} + C_2^\infty e^{\delta z}\right]\sin nky\, e^{i(mkx-\omega t)}. \tag{3.43}$$

In (3.34)–(3.43), the C_i^∞ coefficients are given by

$$C_0^\infty = \frac{-\lambda[\mu(\delta-k)^2 - \delta(\delta-2k)]}{k(\delta-k)(\delta-\delta\mu+k\mu+k\lambda)}, \tag{3.44}$$

$$C_1^\infty = \frac{\delta-\delta\mu+k\mu}{\delta-\delta\mu+k\mu+k\lambda}, \tag{3.45}$$

$$C_2^\infty = \frac{k\lambda}{(\delta-k)(\delta-\delta\mu+k\mu+k\lambda)}. \tag{3.46}$$

The solutions presented above can be found in [12].

3.4.3 Soil Response in a Porous Seabed of Finite Thickness

Employing the solution procedure outlined in Sect. 3.4.1, the soil displacements and pore pressure are expressed with eight unknown coefficients ($a_1 - a_8$) as follows:

$$U_1 = m\{(a_1 + a_2 z)\, e^{kz} + (a_3 + a_4 z)\, e^{-kz} + a_5\, e^{\delta z} + a_6\, e^{-\delta z}\}, \tag{3.47}$$

$$V_1 = \{(a_7 + n a_2 z)\, e^{kz} + (a_8 + n a_4 z)\, e^{-kz} + n a_5\, e^{\delta z} + n a_6\, e^{-\delta z}\}, \tag{3.48}$$

$$\begin{aligned} W_1 = -i\Bigg\{&\left[a_1 + \left(\frac{1+2\lambda}{k} - z\right)a_2\right] e^{kz} \\ &- \left[a_3 + \left(\frac{1+2\lambda}{k} + z\right)a_4\right] e^{-kz} + \delta\big(a_5\, e^{\delta z} - a_6\, e^{-\delta z}\big)\Bigg\}, \end{aligned} \tag{3.49}$$

and

$$\begin{aligned} P_1 = \frac{-2iG}{1-2\mu}\{&(1-\lambda-2\mu)\big(a_2\, e^{kz} - a_4\, e^{-kz}\big) \\ &+ (1-\mu)\big(\delta^2 - k^2\big)\big(a_5\, e^{\delta z} + a_6\, e^{-\delta z}\big)\}, \end{aligned} \tag{3.50}$$

where δ and λ are given by (3.27) and (3.33), respectively.

The eight unknown coefficients a_j ($j = 1$–8) can be determined using boundary conditions of (3.19)–(3.21), which include eight equations. After combining the solution of the reflected and incident wave components, the complete expressions for soil displacements are given as

$$\begin{aligned} u = \frac{imp_o}{2Gk}\{&(C_1 + C_2 kz)\, e^{kz} + (C_3 + C_4 kz)\, e^{-kz} \\ &+ k^2\big(C_5\, e^{\delta z} + C_6\, e^{-\delta z}\big)\}\cos nky\; e^{i(mkx-\omega t)}, \end{aligned} \tag{3.51}$$

$$\begin{aligned} v = \frac{-np_o}{2Gk}[&(C_1 + C_2 kz)\, e^{kz} + (C_3 + C_4 kz)\, e^{-kz} \\ &+ k^2\big(C_5\, e^{\delta z} + C_6\, e^{-\delta z}\big)]\sin nky\; e^{i(mkx-\omega t)}, \end{aligned} \tag{3.52}$$

$$\begin{aligned} w = \frac{p_o}{2Gk}\{&\big[C_1 - (1+2\lambda - kz)C_2\big] e^{kz} - \big[C_3 + (1+2\lambda + kz)C_4\big] e^{-kz} \\ &+ k\delta\big(C_5\, e^{\delta z} - C_6\, e^{-\delta z}\big)\}\cos nky\; e^{i(mkx-\omega t)}. \end{aligned} \tag{3.53}$$

The wave-induced pore pressure is given by

$$\begin{aligned} p = \frac{p_o}{1-2\mu}\{&(1-\lambda-2\mu)\big(C_2\, e^{kz} - C_4\, e^{-kz}\big) \\ &+ (1-\mu)\big(\delta^2 - k^2\big)\big(C_5\, e^{\delta z} + C_6\, e^{-\delta z}\big)\}\cos nky\; e^{i(mkx-\omega t)}. \end{aligned} \tag{3.54}$$

Again, from the stress-strain relationship, the effective normal stresses can be expressed as

$$\sigma_x' = -p_o\left\{\left[m^2(C_1 + C_2kz) + \frac{2\mu\lambda}{1-2\mu}C_2\right]e^{kz}\right.$$
$$+\left[m^2(C_3 + C_4kz) - \frac{2\mu\lambda}{1-2\mu}C_4\right]e^{-kz}$$
$$\left.+\left[k^2m^2 - \frac{\mu(\delta^2-k^2)}{1-2\mu}\right]\left(C_5\,e^{\delta z} + C_6\,e^{-\delta z}\right)\right\}\cos nky\; e^{i(mkx-\omega t)}, \qquad (3.55)$$

$$\sigma_y' = -p_o\left\{\left[n^2(C_1 + C_2kz) + \frac{2\mu\lambda}{1-2\mu}C_2\right]e^{kz}\right.$$
$$+\left[n^2(C_3 + C_4kz) - \frac{2\mu\lambda}{1-2\mu}C_4\right]e^{-kz}$$
$$\left.+\left[k^2n^2 - \frac{\mu(\delta^2-k^2)}{1-2\mu}\right]\left(C_5\,e^{\delta z} + C_6\,e^{-\delta z}\right)\right\}\cos nky\; e^{i(mkx-\omega t)}, \qquad (3.56)$$

$$\sigma_z' = p_o\left\{\left[C_1 + C_2kz - \frac{2\lambda(1-\mu)}{1-2\mu}C_2\right]e^{kz} + \left[C_3 + C_4kz - \frac{2\lambda(1-\mu)}{1-2\mu}C_4\right]e^{-kz}\right.$$
$$\left.+\frac{1}{1-2\mu}\left[\delta^2(1-\mu) - k^2\mu\right]\left(C_5\,e^{\delta z} + C_6\,e^{-\delta z}\right)\right\}\cos nky\; e^{i(mkx-\omega t)}, \qquad (3.57)$$

and the shear stresses are given by

$$\tau_{xz} = imp_o\left\{\left[C_1 + (kz-\lambda)C_2\right]e^{kz} - \left[C_3 + (kz+\lambda)C_4\right]e^{-kz}\right.$$
$$\left.+k\delta\left(C_5\,e^{\delta z} - C_6\,e^{-\delta z}\right)\right\}\cos nky\; e^{i(mkx-\omega t)}, \qquad (3.58)$$
$$\tau_{yz} = -np_o\left\{\left[C_1 + (kz-\lambda)C_2\right]e^{kz} - \left[C_3 + (kz+\lambda)C_4\right]e^{-kz}\right.$$
$$\left.+k\delta\left(C_5\,e^{\delta z} - C_6\,e^{-\delta z}\right)\right\}\sin nky\; e^{i(mkx-\omega t)}, \qquad (3.59)$$
$$\tau_{xy} = -imnp_o\left\{(C_1 + C_2kz)\,e^{kz} + (C_3 + C_4kz)\,e^{-kz}\right.$$
$$\left.+k^2\left(C_5e^{\delta z} + C_6\,e^{\delta z}\right)\right\}\sin nky\; e^{i(mkx-\omega t)}, \qquad (3.60)$$

where the complete expressions for the six resulting coefficients, C_1 to C_6, are systematically presented in Sect. 3.8. The analytical solutions presented in this section can be found in [10, 17].

As pointed out by Sumer [39], there were two typos in the coefficients (C_{11} and C_{32}) originally presented in [10]. Although the influences of these typos on the numerical results are rather minor, these have been corrected in the coefficients listed in Sect. 3.8.

3.4.4 Soil Response in a Layered Seabed

Employing the solution procedure outlined in Sect. 3.4.1, and the principle of superposition to incorporate the solutions produced by incident and reflected wave components, a general solution for the excess pore pressure and soil displacements in the jth sub-layer of a N-layered soil column may be given by

$$\begin{Bmatrix} u_j \\ v_j \\ w_j \\ p_j \end{Bmatrix} = \sum_{s=1}^{2} \begin{Bmatrix} U_{js} \\ V_{js} \\ W_{js} \\ P_{js} \end{Bmatrix} e^{i(mkx+(-1)^{s-1}nky-\omega t)}, \tag{3.61}$$

where the subscript "s" on the right-hand-side of (3.61) represents the contribution from the incident ($s=2$) or reflected ($s=1$) wave components. The solutions of u_j for these two parts render the final form for the governing equations (3.28).

Consequently, the general solutions for the soil displacements and pore pressure, including $8N$ unknown coefficients for the jth sub-layer in a N-layered seabed, are given as

$$u_j = \sum_{s=1}^{2} m\{(a_{js1}+a_{js2}z)\,e^{kz} + (a_{js3}+a_{js4}z)\,e^{-kz} + a_{js5}\,e^{\delta_j z} + a_{js6}\,e^{-\delta_j z}\}\,e^{i(mkx+(-1)^{s-1}nky-\omega t)}, \tag{3.62}$$

$$v_j = \sum_{s=1}^{2} \{(a_{js7}+(-1)^{s-1}na_{js2}z)\,e^{kz} + (a_{js8}+(-1)^{s-1}na_{js4}z)\,e^{-kz} + (-1)^{s-1}n(a_{js5}\,e^{\delta_j z} + a_{js6}\,e^{-\delta_j z})\}e^{i(mkx+(-1)^{s-1}nky-\omega t)}, \tag{3.63}$$

$$w_j = \sum_{s=1}^{2} \Bigg\{\left[-m^2 a_{js1} - (-1)^{s-1}na_{js7} + \frac{1+2\lambda_j-kz}{k}a_{js2}\right]e^{kz} + \left[-m^2 a_{js3} - (-1)^{s-1}na_{js8} + \frac{1+2\lambda_j-kz}{k}a_{js4}\right]e^{-kz} - \frac{\delta_j}{k}(a_{js5}\,e^{\delta_j z} + a_{js6}\,e^{-\delta_j z})\Bigg\}\,e^{i(mkx+(-1)^{s-1}nky-\omega t)}, \tag{3.64}$$

where δ_j and λ_j are given by (3.27) and (3.33), respectively, but now with subscript j.

Similarly, the effective normal stresses in the jth sub-layer can be expressed as

$$\sigma'_{x,j} = \sum_{s=1}^{2} 2iG_j\Bigg\{\left[km^2(a_{js1}+a_{js2}z) + \frac{2\lambda_j\mu_j}{1-2\mu_j}a_{js2}\right]e^{kz}$$

$$
+\left[km^2(a_{js3}+a_{js4}z)-\frac{2\lambda_j\mu_j}{1-2\mu_j}a_{js4}\right]e^{-kz}
$$

$$
+\left[km^2-\frac{\mu_j(\delta_j^2-k^2)}{k(1-2\mu_j)}\right]\left(a_{js5}\,e^{\delta z}+a_{js6}\,e^{-\delta z}\right)\Bigg\}\,e^{i(mkx+(-1)^{s-1}nky-\omega t)}, \tag{3.65}
$$

$$
\sigma'_{y,j}=\sum_{s=1}^{2}2iG_j\Bigg\{\left[(-1)^{s-1}kn\big(a_{js7}+kn^2a_{js2}z\big)+\frac{2\lambda_j\mu_j}{1-2\mu_j}a_{js2}\right]e^{kz}
$$

$$
+\left[(-1)^{s-1}kn\big(a_{js8}+kn^2a_{js2}z\big)-\frac{2\lambda_j\mu_j}{1-2\mu_j}a_{js2}\right]e^{-kz}
$$

$$
+\left[kn^2-\frac{\mu_j(\delta_j^2-k^2)}{k(1-2\mu_j)}\right]\left(a_{js5}\,e^{\delta z}+a_{js6}\,e^{-\delta z}\right)\Bigg\}e^{i(mkx+(-1)^{s-1}nky-\omega t)}, \tag{3.66}
$$

$$
\sigma'_{z,j}=\sum_{s=1}^{2}2iG_j\Bigg\{\left[k\big(m^2a_{js1}+(-1)^{s-1}na_{js7}+a_{js2}z\big)+\frac{2\lambda_j(1-\mu_j)}{1-2\mu_j}a_{js2}\right]e^{kz}
$$

$$
+\left[k\big(m^2a_{js3}+(-1)^{s-1}na_{js8}+a_{js4}z\big)+\frac{2\lambda_j(1-\mu_j)}{1-2\mu_j}a_{js4}\right]e^{-kz}
$$

$$
-\frac{1}{1-2\mu_j}\left[\delta_j^2(1-\mu_j)-k^2\mu_j\right]
$$

$$
\times\left(a_{js5}\,e^{\delta_j z}+a_{js6}\,e^{-\delta_j z}\right)\Bigg\}e^{i(mkx+(-1)^{s-1}nky-\omega t)}, \tag{3.67}
$$

and the shear stresses are given by

$$
\tau_{xz,j}=\sum_{s=1}^{2}mG_j\big\{\big[ka_{js1}+m^2ka_{js2}+(-1)^{s-1}nka_{js7}+2(kz-\lambda_j)a_{js2}\big]e^{kz}
$$

$$
-\big[ka_{js3}+m^2ka_{js3}+(-1)^{s-1}nka_{js8}+2(kz+\lambda_j)a_{js4}\big]e^{-kz}
$$

$$
+2\delta_j\big(a_{js5}\,e^{\delta_j z}-a_{js6}\,e^{-\delta_j z}\big)\big\}\,e^{i(mkx+(-1)^{s-1}nky-\omega t)}, \tag{3.68}
$$

$$
\tau_{yz,j}=\sum_{s=1}^{2}G_j\big\{\big[ka_{js1}+n^2ka_{js7}+(-1)^{s-1}m^2nka_{js2}
$$

$$
+(-1)^{s-1}2n(kz-\lambda_j)a_{js2}\big]e^{kz}
$$

$$
-\big[ka_{js3}+n^2ka_{js8}+(-1)^{s-1}m^2nka_{js3}+(-1)^{s-1}2n(kz+\lambda_j)a_{js4}\big]e^{-kz}
$$

$$
+(-1)^{s-1}2n\delta_j\big(a_{js5}\,e^{\delta_j z}-a_{js6}\,e^{-\delta_j z}\big)\big\}\,e^{i(mkx+(-1)^{s-1}nky-\omega t)}, \tag{3.69}
$$

$$\tau_{xy,j} = \sum_{s=1}^{2} imkG_j\{\left[a_{js7} + (-1)^{s-1}na_{js1} + (-1)^{s-1}2na_{js2}z\right]e^{kz}$$
$$+ \left[a_{js8} + (-1)^{s-1}na_{js3} + (-1)^{s-1}2na_{js4}z\right]e^{-kz}$$
$$+ (-1)^{s-1}2n\left(a_{js5}e^{\delta_j z} + a_{js6}e^{-\delta_j z}\right)\}e^{i(mkx+(-1)^{s-1}nky-\omega t)}. \tag{3.70}$$

In the equations above, a $8N \times 8N$ symbolic solution matrix **F** is established for the $8N$ unknown coefficients for a N-layered soil seabed. The unknown coefficients can be obtained numerically from the matrix form

$$\mathbf{F} \bullet \mathbf{A} = \mathbf{B}, \tag{3.71}$$

where **A** is a column matrix containing the unknown a_{ij} coefficients within each sub-layer, **B** is also a column matrix representing the resultant boundary values to be satisfied, such that

$$\mathbf{A} = \{a_{js1}, a_{js2}, a_{js3}, a_{js4}, a_{js5}, a_{js6}, a_{js7}, a_{js8}\}^T_{1\times 8N}, \quad \text{for } j = 1\text{–}N, \tag{3.72}$$

$$\mathbf{B} = \{p_o, 0, 0, \ldots, 0, 0\}^T_{8N\times 1}, \tag{3.73}$$

and

$$\mathbf{F} = \{f_{ij}\}^T_{8N\times 8N}, \quad \text{for } i = 1\text{–}N, j = 1\text{–}N. \tag{3.74}$$

The coefficient matrix **F** is available in [11]. After solving the matrix numerically and inserting the unknown variables into the relevant equations, numerical results for pore pressure, soil displacements and effective stresses can be further computed.

3.4.5 *Limiting Two-Dimensional Conditions*

Since no exact solutions have been presented explicitly for a porous seabed of finite thickness for a two-dimensional wave condition, it is desirable to present such solutions for practical application. Variation in the incident angle θ renders the two-dimensional limiting cases, i.e. $\theta = 0°$ for standing waves and $\theta = 90°$ for progressive waves (Fig. 3.1). The present general three-dimensional solutions for a seabed of finite thickness under hydraulically anisotropic and unsaturated conditions can be simplified to yield the solutions for the limiting two-dimensional cases. For convenience, only the solution for a seabed of finite thickness is presented here.

3.4.5.1 Progressive Waves

Upon substituting $m = 1$ and $n = 0$ (for $\theta = 90°$) into (3.27) and (3.33), the expressions for δ and λ become

$$\delta^2 = k^2\frac{K_x}{K_z} - \frac{i\omega\gamma_w}{K_z}\left(n'\beta + \frac{1-2\mu}{2G(1-\mu)}\right), \tag{3.75}$$

$$\lambda = \frac{(1-2\mu)\{k^2(1-\frac{K_x}{K_z}) + \frac{i\omega\gamma_w n'\beta}{K_z}\}}{k^2(1-\frac{K_x}{K_z}) + \frac{i\omega\gamma_w}{K_z}(n'\beta + \frac{1-2\mu}{G})}, \tag{3.76}$$

and the soil displacements in a porous seabed of finite thickness are

$$u = \frac{ip_o}{2Gk}\big[(C_1 + C_2 kz)\,e^{kz} + (C_3 + C_4 kz)\,e^{-kz} + k^2\big(C_5\,e^{\delta z} + C_6\,e^{-\delta z}\big)\big]\,e^{i(kx-\omega t)}, \tag{3.77}$$

$$w = \frac{p_o}{2Gk}\{\big[C_1 - (1+2\lambda - kz)C_2\big]e^{kz} - \big[C_3 + (1+2\lambda+kz)C_4\big]e^{-kz} + k\delta\big(C_5\,e^{\delta z} - C_6\,e^{-\delta z}\big)\}\,e^{i(kx-\omega t)}. \tag{3.78}$$

The wave-induced pore pressure is given by

$$p = \frac{p_o}{1-2\mu}\{(1-\lambda-2\mu)\big(C_2\,e^{kz} - C_4\,e^{-kz}\big) + (1-\mu)\big(\delta^2 - k^2\big)\big(C_5\,e^{\delta z} + C_6\,e^{-\delta z}\big)\}\,e^{i(kx-\omega t)}, \tag{3.79}$$

and the effective normal and shear stresses are

$$\sigma_x' = -p_o\Bigg\{\Bigg[(C_1 + C_2 kz) + \frac{2\mu\lambda}{1-2\mu}C_2\Bigg]e^{kz} + \Bigg[(C_3 + C_4 kz) - \frac{2\mu\lambda}{1-2\mu}C_4\Bigg]e^{-kz} + \Bigg[k^2 - \frac{\mu(\delta^2-k^2)}{1-2\mu}\Bigg]\Big(C_5\,e^{\delta z} + C_6\,e^{-\delta z}\Big)\Bigg\}\,e^{i(kx-\omega t)}, \tag{3.80}$$

$$\sigma_z' = p_o\Bigg\{\Bigg[C_1 + C_2 kz - \frac{2\lambda(1-\mu)}{1-2\mu}C_2\Bigg]e^{kz} + \Bigg[C_3 + C_4 kz - \frac{2\lambda(1-\mu)}{1-2\mu}C_4\Bigg]e^{-kz} + \frac{1}{1-2\mu}\big[\delta^2(1-\mu) - k^2\mu\big]\Big(C_5\,e^{\delta z} + C_6\,e^{-\delta z}\Big)\Bigg\}\,e^{i(kx-\omega t)}, \tag{3.81}$$

$$\tau_{xz} = ip_o\{\big[C_1 + (kz-\lambda)C_2\big]e^{kz} - \big[C_3 + (kz+\lambda)C_4\big]e^{-kz} + k\delta\big(C_5\,e^{\delta z} - C_6\,e^{-\delta z}\big)\}\,e^{i(kx-\omega t)}, \tag{3.82}$$

while σ_y', τ_{yz} and τ_{xy} are equal to zero. The six C_i coefficients are listed in Sect. 3.8.

Since (3.77)–(3.82) presented in this study for soil response in a seabed of finite thickness under the action of a two-dimensional progressive wave are new, it is necessary to verify them with other results derived directly for two-dimensional cases, such as [49], which were developed semi-analytically with all unknown coefficients to be calculated numerically from a 6×6 matrix. Comparison with these and other theoretical and experimental results available is presented in Sect. 3.5.

3.4.5.2 Standing Waves

From the three-dimensional general solutions developed in this study, the following expressions may be deduced for the case of standing waves. Substitution of $m = 0$

and $n = 1$ (for $\theta = 0°$) into (3.27) and (3.33), the δ and λ parameters become

$$\delta^2 = k^2 \frac{K_y}{K_z} - \frac{i\omega\gamma_w}{K_z}\left(n'\beta + \frac{1-2\mu}{2G(1-\mu)}\right), \tag{3.83}$$

$$\lambda = \frac{(1-2\mu)\{k^2(1-\frac{K_y}{K_z}) + \frac{i\omega\gamma_w n'\beta}{K_z}\}}{k^2(1-\frac{K_y}{K_z}) + \frac{i\omega\gamma_w}{K_z}[n'\beta + \frac{1-2\mu}{G}]}, \tag{3.84}$$

and the soil displacements in a porous seabed are given by

$$v = \frac{-p_o}{2Gk}\big[(C_1 + C_2 kz)\,e^{kz} + (C_3 + C_4 kz)\,e^{-kz} + k^2\big(C_5\,e^{\delta z} + C_6\,e^{-\delta z}\big)\big] \sin ky\; e^{-i\omega t}, \tag{3.85}$$

$$w = \frac{p_o}{2Gk}\big\{\big[C_1 - (1+2\lambda - kz)C_2\big]e^{kz} - \big[C_3 + (1+2\lambda+kz)C_4\big]e^{-kz} + k\delta\big(C_5\,e^{\delta z} - C_6\,e^{-\delta z}\big)\big\} \cos ky\; e^{-i\omega t}. \tag{3.86}$$

The wave-induced pore pressure is given by

$$p = \frac{p_o}{1-2\mu}\big\{(1-\lambda-2\mu)\big(C_2\,e^{kz} - C_4\,e^{-kz}\big) + (1-\mu)\big(\delta^2 - k^2\big)\big(C_5\,e^{\delta z} + C_6\,e^{-\delta z}\big)\big\} \cos ky\; e^{-i\omega t}, \tag{3.87}$$

and the effective normal stresses are

$$\sigma'_y = -p_o\left\{\left[(C_1 + C_2 kz) + \frac{2\mu\lambda}{1-2\mu}C_2\right]e^{kz} + \left[(C_3 + C_4 kz) - \frac{2\mu\lambda}{1-2\mu}C_4\right]e^{-kz} + \left[k^2 - \frac{\mu(\delta^2-k^2)}{1-2\mu}\right]\big(C_5\,e^{\delta z} + C_6\,e^{-\delta z}\big)\right\} \cos ky\; e^{-i\omega t}, \tag{3.88}$$

$$\sigma'_z = p_o\left\{\left[C_1 + C_2 kz - \frac{2\lambda(1-\mu)}{1-2\mu}C_2\right]e^{kz} + \left[C_3 + C_4 kz - \frac{2\lambda(1-\mu)}{1-2\mu}C_4\right]e^{-kz} + \frac{1}{1-2\mu}\big[\delta^2(1-\mu) - k^2\mu\big]\big(C_5\,e^{\delta z} + C_6\,e^{-\delta z}\big)\right\} \cos ky\; e^{-i\omega t}, \tag{3.89}$$

and the shear stresses are given by

$$\tau_{yz} = -p_o\big\{\big[C_1 + (kz-\lambda)C_2\big]e^{kz} - \big[C_3 + (kz+\lambda)C_4\big]e^{-kz} + k\delta\big(C_5\,e^{\delta z} - C_6\,e^{-\delta z}\big)\big\} \sin ky\; e^{-i\omega t}, \tag{3.90}$$

while σ'_x, τ_{xz} and τ_{xy} are equal to zero. Again, the six C_i coefficients are listed in Sect. 3.8.

3.4.6 A Special Case: Fully Saturated Isotropic Seabed of Infinite Thickness

For a seabed of infinite thickness in hydraulically isotropic conditions ($K_x = K_y = K_z$), subject to a short-crested wave, (3.27) is reduced to

$$\delta^2 = k^2 - \frac{i\omega\gamma_w}{K_z}\left(n'\beta + \frac{1-2\mu}{2G(1-\mu)}\right). \tag{3.91}$$

Equation (3.91) is the same as λ' in [49] (equation (14) in his paper), in which the second part on the right hand side of (3.91) is equivalent to his $\omega c/\lambda$. Physically, this parameter was described as the ratio of the diffusion of water in soil to the wave diffusion, ω/k^2.

For a soil matrix under fully saturated conditions, (3.91) can be further simplified, in which $G\beta \to 0$ is generally assumed, such that,

$$\delta^2 = k^2 - \frac{i\omega\gamma_w}{K_z}\frac{1-2\mu}{2G(1-\mu)}. \tag{3.92}$$

The δ in (3.92) may be interpreted as the modified wave number due to wave action on a saturated and isotropic seabed. Under this specific condition, parameter $\lambda = 0$ in (3.33).

Under saturated and isotropic conditions, final expressions for soil response are obtainable simply by inserting $\lambda = 0$ into (3.77)–(3.82) for the limiting progressive waves, and into (3.85)–(3.90) for the standing waves. Moreover, the C_j and C_{ij} coefficients in Sect. 3.8 can also be simplified systematically using the first six non-zeros B_i coefficients only. This is because the remaining 18 B_j coefficients (for $j = 7$–24) have no contribution to the C_{ij} coefficients, as they are multiplied by the parameter $\lambda = 0$ (Sect. 3.8).

For a fully saturated and isotropic seabed of infinite thickness, the wave-induced soil displacements and pore pressure can be obtained by reduction from (3.34)–(3.37) as follows

$$\frac{2Gku}{p_o} = imkz\, e^{kz}\cos nky\; e^{i(mkx-\omega t)}, \tag{3.93}$$

$$\frac{2Gkv}{p_o} = -nkz\, e^{kz}\sin nky\; e^{i(mkx-\omega t)}, \tag{3.94}$$

$$\frac{2Gkw}{p_o} = (kz-1)\, e^{kz}\cos nky\; e^{i(mkx-\omega t)}, \tag{3.95}$$

and

$$p = p_o\, e^{kz}\cos nky\; e^{i(mkx-\omega t)}. \tag{3.96}$$

The effective normal stresses are

$$\sigma'_x = -m^2 p_o kz\, e^{kz} \cos nky\; e^{i(mkx-\omega t)}, \tag{3.97}$$

$$\sigma'_y = -n^2 p_o kz\, e^{kz} \cos nky\; e^{i(mkx-\omega t)}, \tag{3.98}$$

$$\sigma'_z = -p_o kz\, e^{kz} \cos nky\; e^{i(mkx-\omega t)}, \tag{3.99}$$

and the components of shear stresses are

$$\tau_{xz} = imp_o kz\, e^{kz} \cos nky\; e^{i(mkx-\omega t)}, \tag{3.100}$$

$$\tau_{yz} = -np_o kz\, e^{kz} \sin nky\; e^{i(mkx-\omega t)}, \tag{3.101}$$

$$\tau_{xy} = -imnp_o kz\, e^{kz} \sin nky\; e^{i(mkx-\omega t)}. \tag{3.102}$$

From (3.93)–(3.102), reduced for a fully saturated isotropic seabed of infinite thickness, it is found that the wave-induced effective stresses and pore pressure depend only on wave characteristics, but being independent of soil properties, because δ and λ are not included in the solutions. These formulae have been widely used in the field of marine geotechnical engineering [14].

3.5 Verification

Since the present three-dimensional analytical solutions are new and relevant theoretical and experimental investigations are not available, it is necessary to verify them against previous experimental data, analytical solutions for two-dimensional waves, and numerical models for two-dimensional and three-dimensional wave conditions.

3.5.1 Comparison with Two-Dimensional Experimental Data

Since experimental investigations for the three-dimensional short-crested wave-induced pore pressure have not been available in the literature, verification of the present theory is restricted to its reduced form of two-dimensional case of a single layer of finite thickness.

Comparisons with five sets of two-dimensional experimental data [23, 44] are demonstrated here. Physical variables in the figures are non-dimensionalized with respect to the amplitude of wave pressure p_o given by (3.20).

Comparison with experimental data of Tsui and Helfrich [44] is presented in Fig. 3.4, together with the analytical solutions of Putnam [29] and Liu [20]. In the figure, experimental data from Tsui and Helfrich [44] in dots "o" for dense sand

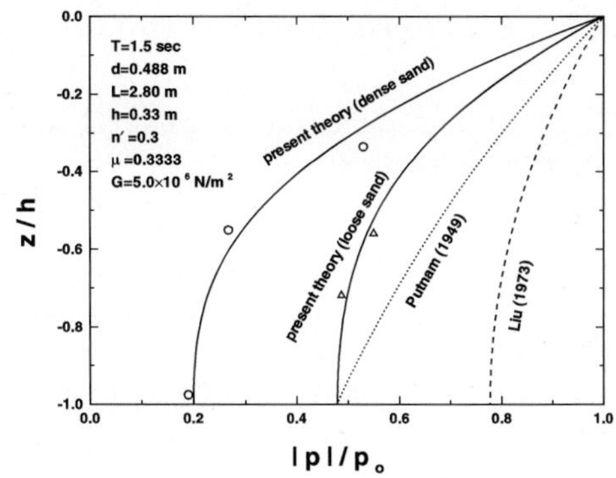

Fig. 3.4 Vertical distributions of the maximum $|p|/p_o$

($K_z = 5.0 \times 10^{-4}$ m/s, $S_r = 0.985$), in "Δ" for loose sand ($K_z = 5.0 \times 10^{-3}$ m/s, $S_r = 0.988$), present theory in solid lines. The solution of Liu [20] in dashed line and Putnam [29] in dotted line. The present theory is in good agreement with the experimental data for both loose and dense sand, with an appropriate value assigned to the degree of saturation S_r (in caption of Fig. 3.4). This implies that the present theory provides a better fit to the wave-induced pore pressures than the previous solutions [20, 29].

Further comparisons using the experimental data of Maeno and Hasegawa [23] are shown in Fig. 3.5, again with the previous theoretical predictions [20, 29]. These figures represent the original experimental data, obtained from different wave conditions. Since the values of the degree of saturation were not mentioned by the original authors, their values are assigned by the writer for the best fit to the data. These figures clearly show that the present solution for soil of finite thickness agrees well with the experimental data recorded and is better than that of [20, 29] (Fig. 3.5). The sensitivity of the degree of saturation in affecting the pore pressure response will be discussed in Sect. 3.6.2.2.

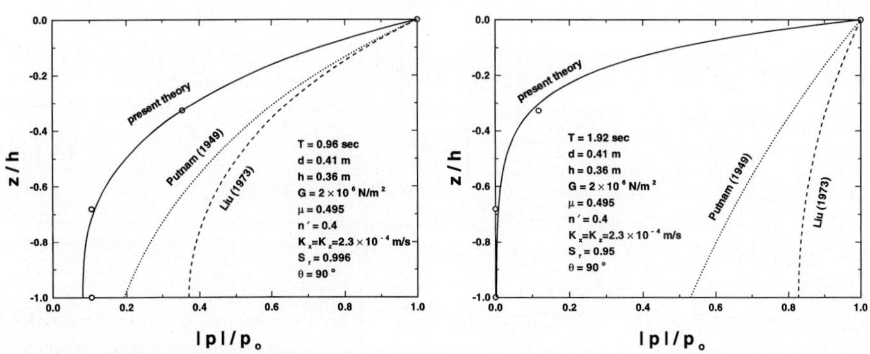

Fig. 3.5 Vertical distributions of the maximum $|p|/p_o$. Experimental data from Maeno and Hasegawa [23] in dots "∘", present theory in *solid line*, the solution of Liu [20] in *dashed line* and Putnam [29] in *dotted line*

3.5.2 Comparison with Two-Dimensional Analytical Solutions

Since the three-dimensional analytical solutions have not been available, the only possible verification is against previous two-dimensional analytical solutions for a seabed of finite thickness. The relevant two-dimensional solutions available include:

- the analytical solution of Madsen [22] for a hydraulically anisotropic seabed of infinite thickness under unsaturated conditions,
- the semi-analytical solution of Yamamoto [49] and the boundary-layer approximation of Mei and Foda [25] for a hydraulically isotropic seabed of finite thickness.

3.5.2.1 Comparison with Madsen [22]

The simplified solution for soil response in a porous seabed of infinite thickness presented in Sect. 3.4.2 is compared with the analytical solution for progressive waves developed by Madsen [22] under hydraulically anisotropic and unsaturated conditions. The mathematical expressions to be discussed are (34)–(42) in [22] and the simplified form of (3.34)–(3.41) (with $\theta = 90°$) for two-dimensional progressive waves given in Sect. 3.4.2. The comparison is carried out on a term-by-term basis, converted to the same notations used in this study.

For convenience, the deduced expressions from the present study for two-dimensional progressive waves (with $\theta = 90°$) are summarized as follows:

$$u = \frac{ip_o}{2G}\left[(C_0^\infty + C_1^\infty z)\, e^{kz} + C_2^\infty e^{\delta z}\right] e^{i(kx-\omega t)}, \tag{3.103}$$

$$w = \frac{p_o}{2G}\left[\left(C_0^\infty - \frac{1+2\lambda}{k}C_1^\infty + C_1^\infty z\right) e^{kz} + \frac{\delta}{k}C_2^\infty e^{\delta z}\right] e^{i(kx-\omega t)}, \tag{3.104}$$

$$p = \frac{p_o}{1-2\mu}\left[(1-2\mu-\lambda)C_1^\infty e^{kz} + \frac{\delta^2-k^2}{k}(1-\mu)C_2^\infty e^{\delta z}\right] e^{i(kx-\omega t)}, \tag{3.105}$$

$$\begin{aligned}\sigma_x' = -p_o\Bigg\{&\left[(C_0^\infty + C_1^\infty z) + \frac{2\mu\lambda}{1-2\mu}C_1^\infty\right] e^{kz}\\ &+\left[k^2 - \frac{\mu(\delta^2-k^2)}{k(1-2\mu)}\right] C_2^\infty e^{\delta z}\Bigg\} e^{i(kx-\omega t)},\end{aligned} \tag{3.106}$$

$$\begin{aligned}\sigma_z' = p_o\Bigg\{&\left[kC_0^\infty + C_1^\infty kz - \frac{2\mu\lambda(1-\mu)}{1-2\mu}C_1^\infty\right] e^{kz}\\ &+\frac{1}{k(1-2\mu)}(\delta^2(1-\mu) - k^2\mu)C_2^\infty e^{\delta z}\Bigg\} e^{i(kx-\omega t)},\end{aligned} \tag{3.107}$$

$$\tau_{xz} = ip_o\left[kC_0^\infty + (kz-\lambda)C_1^\infty\right] e^{kz} + \delta C_2^\infty e^{\delta z}\big] e^{i(mkx-\omega t)}, \tag{3.108}$$

where C_0^∞, C_1^∞ and C_2^∞ are given by (3.44)–(3.46).

It is worth noting that the symbols k and ω are the same in both approaches, but Madsen's δ (his equation (33)) is equal to the λ in (3.76), and his k^*k is equal to δ in (3.75). Thus k^* in Madsen's equation (25) is equal to δ/k in the present study. While Madsen's solution has eight C coefficients (i.e. C_0–C_7 in his equations (39a)–(39h)), with a common denominator D given by his equation (39i), only three coefficients (i.e., C_0^∞, C_1^∞ and C_2^∞, given by (3.44)–(3.46)) are used in this study. Madsen's C coefficients are tentatively referred to as C_0^*–C_7^* in the following comparison to avoid confusion with the C coefficients used in the present study.

The expression for the horizontal soil displacement u obtained by Madsen [22] in dimensional form (his equation (4)) was given by

$$u = \frac{ip_{bm}}{2Gk}\left(C_0^* e^{kz} + C_1^* kz\, e^{kz} - \frac{1}{k^*} C_5^* e^{k^*kz}\right) e^{i(kx-\omega t)}. \tag{3.109}$$

Equation (3.109) is to be compared with (3.103) of this study. There are three terms to be verified, the collective coefficients for e^{kz}, ze^{kz} and $e^{\delta z}$, with $\delta = k^*k$ as mentioned above.

Firstly, the collective coefficient for e^{kz} in (3.109) is converted to

$$\frac{ip_{bm}}{2Gk} C_0^* = \frac{i\lambda p_o[(1-\mu)(2k\delta - \delta^2) + k^2\mu]}{2Gk[k\delta(1-\lambda-2\mu) - (1-\mu)\delta^2 + (\lambda+\mu)k^2]}, \tag{3.110}$$

where $p_{bm} = p_o$. Expanding and arranging the term on the right hand side of (3.110), renders

$$\frac{i\lambda p_o[\mu(\delta-k)^2 - \delta(\delta-2k)]}{-2Gk(\delta-k)(\delta-\delta\mu+k\mu+k\lambda)} = \frac{i\lambda p_o}{2G} C_0^\infty. \tag{3.111}$$

This agrees with the first part on the right-hand side of (3.103), where C_0^∞ is given by (3.44).

Similarly, the collective coefficient for ze^{kz} in (3.109) is transformed to

$$\begin{aligned} \frac{ip_{bm}}{2G} C_1^\infty &= \frac{ip_o[(1-2\mu)k\delta - (1-\mu)\delta^2 + k^2\mu]}{2G[k\delta(1-\lambda-2\mu) - (1-\mu)\delta^2 + k^2(\lambda+\mu)]} \\ &= \frac{ip_o}{2G}\frac{(\delta-\delta\mu+k\mu)}{(\delta-\delta\mu+k\mu+k\lambda)} \\ &= \frac{ip_o}{2G} C_1^\infty. \end{aligned} \tag{3.112}$$

Equation (3.112) is the same as the second part on the right-hand side of (3.103), where C_1^∞ is given by (3.45).

Likewise, the collective coefficient for $e^{\delta z}$ in (3.109) becomes

$$\frac{ip_o}{2Gkk^*} C_5^* = -\frac{ip_o}{2Gk}\frac{k^2\lambda}{[k\delta(1-\lambda-\mu) - \delta^2(1-\mu) + k^2(\lambda+\mu)]}, \tag{3.113}$$

which can be shown as

$$-\frac{ip_o}{2G}\frac{k\lambda}{(k-\delta)[\delta(1-\mu)+k(\lambda+\mu)]}=\frac{ip_o}{2G}C_2^{\infty}, \tag{3.114}$$

which conforms to the equivalent part in (3.103).

The effective normal stress in the vertical direction, σ_z', was given by Madsen [22] as

$$\sigma_z' = p_{bm}\left[-C_0^* e^{kz} - C_1^* kz\, e^{kz} + C_4^* e^{k^*kz} + \frac{2(1-\mu)}{1-2\mu}C_3^* e^{kz}\right]e^{i(kx-\omega t)}, \tag{3.115}$$

which contains two other C^* coefficients, C_3^* and C_4^*. Verification of the collective coefficient associated with the terms e^{kz} and kze^{kz} can be found from (3.112). It is noted that a negative sign was attached to the definition of the effective stresses and strains in [22], due to a different sign notation used.

The coefficient to the third term, $e^{\delta z} = e^{k^*kz}$, within the brackets of (3.115) can be converted to

$$\begin{aligned} C_4^* &= \frac{\lambda}{1-2\mu}\frac{[(1-\mu)\delta^2-k^2\mu]}{[k\delta(1-\lambda-2\mu)-\delta^2(1-\mu)+k^2(\lambda+\mu)]} \\ &= -\frac{k\lambda}{k(1-2\mu)}\frac{[(1-\mu)\delta^2-k^2\mu]}{(\delta-k)[\delta(1-\mu)+k(\lambda+\mu)]} \\ &= -\frac{k\lambda}{k(1-2\mu)}C_1^{\infty}, \end{aligned} \tag{3.116}$$

and the fourth part of the coefficient to e^{kz} in (3.115) is converted to

$$\begin{aligned} \frac{2(1-\mu)}{1-2\mu}C_3^* &= \frac{2\lambda(1-\mu)}{(1-2\mu)}\frac{[k\delta(1-2\mu)-\delta^2(1-\mu)+k^2\mu]}{[k\delta(1-\lambda-2\mu)-\delta^2(1-\mu)+k^2(\lambda+\mu)]} \\ &= \frac{2\lambda(1-\mu)}{(1-2\mu)}\frac{(k-\delta)(\delta-\delta\mu+k\mu)}{[(\delta-k)(\delta(1-\mu)+k(\lambda+\mu))]} \\ &= -\frac{2\lambda(1-\mu)}{1-2\mu}C_1^{\infty}. \end{aligned} \tag{3.117}$$

The above equations agree with the expressions of σ_z' in (3.107).

Finally, the expression for the pore pressure given by Madsen's equation (37) is

$$p = p_{bm}\left(-C_6^* e^{k^*kz} + C_7^* e^{kz}\right)e^{i(kx-\omega t)}. \tag{3.118}$$

Equation (3.118) has two additional C^* coefficients, which yield

$$C_6^* = \cdots = \frac{(1-\mu)(\delta^2-k^2)}{k}C_2^{\infty}, \tag{3.119}$$

and

$$-C_7^* = \frac{(1-\lambda-2\mu)}{1-2\mu}C_1^* = \cdots = \frac{(1-\lambda-2\mu)}{1-2\mu}C_1^{\infty}, \tag{3.120}$$

respectively. The resultant expressions are exactly the same as that given by (3.105).

The solution for the horizontal effective normal stress, σ'_x, which was given by (34) in [22], contains coefficients C_2^*, which can be shown to match that required in (3.106).

Thus, the comparison may be considered as complete.

3.5.2.2 Comparison with Yamamoto [49] and Mei and Foda [25]

To facilitate the comparison, the results of Yamamoto [49] are computed from his 6×6 coefficient matrix, and that of Mei and Foda [25] from their final equations. The comparison is made using the reduced form of the limiting two-dimensional progressive waves ($\theta = 90°$), i.e., (3.77)–(3.82) for fully saturated and partially saturated seabeds. The wave conditions used in the first example is for a 15 s progressive wave in a water depth $d = 70$ m (wave length $L = 311.59$ m). Fine and coarse sands are used, with permeability coefficients of $K_z = 10^{-4}$ m/s and 10^{-2} m/s, respectively.

For a fully saturated and isotropic seabed of finite thickness, $h = 25$ m and $G\beta = 0$, this data set represents soil-wave conditions for the North Sea [4]. The vertical distribution of the non-dimensional pore pressure $|p|/p_o$ and effective stresses, $|\sigma'_x|/p_o$, $|\sigma'_z|/p_o$ and $|\tau_{xz}|/p_o$ are drawn as a function of z/h in Fig. 3.6. It is found that the simplified exact solutions for the limiting two-dimensional progressive waves are in complete agreement with the results of [49] for both the fine and coarse sands. It also agrees with the computed results of [25] for fine sand (Fig. 3.6), but they deviate slightly from it for coarse sand (Fig. 3.6).

The wave and soil conditions employed in the second example are the same as that for producing Fig. 3.6b, except now for a partially saturated condition at $S_r = 0.932$ ($G\beta = 1.0$). For fine sand under an unsaturated condition, the normalized pore pressure and effective stresses against z/h from [18] are in almost complete agreement with the exact solutions developed in this study (Fig. 3.6b). However, their results for coarse sand (the circle "o" in Fig.3.6) also deviate slightly from the exact solution, similar to that for a fully saturated condition.

Although the results from the boundary-layer approximation proposed by Mei and Foda [25] differ slightly from the present exact solution near the seabed bottom, their solutions provide much simpler expressions than the present theory. Since the wave-induced seabed instability only occurs near the seabed surface [15], and the results of boundary-layer approximation are almost identical to the exact solutions in these regions, the solution of [25] may be more readily accessible for engineering applications.

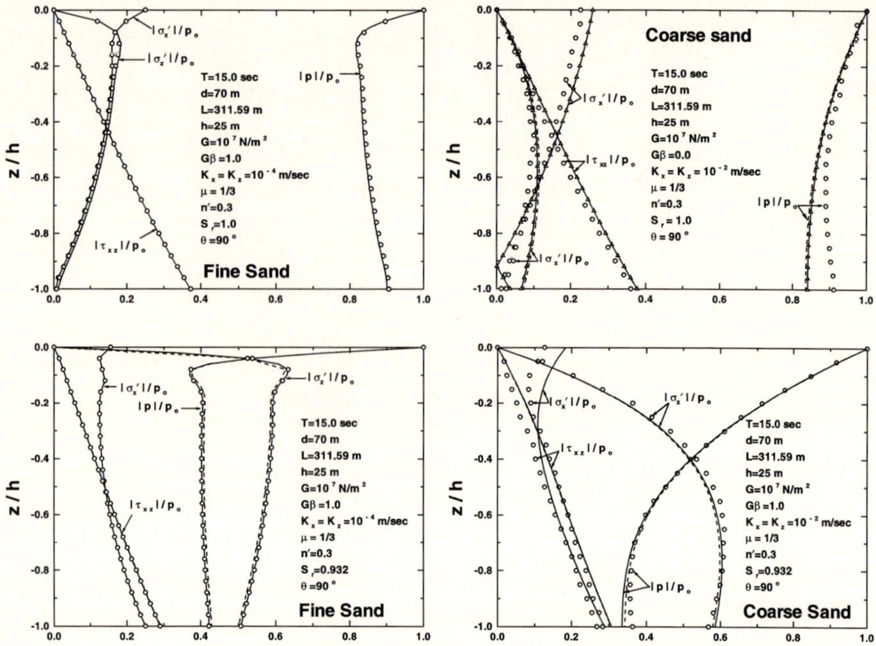

(a) Fine sand (b) Coarse sand

Fig. 3.6 Vertical distribution of the maximum pore pressure and effective stresses versus z/h for a hydraulically isotropic seabed. Calculated results from [25] in "o", [40, 41] in *dashed lines*, and present theory in *solid lines*

3.5.3 Comparison with Numerical Model [18, 40, 41]

In addition to the comparisons with other two-dimensional analytical solutions and experimental data, verifications against the results of numerical models are now presented. Two **FEM** (Finite Element Model) numerical approaches, proposed by [18, 40, 41], are used for the numerical verification.

In Fig. 3.6, the results of a two-dimensional numerical model (finite element method) of [40, 41] are also included for saturated fine and coarse sands. Complete agreement was found in the results of the present study, Yamamoto [49] and Thomas [40, 41], for both sand grain sizes.

From the three-dimensional numerical model [18], results for the two-dimensional progressive waves condition can be deduced (i.e., [40, 41]), as shown in Fig. 3.6 (in dashed lines). Further comparisons between the present theory and the numerical model [18] for a three-dimensional short-crested wave are also depicted in Fig. 3.7. The numerical values of the wave-induced pore pressure and the vertical effective normal stress are also tabulated in Table 3.1. At least from these examples, it is clear that the present solution agrees well with the three-dimensional numerical finite element model of [18].

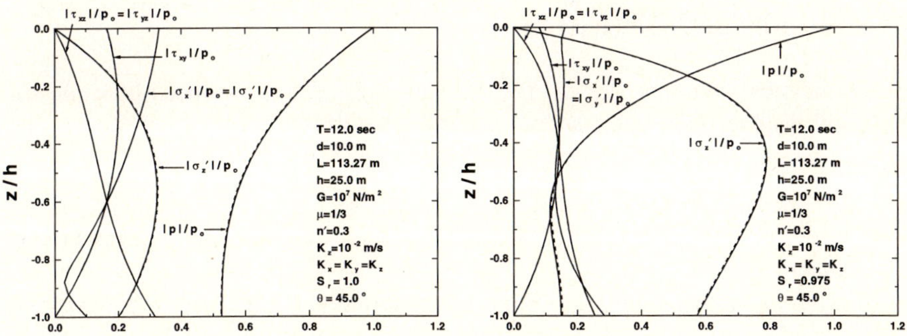

(a) full saturated seabed (b) unsaturated seabed

Fig. 3.7 Vertical distributions of the maximum pore pressure and effective stresses versus z/h for a coarse sand. Results from [41] in *dashed lines*, and present theory in *solid lines*

Table 3.1 Comparison of the present solution and a three-dimensional numerical model[a]

$z(m)$	Present solution		3-D numerical model[b]	
	$\lvert p\rvert/p_o$	$\lvert\sigma_z'\rvert/p_o$	$\lvert p\rvert/p_o$	$\lvert\sigma_z'\rvert/p_o$
0.0	1.0000	0.0000	1.0000	0.0000
−1.0	0.8675	0.1716	0.8676	0.1717
−2.0	0.7514	0.3153	0.7515	0.3151
−3.0	0.6494	0.4339	0.6495	0.4335
−4.0	0.5596	0.5305	0.5597	0.5297
−5.0	0.4808	0.6077	0.4809	0.6066
−6.0	0.4115	0.6680	0.4116	0.6667
−7.0	0.3509	0.7138	0.3511	0.7121
−8.0	0.2981	0.7469	0.2984	0.7451
−9.0	0.2525	0.7694	0.2529	0.7673
−10.0	0.2136	0.7829	0.2143	0.7805
−11.0	0.1812	0.7888	0.1822	0.7862
−12.0	0.1552	0.7885	0.1567	0.7857
−13.0	0.1357	0.7831	0.1377	0.7802
−14.0	0.1227	0.7736	0.1252	0.7706
−15.0	0.1156	0.7610	0.1187	0.7578
−16.0	0.1136	0.7459	0.1172	0.7426
−17.0	0.1156	0.7290	0.1194	0.7256
−18.0	0.1198	0.7108	0.1238	0.7073
−19.0	0.1253	0.6917	0.1294	0.6882
−20.0	0.1311	0.6721	0.1353	0.6686
−21.0	0.1367	0.6523	0.1409	0.6486
−22.0	0.1414	0.6323	0.1456	0.6286
−23.0	0.1450	0.6123	0.1493	0.6086
−24.0	0.1473	0.5924	0.1516	0.5887
−25.0	0.1481	0.5725	0.1524	0.5687

[a]The input data are the same as the legend of Fig. 3.7
[b]The results are calculated from [18]

3.6 Results and Discussion

The analytical solutions derived for the short-crested wave induced pore pressure (p), soil displacements (u, v and w) and effective stresses (σ'_x, σ'_y, σ'_z, τ_{xz}, τ_{yz} and τ_{xy}), contain three groupings of parameters:

- *wave parameters*: these being wave height H_s, period T (explicitly in frequency ω), wave obliquity θ, and relative water depth d/L (in terms of wave number k);
- *soil parameters*: including hydraulic permeabilities (K_x, K_y and K_z), porosity n', Poisson's ratio μ, degree of saturation S_r (in term of compressibility β) and shear modulus G;
- *a combined obliquity-permeability parameter*: appearing in the final governing equation (3.26) and also in the collective parameters δ and λ given by (3.27) and (3.33), respectively.

3.6.1 Effect of Wave Characteristics

In a short-crested wave system, one of the main variables to determine water particle motion and pressure fluctuations at the bed surface is that of the angle between the wave components. In the case of wave reflection for a structure, this wave obliquity (θ) is measured between the orthogonals and the normal to the wall, or alternatively between crest alignments and the wall itself. The other major distinctive variable is the depth ratio (d/L), which for any depth is fixed by the wave period (T). The influence of θ and d/L on various soil parameters is now examined.

3.6.1.1 Wave Obliquity

The vertical distributions of the maximum values of $|2Gkw|/p_o$, $|p|/p_o$ and $|\sigma'_z|/p_o$ versus z/L can be calculated from (3.95), (3.96) and (3.97) for a fully saturated, hydraulically isotropic seabed of infinite thickness. They are found to be independent of the wave obliquity θ (Fig. 3.8a). However, it is worth noting that this conclusion is based on the assumption of using a linear wave theory. If a non-linear wave theory is employed in the seabed surface boundary conditions, the wave-induced $|2Gkw|/p_o$, $|p|/p_o$ and $|\sigma'_z|/p_o$ are affected slightly by the wave obliquity θ [18]. The pore pressure $|p|/p_o$ and $|2Gkw|/p_o$ decrease as z/L increases. On the other hand, the vertical effective normal stress $|\sigma'_z|/p_o$ increases as z/L increases near the seabed surface ($-0.159 \le z/L \le 0$), and then decreases as z/L increases.

Figures 3.8b–3.8d illustrate the vertical distributions of other soil response parameters against z/L for various wave obliquities θ. It is found that the maximum values of $|\sigma'_x|/p_o$ increase as the incident angle θ increases and the general patterns are similar (Fig. 3.8b). The curves for $|\sigma'_y|/p_o$ are also presented in Fig. 3.8b. The values are complimentary to those for $|\sigma'_x|/p_o$. For example, the curve of $\theta = 30°$

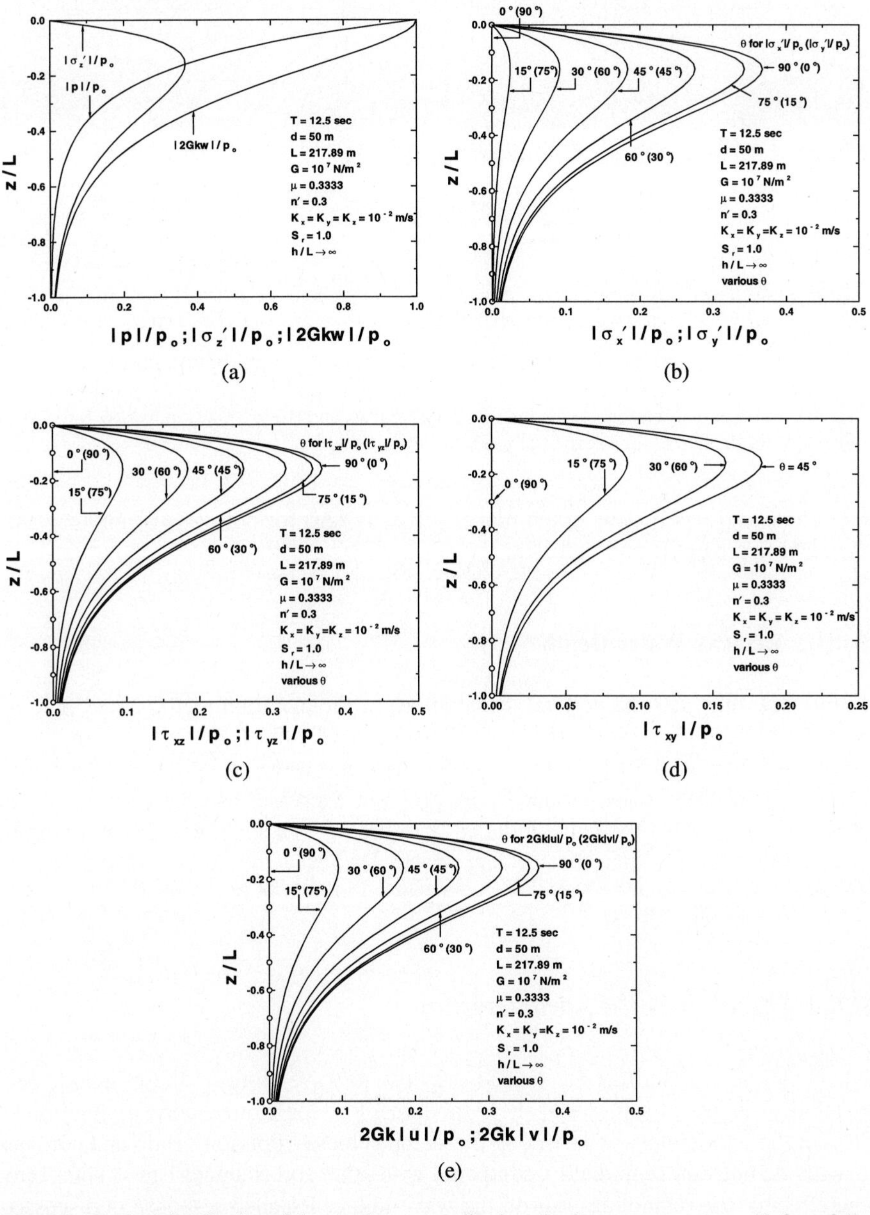

Fig. 3.8 Vertical distribution of the maximum amplitudes of soil response versus z/L for a fully saturated and hydraulically isotropic seabed of infinite thickness

for $|\sigma_x'|/p_o$ is also for $\theta = 60°$ for $|\sigma_y'|/p_o$. Similar trends are also observed in Fig. 3.8c for $|\tau_{xz}|/p_o$ and $|\tau_{yz}|/p_o$, and in Fig. 3.8e for $|2Gku|/p_o$ and $|2Gkw|/p_o$.

For the curves of $|\tau_{xy}|/p_o$ (Fig. 3.8d), the maximum values occur at $\theta = 45°$, while the curves of $\theta = 30°$ and $\theta = 60°$ are identical, and similarly for curves of

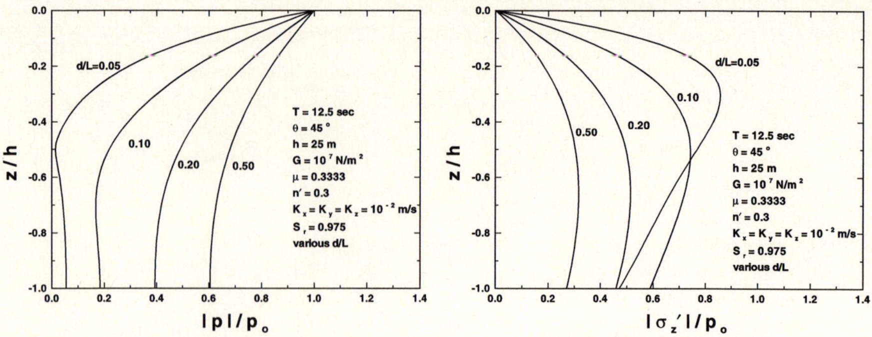

(a) $|p|/p_o$ (b) $|\sigma_z'|/p_o$

Fig. 3.9 Vertical distributions of the maximum (**a**) $|p|/p_o$ and (**b**) $|\sigma_z'|/p_o$ versus z/h for various relative water depth d/L in a seabed of finite thickness

$\theta = 15°$ and $\theta = 75°$. It is noted that $|\tau_{xy}|/p_o$ is zero for both $\theta = 0°$ and $\theta = 90°$ throughout the entire soil column.

3.6.1.2 Relative Water Depth

Figure 3.9 illustrates the vertical distributions of the maximum $|p|/p_o$ and $|\sigma_z'|/p_o$ versus z/h for four typical values of relative water depth (d/L), ranging from 0.05 (shallow water) to 0.5 (deep water) under the same wave period ($T = 12.5$ s). As shown in Fig. 3.9a, wave-induced pore pressure decreases as relative water depth d/L decreases. Also the wave-induced pore pressure in shallow water is found to reduce more rapidly than that in deep water. However, the maximum $|\sigma_z'|/p_o$ increases as relative water depth d/L decreases, as shown in Fig. 3.9b.

3.6.2 Effect of Soil Characteristics

Many soil variables affect the seabed response to wave motion. These include permeability, degree of saturation, shear modulus, Poisson's ratio, porosity and thickness of the sedimentary bed. Among these, the values of Poisson's ratio and porosity usually do not vary remarkably, compared with other soil characteristics. Thus, only the effect of the remaining four on the wave-induced seabed response is discussed here.

3.6.2.1 Soil Permeability

The permeability of a soil is a measure of how rapidly fluid is transmitted through the voids between grains. The vertical permeability of the soil K_z varies from

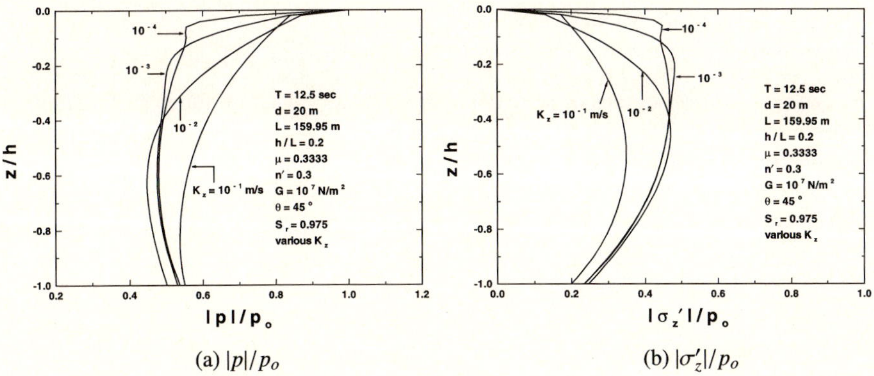

Fig. 3.10 Vertical distribution of the maximum (**a**) $|p|/p_o$ and (**b**) $|\sigma_z'|/p_o$ versus z/h for various permeability K_z in a seabed of finite thickness

10^{-1} m/s for gravel to 10^{-6} m/s for silts and clays [2]. The former may be considered as permeable and the latter as semi-permeable. In this study, the typical value of permeability is taken as 10^{-1} m/s for gravel, 10^{-2} m/s for coarse sand and 10^{-4} m/s for fine sand, unless specified otherwise.

The wave-induced seabed response is affected significantly by the vertical permeability of the soil (Fig. 3.10). For instance, for the case with $K_z = 10^{-4}$ m/s (fine sand), pore pressure attenuates very rapidly within the top 5 % of its thickness, where a minimum value is reached. Then, a slight increase in magnitude with depth is observed, and finally decreases slightly with depth again. On the other hand, the results for $K_z = 10^{-2}$ m/s (coarse sand) indicate a relatively slow decrease in the magnitude of pore pressure versus soil depth, and with less reduction in pore pressure at $K_z = 10^{-1}$ m/s (gravel).

For the vertical effective normal stress $|\sigma_z'|/p_o$, it increases as the relative soil depth z/h increases in the region near the seabed surface (Fig. 3.10b). It reaches a maximum value at a z/h, then decreases as z/h increases further. All curves with different values of permeability display similar trends, despite their maxima being found at different z/h. The curve for fine sand ($K_z = 10^{-4}$ m/s) has a swift increase at small z/h values than for coarse sand ($K_z = 10^{-2}$ m/s), as seen in the figure.

3.6.2.2 Degree of Saturation

It is not uncommon to find air/gas within marine sediments. For example, samples taken from Mississippi Delta sediment equilibrate when exposed to atmospheric pressure has the degrees of saturation generally between 75 % and 95 % [5]. It is believed that most marine sediments have degrees of saturation very close to unity, implying nearly full saturation [28]. However, it is rare that full saturation can be attained in field or laboratory conditions, except for an ideal condition.

Generally speaking, the mechanical behavior at various degrees of saturation in partially saturated soils depends on the micro-mechanical structures of the three phase-soil skeletons, liquid (water) and gas (air). Four different types of microstructure of partially saturated soils were distinguished by [28, 48]. They are:

- Type A: At low degrees of saturation, the gas phase is continuous and menisci of liquid adhere to most grain boundaries.
- Type B: At higher degrees of saturation, the liquid phase is continuous in part of the sample, whereas in the remaining part gas phase remains continuous.
- Type C: As the degree of saturation is further increased, the liquid phase becomes continuous while the gas phase becomes discontinuous in the form of bubbles embedded in the liquid phase.
- Type D: It is also called *gassy soil*. In this type, the gas phase is discontinuous in the form of discrete bubbles that are entrapped in a fully saturated soil matrix. Such a structure is formed when the gas (produced by decomposition of organic matter) pushes against the soil skeleton, creating gas voids of a size that is much larger than the average particle size.

It is worth noting that the microstructure of soil may change from one type to the other during the process of deformation [28].

Among these, types C and D are commonly observed in marine sediments [28]. The structure of type D has been studied by [27, 40, 41, 47]. For type C, Verruijt [46] and Madsen [22] has approached problems by calculating the compressibility of air-water mixture assuming that pore water pressure and the air pressure are the same. Each fluid phase is assumed to be homogeneous in an average sense. Therefore, it is treated as a separate physical continuum described by a unique constitutive equation. This microstructure corresponds to a high degree of saturation (approximately more than 90 %). More detailed information regarding different types of unsaturated soil can refer to [28]. In this study, only type C soil is considered when partial saturation is addressed.

The relationships of the maximum pore pressure ($|p|/p_o$) and vertical effective stress ($|\sigma_z'|/p_o$) versus relative depth (z/h) are depicted in Fig. 3.11, for various values of degrees of saturation (S_r) with $K_z = 10^{-2}$ m/s (coarse sand) and $h =$ 25 m. These figures clearly show that $|p|/p_o$ increases as the degree of saturation S_r increases, while $|\sigma_z'|/p_o$ decreases as S_r increases. As seen in Fig. 3.11a, the maximum of $|p|/p_o$ at $S_r = 0.975$ is almost 80 % of that at $S_r = 1.0$ (i.e., fully saturated) at the rigid bottom. On the other hand, the maximum $|\sigma_z'|/p_o$ at $S_r =$ 0.975 is about three times of that at $S_r = 1.0$ (Fig. 3.11b). This means that the degree of saturation may affect significantly the wave-induced soil response. Thus, geotechnical and coastal engineers should note that: *the accuracy in determining the degree of saturation in the field and laboratory experiments has an important role in the analysis of the wave-induced seabed response*.

Since the degree of saturation affects the wave-induced soil response significantly, its evaluation may raise an interesting subject in geotechnical engineering. Based on the present analytical solution, it may become possible to back-calculate the degree of saturation by measuring the soil displacements at the seabed surface. This concept is similar to **BSMP** (Bottom Shear Modulus Profile), which was proposed by Yamamoto's group [1, 50, 52]. However, the validity of this concept re-

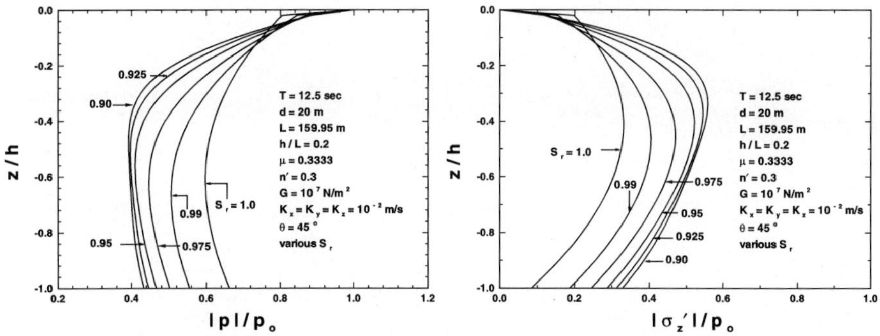

(a) $|p|/p_o$ (b) $|\sigma_z'|/p_o$

Fig. 3.11 Vertical distributions of the maximum (**a**) $|p|/p_o$ and (**b**) $|\sigma_z'|/p_o$ versus z/h for various degrees of saturation S_r in a seabed of finite thickness

quires more comprehensive verification from laboratory experiments and field measurements in the future.

3.6.2.3 Shear Modulus

The shear modulus of soil (G) is defined as the proportional coefficient in the shear stress-shear strain relationship. It varies from about 5×10^8 N/m^2 for gravel or very dense sand to 10^6 N/m^2 for very fine sand or silt and clay [51]. Since shear modulus of a soil has been recognized as one of the dominant soil properties in the literature of soil mechanics, it is beneficial to examine its effect on the wave-induced pore pressure and effective stresses.

It has been proved in Sect. 3.4.6 that the wave-induced pore pressure and effective stresses only depend on wave conditions in a saturated and hydraulically isotropic seabed of infinite thickness (also see [12, 51]). However, this conclusion is invalid for a saturated and hydraulically isotropic seabed of finite thickness [17]. That is, the wave-induced soil response depends on both wave and soil characteristics. For example, the wave-induced pore pressure and vertical effective normal stress in a saturated and hydraulically isotropic seabed of coarse sand are affected by the shear modulus G as demonstrated in Fig. 3.12a. The influence of shear modulus on $|p|/p_o$ and $|\sigma_z'|/p_o$ increases as the degree of saturation decreases. This can be observed from the results in Fig. 3.12.

3.6.2.4 Seabed Thickness

Using the same wave and soil conditions as Gatmiri [8], the effect of seabed thickness on the vertical distribution of pore pressure is now examined. Considering a two-dimensional progressive wave with period of 7 s in a water depth $d = 3.7$ m

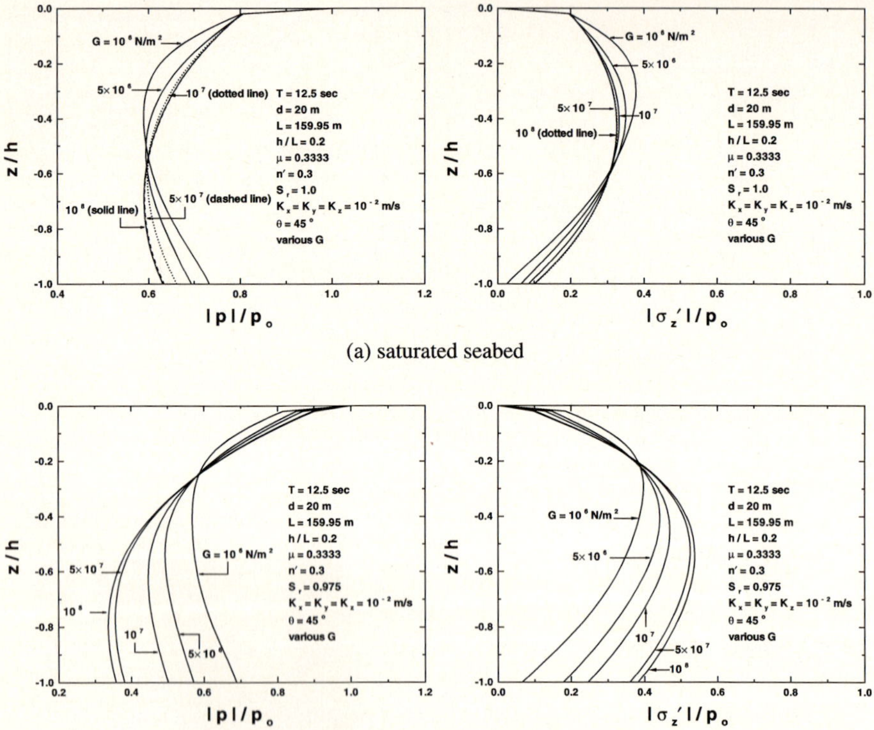

Fig. 3.12 Vertical distribution of the maximum $|p|/p_o$ and $|\sigma_z'|/p_o$ versus z/h for various shear modulus G in a seabed of finite thickness.

(or wave length $L = 40$ m), the porous seabed has fine sand with permeability $K_z = 5 \times 10^{-4}$ m/s, (including Poisson's ratio $\mu = 1/3$, porosity $n' = 0.3$), and shear modulus $G = 10^7$ N/m^2. The condition calculated is for a fully saturated and isotropic soil, hence $\lambda = 0$.

As seen in Fig. 3.13a, the present analytical solution renders a set of similar distribution profiles for pore pressure versus h/L ranging from 0.1 to 1.0. Each h/L curve stops at its rigid impermeable bottom. The curve for $h/L = 1.0$ approaches the result for soil of infinite thickness ($h/L \to -\infty$, in dashed line). However, the numerical results of [8] (in dotted lines) indicate that the $|p|/p_o$ value for $h/L = 0.2$ is greater than that for $h/L = 0.1$, and that of $h/L = 0.5$ greater than 0.4. These are inconsistent in the trend for pore pressure distribution versus h/L, compared with the present analytical solution (in solid lines). It seems difficult to provide a reason for the *jumping order* in pore pressure for these h/L values. Although there is no experimental evidence to support a maximum pore pressure for $h/L = 0.1$ being greater than for $h/L = 0.2$, and of $h/L = 0.4$ being greater than 0.5, the ordering in magnitude for $|p|/p_o$ versus h/L presented in this study would appear to be more consistent.

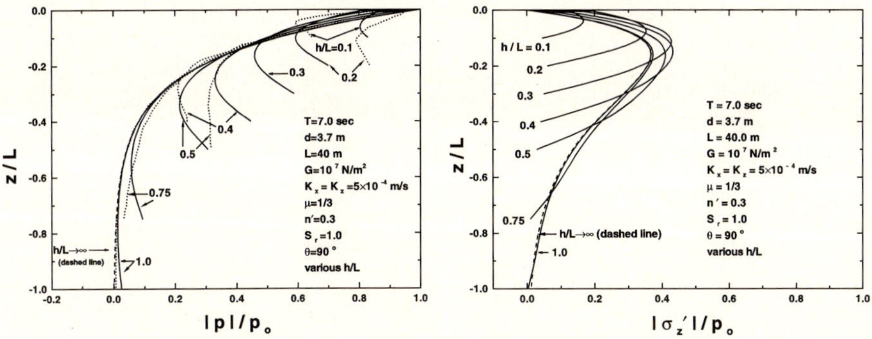

(a) saturated seabed

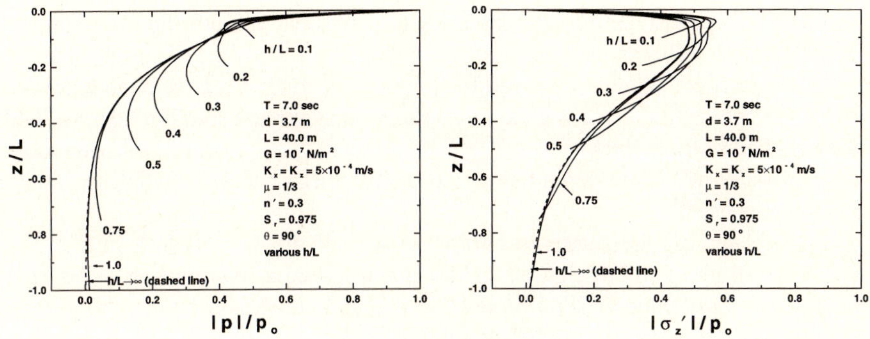

(b) unsaturated seabed

Fig. 3.13 Vertical distributions of the maximum $|p|/p_o$ versus z/h for various seabed thickness in a porous seabed of finite thickness. Results from [8] in *dotted lines*, the present solution in *solid lines* for finite thickness, and in *dashed line* for infinite thickness ($\theta = 90°$)

A possible error in the results of [8] may have stemmed from the boundary conditions used. The lateral boundaries at $x/L = 1/4, 3/4$ (i.e., $x/L = 0, 1$ in his paper), "$v = 0,\ p = 0$ and u free", were used in his model. However, it has been proved that there is a phase lag in soil response in a fully saturated seabed of finite thickness [17]. This implies that the lateral boundary conditions, $v = 0$ and $p = 0$, are invalid in a porous seabed of finite thickness. Thus, the numerical results of [8, 9] seem doubtful.

Under the same wave and soil conditions used in Fig. 3.13a, except now for a partially unsaturated soil of $S_r = 0.975$, the values $|p|/p_o$ at the seabed bottom vary from 0.5 for $h/L = 0.1$ to 0.02 for $h/L = 1$. However, the pore pressure is almost the same for various h/L near the seabed surface ($-0.02 \le z/L \le 0$) in Fig. 3.13b.

Figure 3.13a illustrates the vertical distribution of the maximum $|\sigma_z'|/p_o$ versus z/L for various values of relative seabed thickness h/L in a fully saturated seabed. Generally speaking, the maximum vertical effective stress increases as h/L increases, when $h/L \le 0.4$ under a saturated condition (Fig. 3.13a). However, it decreases as h/L increases, when $h/L > 0.4$. Similar trend can also be observed from Fig. 3.13b for an unsaturated seabed.

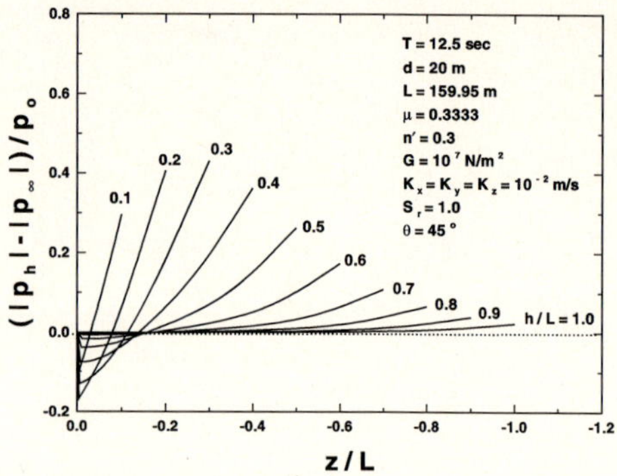

Fig. 3.14 Relative difference in pore pressure calculated from solutions for soil in infinite and finite thickness $(|p_h| - |p_\infty|)/p_o$ versus z/L for various h/L in a saturated coarse sand

Practically, most laboratory experiments and field monitoring programs for wave-induced soil response have been conducted on a seabed of finite thickness, and the analytical solutions derived for infinite soil thickness have been taken as a reference against the experimental data. With the analytical solution now available for a seabed of finite thickness, it is possible to demonstrate the percentage error of pore pressure in a finite depth below the seabed surface if the solution for infinite thickness is used.

From observing the curves of $|p|/p_o$ versus z/L for various h/L in Fig. 3.13, a practical limit of soil thickness, $h/L = 1$, considered as *infinitive* ($h \to \infty$), may be taken as a reference. This implies that if seabed thickness h is slightly greater than the prevailing wave length L, the soil thickness may be theoretically regarded as infinite, for which the mathematical solution for soil of infinite thickness may be used. However, if $h/L < 1$, as shown in Fig. 3.14, the use of infinite thickness solution for soil of finite thickness may lead to error, either under- or over-estimating the pore pressure.

An example demonstrating the possible relative error in pore pressure for a seabed of finite thickness using the solution for infinite thickness is now given for a fully saturated and isotropic seabed. Under the wave and soil conditions specified in Fig. 3.14, the relative differences in pore pressures $(|p_h| - |p_\infty|)/p_o$ are plotted versus z/L for various h/L. In this figure, p_∞ denotes the pore pressure using the solution for soil of infinite thickness, such as (3.37), and p_h is the pore pressure for soil of finite thickness given by (3.54). This results in either positive or negative sign, respectively, for any depth immediately below the seabed surface, implying either under-estimation or over-estimation of the pore pressure for soil of finite thickness.

Following the curve of a particular h/L in Fig. 3.14, it is observed that p_∞ is greater than p_h within an upper layer of soil matrix immediately beneath the seabed surface ($z = 0$) to a specific level of z/L, where $|p_h| = |p_\infty|$. At points further below this critical depth, $|p_\infty|$ values become less than $|p_h|$. The relative errors are generally greater for larger z/L for any h/L curve, indicating that the infinite solution under-estimates the pore pressure in soil of finite thickness, except for locations near the seabed surface. Similarly, inaccuracy may also be generated for other soil-response quantities, such as effective stresses and soil displacements, if the solution for soil of infinite thickness is used in a seabed of finite thickness.

For an unsaturated seabed, the calculated relative differences in pore pressure are smaller than those demonstrated above for soil in saturated and isotropic conditions. This can be observed by comparing Figs. 3.13a and 3.13b, showing the relatively smaller range in $(|p_h| - |p_\infty|)/p_o$ between a specific h/L value and $h \to \infty$ in the latter. Therefore, it is relatively safe to employ the analytical solution of infinite thickness for a seabed of finite thickness, if the seabed is in an unsaturated condition.

3.6.3 Effect of a Combined Obliquity-Permeability Parameter

Almost all non-cohesive soils in their natural state are hydraulically anisotropic, in which their permeabilities in the horizontal direction (K_x and K_y) are usually several times different from its vertical permeability, K_z. A new parameter, combining obliquity-permeability parameters *KMN*, appears only in the solutions of the *short-crested* wave-induced soil response [12]. It exists in the collective parameters δ and λ, which in turn appear in the C_i and B_i coefficients (see Sect. 3.8) for the solution of finite thickness. Therefore, it is expected that this *KMN* parameter may affect the result of soil response.

Figure 3.15 shows the maximum pore pressure ($|p|/p_o$) and vertical effective normal stress ($|\sigma_z'|/p_o$) versus relative soil depth (z/h) as a function of the *KMN* parameter. To understand the general trends of the influence of the *KMN* parameter, the values of *KMN* have been chosen from 0.1 to 100, which cover possible values of permeabilities in the field. The maximum pore pressure decreases slightly as *KMN* increases when *KMN* is less than 10 (Fig. 3.15a). However, an opposite trend is observed for $|\sigma_z'|/p_o$ (Fig. 3.15b). With the large range in difference for *KMN* from 10 to 100, its effect on pore pressure and vertical effective normal stress appears to be rather significant.

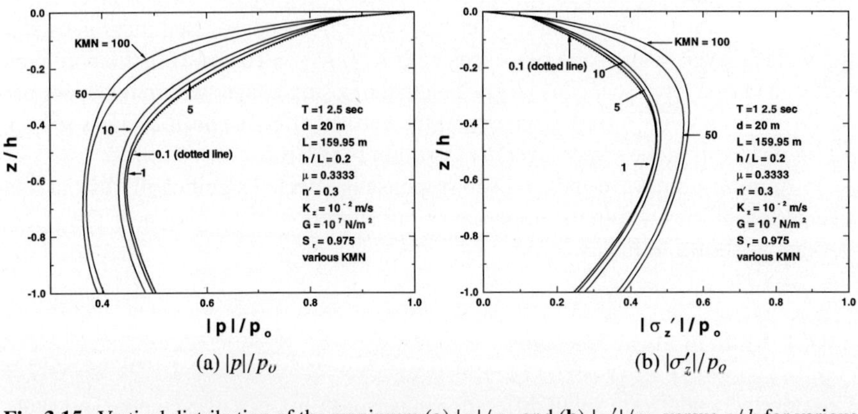

(a) $|p|/p_o$ (b) $|\sigma_z'|/p_o$

Fig. 3.15 Vertical distribution of the maximum (**a**) $|p|/p_o$ and (**b**) $|\sigma_z'|/p_o$ versus z/h for various combined factor *KMN* in an unsaturated seabed of finite thickness

3.6.4 Effect of a Top Layer

In general, natural seabeds consist of multiple layers with different soil properties. For example, in the Ekofisk oil field, the soil profile has an upper layer of 75 m thick consisting of a mixture of sand and clay, followed by clay further down [4]. Furthermore, it is common to place concrete armour block or gravel on the seabed of finer material in the vicinity of marine structures for seabed protection. Thus, the soil matrix may be considered as a multiple-layered medium in both the natural and artificial seabeds.

It has been reported that the wave-induced seabed instability, such as liquefaction and shear failure, is more likely to occur in the immediate vicinity of the seabed surface [54]. Thus, the soil characteristics of a top layer is expected to play an important role on the analysis of the seabed instability. In this section, the effects of a top layer on the wave-induced soil response will be investigated.

3.6.4.1 Ratio of Permeabilities

As reported in the literature [32], the permeability of a top layer may not affect the pore pressure in a porous seabed consisting of a cover layer and a sandy sublayer below. However, a seemingly unrealistic assumption was made in their study, that excess pore pressure was assumed to be independent of the soil permeability ([32]). Since soil permeability has been proved to play a significant role on the wave-induced soil response in a single layer as the numerical results presented in this study, it is worthwhile to re-examine its effect in a multiple-layered seabed.

With the layered solution presented in Sect. 3.4.4, the following comparison becomes possible. Compared with the results of a single layer (i.e., $K_{z1}/K_{z2} = 1$), pore pressure decreases and the vertical effective normal stress increases more rapidly when K_{z1}/K_{z2} is much smaller than unity (Fig. 3.16a), where K_{z1} is the permeability for the top layer, indicated as the region above the dotted line in the figure. The reduction in pore pressure is attributed to the damping by the surface layer with a soil of much smaller permeability. On the contrary, if a coarser sediment is used as the surface layer, such as the two cases with $K_{z1}/K_{z2} = 10$ and 100, the pore pressure of a two-layered seabed is larger than that of a single layer within the upper part of the top layer (Fig. 3.16a). Conversely, the vertical effective normal stress $|\sigma_z'|/p_o$ will reduce within the upper layer, as shown in Fig. 3.16b.

In summary, the wave-induced soil response is affected significantly by the relative permeabilities within the upper layer, but slightly in the region near the seabed bottom, as shown in Fig. 3.16.

3.6.4.2 Ratio of Shear Modulus

Considering a relatively small difference in permeability in a two-layered soil skeleton, such as a thin layer of fine sand ($K_z = 10^{-4}$ m/s, $h_1/h = 0.1$ for $h_1 = 5$ m and

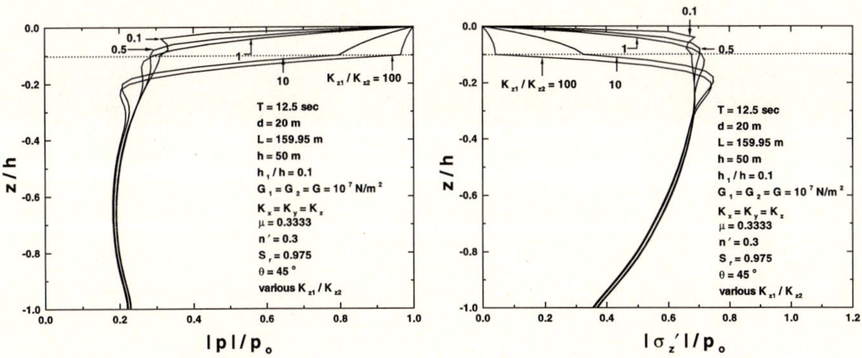

(a) $|p|/p_o$ (b) $|\sigma_z'|/p_o$

Fig. 3.16 Vertical distribution of the maximum (**a**) $|p|/p_o$ and (**b**) $|\sigma_z'|/p_o$ versus z/h for various K_{z1}/K_{z2} within a layered seabed

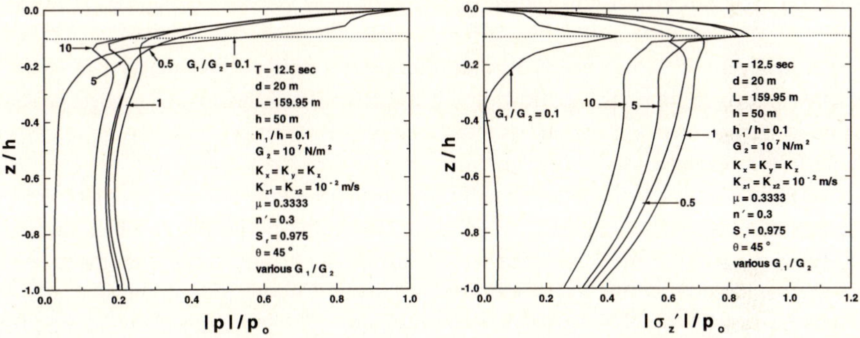

(a) $|p|/p_o$ (b) $|\sigma_z'|/p_o$

Fig. 3.17 Vertical distribution of the maximum (**a**) $|p|/p_o$ and (**b**) $|\sigma_z'|/p_o$ versus z/h for various G_1/G_2 within a layered seabed

$h = 50$ m) over a thicker layer of coarser sand ($K_z = 10^{-3}$ m/s, $h_2/h = 0.9$ for $h_2 = 45$ m and $h = 50$ m) of finite thickness. As shown in Fig. 3.17a, the maximum amplitude of pore pressure, $|p|/p_o$, increases as the relative shear modulus G_1/G_2 decreases (G_1 and G_2 denotes the shear modulus of top and bottom layers, respectively) within the top layer. However, the vertical effective normal stress $|\sigma_z'|/p_o$ increases as G_1/G_2 increases (Fig. 3.17b). It is worth noting that the relative shear modulus G_1/G_2 affects the wave-induced soil response slightly near the seabed surface when G_1/G_2 is greater than unity.

Conversely, the influence of relative shear modulus G_1/G_2 on the soil response in the bottom layer becomes more significant. The pore pressure of a two-layered seabed is larger than that of a single layer, but the vertical effective normal stress becomes smaller (Figs. 3.17a–b).

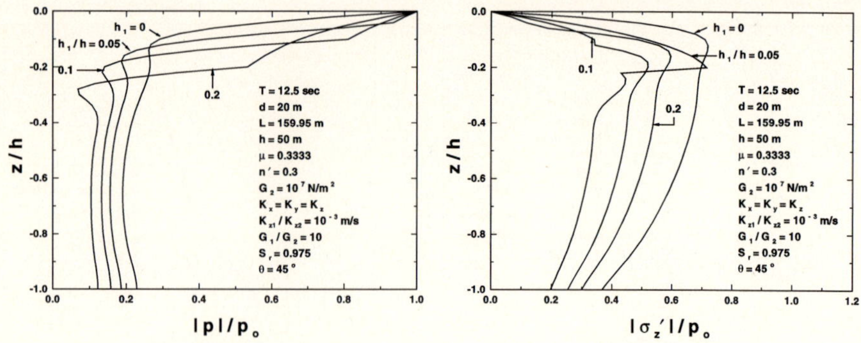

(a) $|p|/p_o$ (b) $|\sigma_z'|/p_o$

Fig. 3.18 Vertical distribution of the maximum (**a**) $|p|/p_o$ and (**b**) $|\sigma_z'|/p_o$versus z/h versus z/h for various h_1/h within a layered seabed

3.6.4.3 Thickness of a Top Layer

The thickness of a top layer is also an important concern in the analysis of seabed instability in engineering applications. It directly affects the cost of seabed protection in the vicinity of a marine structure. Thus, it is necessary to examine its effect on the soil response.

Considering a single layer of medium sand (with permeability $K_{z1} = 10^{-3}$ m/s) and a two-layered seabed ($K_{z1}/K_{z2} = 10$, where the permeability of the bottom layer is $K_z = 10^{-3}$ m/s) with various values of h_1/h. This represents the case of a coarser material on the top of a finer sandy layer. Compared with the single layer with $K_{z1} = 10^{-3}$ m/s, the pore pressure increases within the upper layer when the thickness h_1 increases, and decreases within the bottom layer. On the other hand, an opposite trend is true for the vertical effective normal stress. The thickness of a top layer affects the wave-induced pore pressure and vertical effective stress significantly near the seabed surface within the upper layer (Fig. 3.18). Once the burial soil depth becomes larger, i.e., near the seabed bottom ($z/h = -1$), the influences of a top layer become insignificant.

3.7 Summary

In this chapter, analytical solutions of the wave-induced soil response in front of a breakwater have been presented for a seabed with uniform permeability throughout the entire depth or within each sub layer. The seabed considered covers infinite to finite thickness and a layered medium. Comprehensive verifications of the present solutions have been performed against previous two-dimensional analytical solutions, experimental data and three-dimensional numerical models. Based on the general solutions developed, the effects of wave and soil characteristics on the wave-induced soil response have been discussed through a parametric study. The effects of the soil characteristics of a top layer have also been investigated.

Based on the analytical solutions of the wave-induced soil response presented in this chapter, the wave-induced seabed instability, including liquefaction and shear failure, is investigated in Chap. 4.

3.8 List of Coefficients B_i and C_i

Coefficients C_1 to C_6 in (3.51)–(3.60) can be expressed as follows:

$$C_i = \frac{D_i}{D_0}, \quad \text{for } i = 1, 2, \ldots, 6, \tag{3.121}$$

where

$$D_j = C_{j0} + C_{j1}e^{-2kh} + C_{j2}e^{-(k+\delta)h} + C_{j3}e^{-4kh} + C_{j4}e^{-2\delta h} + C_{j5}e^{-2(k+\delta)h} + C_{j6}e^{-(3k+\delta)h} + C_{j7}e^{-(4k+2\delta)h}, \quad \text{for } j = 0, 1, 2, \ldots, 6. \tag{3.122}$$

The common denominator D_0 can be calculated upon inserting the following coefficients into (3.122):

$$C_{00} = (\delta - k)^2(\delta - \delta\mu + k\mu + k\lambda)(B_1 + \lambda B_7), \tag{3.123}$$

$$C_{01} = -2\delta\{(k^2\mu - \delta^2 + \delta^2\mu)^2 + k^4(1-2\mu)^2 + 2k^2h^2(1-\mu)^2(\delta^2 - k^2)^2\} + 4k^2h(\delta^4 - k^4)(1-2\mu) + \lambda B_8, \tag{3.124}$$

$$C_{02} = -8\delta k^2(1 - \lambda - 2\mu)[kh(\delta^2 - k^2)(1-\mu) - \delta^2(1-\mu) + k^2\mu + \lambda B_9], \tag{3.125}$$

$$C_{03} = (\delta + k)^2(\delta - \delta\mu - k\mu k\lambda)(B_2 + \lambda B_{10}), \tag{3.126}$$

$$C_{04} = C_{03}, \tag{3.127}$$

$$C_{05} = C_{01} - 8k^2h(\delta^4 - k^4)(1-\mu)(1 - \lambda - 2\mu), \tag{3.128}$$

$$C_{06} = C_{02} + 16\delta k^3h(\delta^2 - k^2)(1-\mu)(1 - \lambda - 2\mu), \tag{3.129}$$

$$C_{07} = C_{00}. \tag{3.130}$$

The final expression of coefficients C_i (for $i = 1$–6) can be obtained from $C_i = D_i/D_0$ using (3.121) in conjunction with the coefficients C_{ij} listed below. Where zero-values C_{ij} coefficients exist, they are indicated collectively at the end of this listing.

$$C_{10} = -\lambda(\delta - k)[-\delta(\delta - 2k)(1-\mu) + k^2\mu](B_1 + \lambda B_7), \tag{3.131}$$

$$C_{11} = 2k^2h(\delta + k)(\delta - \delta\mu - k\mu)B_3 + \lambda(B_{11} + B_{12}), \tag{3.132}$$

$$C_{14} = \lambda(\delta + k)\big[\delta(\delta + 2k)(1 - \mu) - k^2\mu\big](B_2 + \lambda B_{10}) \tag{3.133}$$

$$C_{15} = 2k^2h(\delta - k)(\delta - \delta\mu + k\mu)B_4 + \lambda(B_{12} - B_{11}), \tag{3.134}$$

$$C_{16} = 2\delta k^2\lambda\big\{(1 - \mu)\big[\delta^2(1 + 2\lambda) + kh\big(\delta^2 - k^2\big)\big] + k^2(2\lambda\mu - 3\lambda - \mu)\big\}, \tag{3.135}$$

$$C_{20} = (\delta - k)^2(\delta - \delta\mu + k\mu)(B_1 + \lambda B_7), \tag{3.136}$$

$$C_{22} = 4\delta k^2(1 - 2\mu)\big[2\delta^2(1 - \mu) - 2k^2\mu - kh(1 - \mu)\big(\delta^2 - k^2\big)\big], \tag{3.137}$$

$$C_{24} = (\delta + k)^2(\delta - \delta\mu - k\mu)(B_2 + \lambda B_{10}), \tag{3.138}$$

$$C_{25} = C_{00} + (\delta - k)\big[(\delta - k)(\delta - \delta\mu + k\mu)B_6 + \lambda(B_{14} - B_{15})\big] \tag{3.139}$$

$$C_{26} = 2\delta k^2\lambda(1 - \mu)\big(\delta^2 - k^2\big), \tag{3.140}$$

$$C_{31} = 2k^2h(\delta - k)(\delta - \delta\mu + k\mu)B_3 + \lambda(B_{11} + B_{12} - 2kB_{17}), \tag{3.141}$$

$$C_{32} = -2\delta^2k\lambda\big[(kh - \lambda)\big(\delta^2 - k^2\big)(1 - \mu) - \delta^2(1 - \mu) + k^2\mu + \lambda B_9\big], \tag{3.142}$$

$$C_{33} = C_{14}, \tag{3.143}$$

$$C_{35} = 2k^2h(\delta + k)(\delta - \delta\mu - k\mu)B_4 + \lambda(B_{12} - B_{11} - 2kB_{18}), \tag{3.144}$$

$$C_{36} = -C_{12} - 4\delta k^2\lambda B_{19}, \tag{3.145}$$

$$C_{41} = C_{25} - 2C_{00} - 2\lambda(\delta - k)(B_{20} + B_{21}), \tag{3.146}$$

$$C_{42} = -C_{26}, \tag{3.147}$$

$$C_{43} = -C_{24}, \tag{3.148}$$

$$C_{45} = C_{21} - 2C_{03} - 2\lambda(\delta + k)(B_{20} - B_{21}), \tag{3.149}$$

$$C_{46} = 4\delta k^2(1 - 2\mu)\big[2k^2\mu - 2\delta^2(1 - \mu) - kh(1 - \mu)\big(\delta^2 - k^2\big)\big] - \lambda B_{16}, \tag{3.150}$$

$$C_{47} = -C_{20}, \tag{3.151}$$

$$C_{50} = \lambda(\delta - k)(B_1 + \lambda B_7), \tag{3.152}$$

$$C_{51} = -4k^2h(1 - 2\mu)B_3 + 2\lambda\big[B_{22} - k^4h(1 - 2\mu)\big], \tag{3.153}$$

$$C_{52} = -2k(1 - \lambda - 2\mu)\big[-kh(\delta - k)(\delta - \delta\mu + k\mu) + \lambda B_{23}\big], \tag{3.154}$$

$$C_{56} = 2k(1 - \lambda - 2\mu)\big[-kh(\delta - k)(\delta - \delta\mu + k\mu) + \lambda B_{24}\big], \tag{3.155}$$

$$C_{62} = -C_{56} + 4k\lambda(1 - \lambda - 2\mu)B_{24}, \tag{3.156}$$

$$C_{64} = -C_{53}, \tag{3.157}$$

$$C_{65} = -4k^2h(1 - 2\mu)B_4 + 2\lambda\big[2k^4h(1 - 2\mu) + B_{22}\big], \tag{3.158}$$

$$C_{66} = -C_{52} - 4k\lambda(1 - \lambda - 2\mu)B_{23}, \tag{3.159}$$

$$C_{67} = C_{50}. \tag{3.160}$$

The zero-vale C_{ij} coefficients are

$$C_{13} = C_{17} = C_{23} = C_{27} = C_{30} = C_{34} = C_{40} = C_{44} = C_{54} \\ = C_{55} = C_{57} = C_{60} = C_{61} = C_{63} = 0. \tag{3.161}$$

The 24 B_{ij} coefficients in (3.123)–(3.161) above are given by

$$B_1 = k^2\mu - (1-\mu)(\delta^2 + \delta k + k^2), \tag{3.162}$$

$$B_2 = -\delta^2 + \delta k - k^2 + \delta^2\mu - \delta k\mu + 2k^2\mu, \tag{3.163}$$

$$B_3 = (\delta^3 h - k^2 - \delta k^2 h)(1-\mu) + k^2\mu, \tag{3.164}$$

$$B_4 = (\delta^3 h + k^2 - \delta k^2 h)(1-\mu) - k^2 h, \tag{3.165}$$

$$B_5 = 2\delta kh(\delta k)(1-\mu), \tag{3.166}$$

$$B_6 = 2\delta kh(\delta + k)(1-\mu), \tag{3.167}$$

$$B_7 = -2\delta(\delta + k)(1-\mu) + k^2, \tag{3.168}$$

$$B_8 = 4k^6 h(1-\mu) + 4\delta k^4\big[(1-4\mu+\mu^2) - \lambda(3-3\mu+\mu^2)\big] \\ + 4\delta^3 k^2(1-\mu)\big[2 + \lambda(3-2\mu)\big] \\ - 4\delta^4 k^2 h(1-\mu) - 4\delta^5(1+\lambda)(1-\mu)^2, \tag{3.169}$$

$$B_9 = -\delta^2(1\mu) + k^2(2-\mu), \tag{3.170}$$

$$B_{10} = -2\delta(\delta - k)(1-\mu) + k^2, \tag{3.171}$$

$$B_{11} = k^5(1-\lambda-2\mu)02k^6 h\mu + 2\delta^4 k(1-\mu)^2(1+2kh+\lambda) \\ - \delta^2 k^3(1-\mu)\big[1 + 2kh(1-2\mu) + k\lambda(5-4\mu)\big], \tag{3.172}$$

$$B_{12} = \delta k^4\big[2kh(1-\mu-mu^2) + \lambda(2+\mu-\mu^2) - (2-6\mu+3\mu^2)\big] \\ - \delta^3 k^2(1-\mu)(2kh+3\lambda+2\mu) + \delta^5(1-\mu)^2(1+2kh+2\lambda), \tag{3.173}$$

$$B_{13} = 2k^4\delta\big[kh(2-5\mu+4\mu^2) - \lambda(5-10\mu+4\mu^2) - \mu(3-4\mu)\big] \\ + k^2\delta^3(1-\mu)\big[(3-4\mu) - 2kh(5-4\mu) + 2\lambda(1-2\mu)\big], \tag{3.174}$$

$$B_{14} = -k^4(1-\lambda-2\mu) - 2\delta^2 k^2(2-\mu)(1-\mu) + 2\delta^4(1-\mu)^2, \tag{3.175}$$

$$B_{15} = \delta k^3(\lambda+\mu)(3-2\mu) + \delta^3 k(1-\mu)(1-2\lambda-2\mu), \tag{3.176}$$

$$B_{16} = -2\delta k^2\big[\delta^2(1-\mu)(1+4\mu) + k^2(3-11\mu+4\mu^2)\big], \tag{3.177}$$

$$B_{17} = k^4\mu(1-\lambda-2\mu) + 2\delta k^4 h(1-\mu-\mu^2) - \delta^2 k^2(1-\mu)\big[1+\lambda(5-4\mu)\big] \\ - 2\delta^3 k^2 h(1-\mu) + 2\delta^4(1+2\lambda)(1-\mu)^2 + 2\delta^5 h(1-\mu)^2, \tag{3.178}$$

$$B_{18} = -k^4\mu(1-\lambda-2\mu) + 2\delta k^4 h\big(1-\mu-\mu^2\big) - \delta^2 k^2(1-\mu)\big[1+\lambda(5-4\mu)\big] - 2\delta^3 k^2 h(1-\mu) - 2\delta^4(1+2\lambda)(1-\mu)^2 + 2\delta^5 h(1-\mu)^2, \quad (3.179)$$

$$B_{19} = -\delta^2(1-\mu)\big[3-4\mu+2\lambda(1-2\mu)\big] + k^2\big[\mu(3-4\mu)+\lambda\big(5-10\mu+4\mu^2\big)\big], \quad (3.180)$$

$$B_{20} = B_{14}, \qquad B_{21} = -B_{15}, \qquad B_{22} = \delta B_{19}, \quad (3.181)$$

$$B_{23} = -\delta(\delta+2k)(1-\mu) + k^2\mu, \qquad B_{24} = -\delta(\delta-2k)(1-\mu) + k^2\mu. \quad (3.182)$$

References

1. Badiey, M., Zen, K., Yamazaki, H., Suzuki, H.: Field and laboratory measurement of shear modulus profile in seabed. Rep. Port Harb. Res. Inst. Jpn. **29**(4), 3–26 (1990)
2. Bear, J., Verruijt, A.: Modeling Groundwater Flow and Pollution. Reidel, Dordrecht (1987)
3. Biot, M.A.: General theory of three-dimensional consolidation. J. Appl. Phys. **26**(2), 155–164 (1941)
4. Bjerrum, J.: Geotechnical problem involved in foundations of structures in the North Sea. Géotechnique **23**(3), 319–358 (1973)
5. Esrig, M.I., Kirby, R.C.: Implication of gas content for predicting the stability of submarine slopes. Mar. Geotechnol. **17**, 58–67 (1977)
6. Fenton, J.D.: Wave forces on vertical walls. J. Waterw. Port Coast. Ocean Eng. **114**(4), 693–718 (1985)
7. Fenton, J.D.: Wave forces on vertical walls. In: Proceedings Australian Conference on Coastal and Ocean Engineering, pp. 315–324 (1985)
8. Gatmiri, B.: A simplified finite element analysis of wave-induced effective stress and pore pressures in permeable sea beds. Géotechnique **40**(1), 15–30 (1990)
9. Gatmiri, B.: Response of cross-anisotropic seabed to ocean waves. J. Geotech. Eng. **118**(9), 1295–1314 (1992)
10. Hsu, J.R.C., Jeng, D.S.: Wave-induced soil response in an unsaturated anisotropic seabed of finite thickness. Int. J. Numer. Anal. Methods Geomech. **18**(11), 785–807 (1994)
11. Hsu, J.R.C., Jeng, D.S., Lee, C.P.: Oscillatory soil response and liquefaction in an unsaturated layered seabed. Int. J. Numer. Anal. Methods Geomech. **19**(12), 825–849 (1995)
12. Hsu, J.R.C., Jeng, D.S., Tsai, C.P.: Short-crested wave-induced soil response in a porous seabed of infinite thickness. Int. J. Numer. Anal. Methods Geomech. **17**(8), 553–576 (1993)
13. Hsu, J.R.C., Tsuchiya, Y., Silvester, R.: Third-order approximation to short-crested waves. J. Fluid Mech. **90**, 179–196 (1979)
14. Ishihara, K., Yamazaki, A.: Analysis of wave-induced liquefaction in seabed deposit of sand. Soil Found. **24**(3), 85–100 (1984)
15. Jeng, D.S.: Wave-induced seabed instability in front of a breakwater. Ocean Eng. **24**(10), 887–917 (1997)
16. Jeng, D.S., Cha, D.H.: Effects of dynamic soil behavior and wave non-linearity on the wave-induced pore pressure and effective stresses in porous seabed. Ocean Eng. **30**(16), 2065–2089 (2003)
17. Jeng, D.S., Hsu, J.R.C.: Wave-induced soil response in a nearly saturated seabed of finite thickness. Géotechnique **46**(3), 427–440 (1996)
18. Jeng, D.S., Lin, Y.S.: Finite element modelling for water waves-soil interaction. Soil Dyn. Earthq. Eng. **15**(5), 283–300 (1996)
19. Jeng, D.S., Lin, Y.S.: Non-linear wave-induced response of porous seabed: A finite element analysis. Int. J. Numer. Anal. Methods Geomech. **21**(1), 15–42 (1997)

20. Liu, P.L.F.: Damping of water waves over porous bed. J. Hydraul. Div. **99**(12), 2263–2271 (1973)
21. Lundgren, H., Lindhardt, J.H.C., Romold, C.J.: Stability of breakwaters on porous foundation. In: Proceeding of 12th International Conference on Soil Mechanics and Foundation Engineering, vol. 1, pp. 451–454 (1989)
22. Madsen, O.S.: Wave-induced pore pressures and effective stresses in a porous bed. Géotechnique **28**(4), 377–393 (1978)
23. Maeno, Y.H., Hasegawa, T.: Evaluation of wave-induced pore pressure in sand layer by wave steepness. Coast. Eng. Jpn. **28**, 31–44 (1985)
24. Marchant, T.R., Roberts, A.J.: Properties of short-crested waves in water of finite depth. J. Aust. Math. Soc. Ser. B, Appl. Math. **29**, 103–125 (1987)
25. Mei, C.C., Foda, M.A.: Wave-induced response in a fluid-filled poro-elastic solid with a free surface-a boundary layer theory. Geophys. J. R. Astron. Soc. **66**, 597–631 (1981)
26. Mynett, A.E., Mei, C.C.: Wave-induced stresses in a saturated poroelastic seabed beneath a rectangular caisson. Géotechnique **32**, 235–248 (1982)
27. Okusa, S.: Wave-induced stress in unsaturated submarine sediments. Géotechnique **35**(4), 517–532 (1985)
28. Pietruszczak, S., Pande, G.N.: Constitutive relations for partially saturated soils containing gas inclusions. J. Geotech. Eng. **122**(1), 50–59 (1996)
29. Putnam, J.A.: Loss of wave energy due to percolation in a permeable sea bottom. Trans. Am. Geophys. Union **30**(3), 349–356 (1949)
30. Roberts, A.J.: Highly nonlinear short-crested water waves. J. Fluid Mech. **135**, 310–321 (1983)
31. Roberts, A.J., Schwartz, L.W.: The calculation of nonlinear short-crested gravity waves. Phys. Fluids **26**(9), 2388–2392 (1983)
32. Seed, H.B., Rahman, M.S.: Wave-induced pore pressure in relation to ocean floor stability of cohesionless soils. Mar. Geotechnol. **3**(2), 123–150 (1978)
33. Silvester, R.: Sediment by-passing across coastal inlets by natural means. Coast. Eng. **9**(4), 327–345 (1985)
34. Silvester, R.: The influence of oblique reflection on breakwaters. In: Proceedings 20th International Conference on Coastal Engineering, ASCE, vol. 3, pp. 2253–2267 (1986)
35. Silvester, R., Hsu, J.R.C.: Scouring due to reflection of oblique waves on breakwaters. In: Proceedings 8th Australian Conference on Coastal and Ocean Engineering, pp. 145–149 (1987)
36. Silvester, R., Hsu, J.R.C.: Sines revisited. J. Waterw. Port Coast. Ocean Eng. **115**(3), 327–344 (1989)
37. Silvester, R., Hsu, J.R.C.: Coastal Stabilization. PTR Prentice-Hall, New Jersey (1993)
38. Smith, A.W., Gordon, A.D.: Large breakwater toe failures. J. Waterways Harbor Coast. Eng. Div. **109**(2), 253–255 (1983)
39. Sumer, B.M.: Liquefaction Around Marine Structures. World Scientific, New Jersey (2013)
40. Thomas, S.D.: A finite element model for the analysis of wave induced stresses, displacements and pore pressure in an unsaturated seabed. I: Theory. Comput. Geotech. **8**(1), 1–38 (1989)
41. Thomas, S.D.: A finite element model for the analysis of wave induced stresses, displacements and pore pressure in an unsaturated seabed. II: Model verification. Comput. Geotech. **17**(1), 107–132 (1995)
42. Tsai, C., Jeng, D.S.: Forces on vertical walls due to obliquely-incident waves. In: Proceedings 22nd International Conference on Coastal Engineering, ASCE, pp. 1742–1754 (1990)
43. Tsai, C.P., Jeng, D.S.: A Fourier approximation for finite amplitude short-crested waves. J. Chin. Inst. Eng. **15**(6), 713–721 (1992)
44. Tsui, Y.T., Helfrich, S.C.: Wave-induced pore pressures in submerged sand layer. J. Geotech. Eng. **109**(4), 603–618 (1983)
45. Ulker, M.B.C., Rahman, M.S., Jeng, D.S.: Wave-induced response of seabed: Various formulations and their applicability. Appl. Ocean Res. **31**(1), 12–24 (2009)
46. Verruijt, A.: Elastic storage of aquifers. In: De Wiest, R.J.M. (ed.) Flow Through Porous Media, pp. 331–376. Academic Press, New York (1969)

47. Wheeler, S.J.: A conceptual model for soils containing large gas bubbles. Géotechnique **38**(3), 389–397 (1988)
48. Wroth, C.P., Houlsby, G.T.: Soil mechanics: Property characterization and analysis procedures. In: Proceedings 11th International Conference on Soil Mechanics and Foundation Engineering, pp. 1–55 (1985)
49. Yamamoto, T.: Wave induced instability seabed. In: Proceedings A.S.C.E. Special Conference, Coastal Sediments'77, pp. 898–913 (1977)
50. Yamamoto, T.: On the response of a coulomb-damped poro-plastic bed to water waves. Mar. Geotechnol. **5**(2), 93–130 (1983)
51. Yamamoto, T., Koning, H., Sellmeijer, H., Hijum, E.V.: On the response of a poro-elastic bed to water waves. J. Fluid Mech. **87**(1), 193–206 (1978)
52. Yamamoto, T., Trevorrow, M.: Experimental verifications of bottom shear modulus profiler (bsmp) method. In: Proceedings of the International Conference on Geotechnical Engineering for Coastal Development—Theory and Practice on Soft Ground (Geot-Coastal 91), vol. 1, pp. 123–128 (1991)
53. Ye, J., Jeng, D.S.: Effects of shear stresses on the wave-induced dynamic response in a porous seabed: Poro-wssi (shear). Acta Mech. Sin. **27**(6), 898–911 (2011)
54. Zen, K., Yamazaki, H.: Field observation and analysis of wave-induced liquefaction in seabed. Soil Found. **31**(4), 161–179 (1991)

Chapter 4
Wave-Induced Seabed Instability

Abstract Based on the analytical solutions presented in Chap. 3, two types of seabed instability, liquefaction and shear failure, were discussed. Among these, Mohr-Coumb criterion was used for the estimation of shear failure, and the concept of excess pore pressure was used for the prediction of wave-induced liquefaction. With the analytical solutions and criteria of seabed instability, a parametric study was conducted to examine the effects of wave and soil characteristics on the wave-induced seabed instability. Finally, an optimal design for seabed protection was suggested.

Keywords Seabed instability · Liquefaction · Shear failure · Seabed protection

4.1 Introduction

Evaluation of seabed stability is an important part of the foundation design for various marine facilities, because some structures have failed due to seabed instability and committed subsidence [5, 10, 11].

Generally speaking, the state of excess pore pressure and diminishing effective stresses within sediments may lead to seabed instability due to either shear failure or liquefaction [8]. The shear stresses at a point within the marine sediment, induced by gravity forces and storm waves, may be significant enough to overcome its shear resistance, thus resulting in seabed instability. This type of seabed instability, referred to as *shear failure*, may produce a horizontal movement of sediment.

It is well known that when excess pore pressure equals the downward effective soil weight, liquefaction may occur. This causes a *quick sand* effect in which external loads cannot be supported. As this occurs, the soil particles in a suspended state may be readily transported as a fluid. Thus, *liquefaction* is an extreme form of seabed instability, which may cause a vertical movement of sediment.

Based on an elastic approach, the criteria for seabed instability, including both shear failure and soil liquefaction, are described in this chapter. Using the analytical solutions for soil response presented in Chap. 3, the wave-induced seabed instability is then investigated. Finally, employing the layered solution, protection of seabeds in front of a breakwater will be discussed in detail.

D.-S. Jeng, *Porous Models for Wave-seabed Interactions*,
DOI 10.1007/978-3-642-33593-8_4,

4.2 Shear Failure

The wave-induced shear stress at a point within the sediment may become large enough to overcome its shearing resistance, thus causing seabed failure. Once this occurs, the soil particles have potential to move horizontally. The actual mode of such instability depends on the spatial distribution of the wave-induced shear failure and the shear strength of the sediment. Conventionally, prediction of failure for soils has been based on Mohr-Coulomb's failure criterion, which remains the most widely used in geotechnical engineering practice. Although other criteria of failure have been suggested in the literature [2, 3], Mohr-Coulomb's criterion is used here because of its simplicity and conservatism.

4.2.1 Principal Stresses

In Chap. 3, only the wave-induced incremental changes in effective stresses and pore pressure within soils from the initial equilibrium have been considered. Thus, the effective normal stresses $\bar{\sigma}_x'$, $\bar{\sigma}_y'$ and $\bar{\sigma}_z'$ in the x-, y- and z-directions are given by

$$\bar{\sigma}_x' = \bar{\sigma}_{xo}' - \sigma_x' = -(\gamma_s - \gamma_w)K_o z - \sigma_x', \tag{4.1}$$

$$\bar{\sigma}_y' = \bar{\sigma}_{yo}' - \sigma_y' = -(\gamma_s - \gamma_w)K_o z - \sigma_y', \tag{4.2}$$

$$\bar{\sigma}_z' = \bar{\sigma}_{zo}' - \sigma_z' = -(\gamma_s - \gamma_w)z - \sigma_z', \tag{4.3}$$

where $\bar{\sigma}_{xo}'$, $\bar{\sigma}_{yo}'$ and $\bar{\sigma}_{zo}'$ are the effective stresses at the initial equilibrium in the x-, y- and z-directions, respectively, while γ_s and γ_w are the unit weights of soil and water, respectively.

In (4.1)–(4.3), K_o is the coefficient of earth pressure at rest. The value of K_o for soils ranges from 0.4 to 1.0 and $K_o = 0.5$ is commonly used for marine sediments [9].

Since the initial shear stresses are zero, the shear stresses, $\bar{\tau}_{xz}$, $\bar{\tau}_{yz}$ and $\bar{\tau}_{xy}$, can be expressed as

$$\bar{\tau}_{xz} = -\tau_{xz}, \qquad \bar{\tau}_{yz} = -\tau_{yz}, \quad \text{and} \quad \bar{\tau}_{xy} = -\tau_{xy}. \tag{4.4}$$

For studying the general stress fields that occur in a complicated boundary value problem, it is convenient to use a principal stress space. This leads to a convenient geometric representation of various failure criteria. The effective principal stresses, σ_1, σ_2 and σ_3, can be expressed as [2]

$$\sigma_1 = \frac{\bar{s}}{\sqrt{3}} + \sqrt{\frac{2}{3}}\bar{t}\sin\left(\zeta - \frac{2\pi}{3}\right), \tag{4.5}$$

$$\sigma_2 = \frac{\bar{s}}{\sqrt{3}} + \sqrt{\frac{2}{3}}\bar{t}\sin\zeta, \tag{4.6}$$

$$\sigma_3 = \frac{\bar{s}}{\sqrt{3}} + \sqrt{\frac{2}{3}}\bar{t}\sin\left(\zeta + \frac{2\pi}{3}\right), \tag{4.7}$$

where

$$\bar{s} = \frac{1}{\sqrt{3}}(\bar{\sigma}'_x + \bar{\sigma}'_y + \bar{\sigma}'_z), \tag{4.8}$$

$$\bar{t} = \sqrt{\frac{(\bar{\sigma}'_x - \bar{\sigma}'_y)^2 + (\bar{\sigma}'_x - \bar{\sigma}'_z)^2 + (\bar{\sigma}'_y - \bar{\sigma}'_z)^2 + 6(\bar{\tau}^2_{xy} + \bar{\tau}^2_{xz} + \bar{\tau}^2_{yz})}{3}}, \tag{4.9}$$

and

$$\zeta = \frac{1}{3}\sin^{-1}\left(\frac{-3\sqrt{6}J_4}{\bar{t}^3}\right), \tag{4.10}$$

in which

$$J_4 = S_x S_y S_z - S_x\bar{\tau}^2_{yz} - S_y\bar{\tau}^2_{xz} - S_z\bar{\tau}^2_{xy} + 2\bar{\tau}^2_{xz}\bar{\tau}^2_{yz}\bar{\tau}^2_{xy}, \tag{4.11}$$

$$S_x = \frac{2\bar{\sigma}'_x - \bar{\sigma}'_y - \bar{\sigma}'_z}{3}, \tag{4.12}$$

$$S_y = \frac{2\bar{\sigma}'_y - \bar{\sigma}'_x - \bar{\sigma}'_z}{3}, \tag{4.13}$$

$$S_z = \frac{2\bar{\sigma}'_z - \bar{\sigma}'_x - \bar{\sigma}'_y}{3}. \tag{4.14}$$

Equations (4.5)–(4.7) ensure that $\sigma_1 \leq \sigma_2 \leq \sigma_3$.

In (4.5)–(4.7), three principal stresses, σ_1, σ_2 and σ_3 satisfy a third-order polynomial equation [3]

$$\sigma^3 - J_1\sigma^2 + J_2\sigma - J_3 = 0, \tag{4.15}$$

where σ denotes the principal stress (namely σ_1, σ_2 and σ_3), and J_1, J_2 and J_3 are given by

$$J_1 = \sigma_1 + \sigma_2 + \sigma_3 = \bar{\sigma}'_x + \bar{\sigma}'_y + \bar{\sigma}'_z, \tag{4.16}$$

$$J_2 = \sigma_1\sigma_2 + \sigma_2\sigma_3 + \sigma_1\sigma_3 = \begin{vmatrix} \bar{\sigma}'_y & \bar{\tau}_{yz} \\ \bar{\tau}_{yz} & \bar{\sigma}'_z \end{vmatrix} + \begin{vmatrix} \bar{\sigma}'_z & \bar{\tau}_{xz} \\ \bar{\tau}_{xz} & \bar{\sigma}'_x \end{vmatrix} + \begin{vmatrix} \bar{\sigma}'_x & \bar{\tau}_{xy} \\ \bar{\tau}_{xy} & \bar{\sigma}'_y \end{vmatrix}, \tag{4.17}$$

$$J_3 = \sigma_1\sigma_2\sigma_3 = \begin{vmatrix} \bar{\sigma}'_x & \bar{\tau}_{xy} & \bar{\tau}_{xz} \\ \bar{\tau}_{xy} & \bar{\sigma}'_y & \bar{\tau}_{yz} \\ \bar{\tau}_{xz} & \bar{\tau}_{yz} & \bar{\sigma}'_z \end{vmatrix}. \tag{4.18}$$

In (4.15), J_1, J_2 and J_3 are so called "invariants of stresses", being invariant parameters for arbitrary co-ordinates.

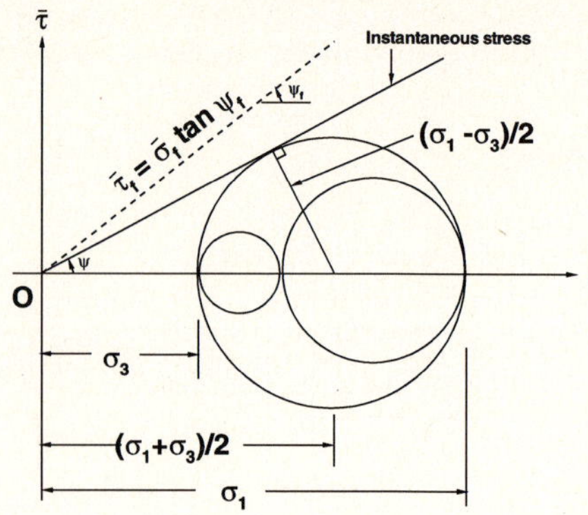

Fig. 4.1 Diagram of Mohr's circle

The stress state at a given location and instant may be expressed by the angle ψ between the tangent from the origin to the instantaneous Mohr's circle and the $\bar{\sigma}$-axis (Fig. 4.1). The *stress angle* ψ is defined by

$$\sin\psi = \frac{\sigma_3 - \sigma_1}{\sigma_3 + \sigma_1} = \frac{\sqrt{3}\bar{t}\cos\zeta}{\bar{t}\sin\zeta - \sqrt{2}\bar{s}}. \tag{4.19}$$

4.2.2 *Mohr-Coulomb's Criterion*

According to Mohr-Coulomb's failure criterion, as shown in Fig. 4.1, the limiting condition in a given soil may be expressed by

$$\bar{\tau}_f = \bar{\sigma}_f \tan\psi_f, \tag{4.20}$$

where ψ_f denotes the angle of internal friction of soil, $\bar{\tau}_f$ and $\bar{\sigma}_f$ represent the shear stress and effective normal stress on the failure plane, respectively. When the stress reaches the failure envelope, the stress angle (ψ) becomes identical to the internal friction angle of soil (ψ_f). Thus, the failure criterion of shear failure at a specific location and time may be defined as

$$\psi \geq \psi_f. \tag{4.21}$$

In general, the value of ψ_f depends on the soil type, for example, 30–35 degrees for sand and 35–40 degrees for small gravels. Based on criterion (4.21), examples showing the distribution of the computed stress angle for gravel and coarse sand are illustrated in Fig. 4.2. If we consider $\psi_f = 30°$ for a sandy seabed, the area above the curve $\psi = 30°$ will fail due to shearing (Fig. 4.2). Similarly, shear failure occurs in the region between the seabed surface and $\psi = 35°$ in a graveled seabed, as shown in Fig. 4.2. A detailed discussion of the wave-induced shear failure will be presented in Sect. 4.4.

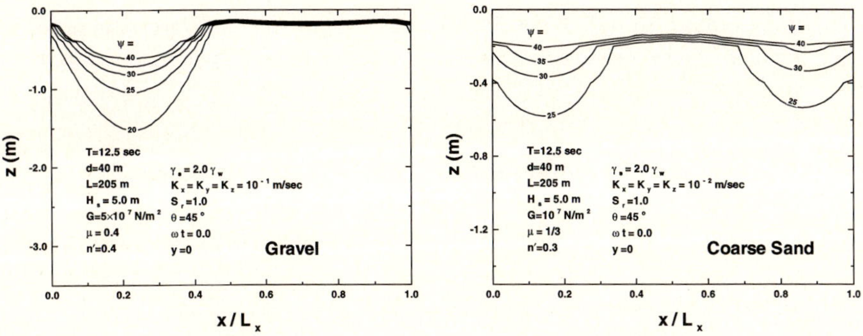

(a) Gravel (b) Coarse sand

Fig. 4.2 Distribution of the computed stress angles in degrees for (**a**) gravel and (**b**) coarse sand

4.3 Soil Liquefaction

The failure mode of liquefaction is quite different from that of a shear failure. Generally speaking, liquefaction is considered as a kind of quick sand or boiling action closely related to seepage flows. The effective stresses and strains generated by waves cause inter-granular slip at grain to grain contacts, and so lead to volumetric compaction. Consequently, the relaxation in soil skeleton transfers effective stresses to the pore water, thus giving rise to the pore water pressure within the soil skeleton. The corresponding reduction in effective stresses leads to a structural rebound in the soil skeleton to absorb the difference in volume due to the increase of pore water pressure. In the extreme case, excess pore pressure may increase until all inter-granular or effective stresses have been eliminated from the system. In this state, no shear resistance of the soil skeleton is expected and liquefaction occurs.

4.3.1 Excess Pore Pressure

An excess pore pressure is defined as the excess component of pressure beyond the hydrostatic pressure. Since the water pressure on the seabed surface varies according to the propagation of waves, it implies that not only the pore pressure but also the hydrostatic pressure changes. The excess pore pressure can be determined by taking into account this variation of the hydrostatic pressure, which is defined by Zen and Yamazaki [15] as

$$\begin{aligned} u_e &= -(P_b - p) \\ &= -p_o \cos nky \cos(mkx - \omega t) \\ &\quad + \frac{p_o}{1-2\mu} \cos nky Re\big\{\big[(1-\lambda-\mu)\big(C_2 e^{kz} - C_4 e^{-kz}\big) \\ &\quad + (1-\mu)\big(\delta^2 - k^2\big)\big(C_5 e^{\delta z} + C_6 e^{-\delta z}\big)\big] \cos nky \, e^{i(mkx-\omega t)}\big\}. \end{aligned} \tag{4.22}$$

The first term on the right-hand side of (4.22) represents the hydrostatic pressure, and the second term denotes the wave-induced pore pressure within the seabed.

4.3.2 Criteria of Liquefaction

It has generally been accepted that, when the vertical effective stress vanishes, the soil will be liquefied. Thus, the soil matrix loses its strength to carry any load and consequently causes seabed instability. However, the mechanisms of the wave-induced soil liquefaction in marine sediments have not been clearly addressed in geotechnical terms at the present stage. The liquefaction is also affected by the state of soil compaction, permeability, the wave-induced cyclic stress as well as the degree of drainage.

To date, several criteria for liquefaction have been suggested. The first is the effective vertical normal stress becomes zero [7], the second is that the mean effective normal stress tends to zero [7, 12] and the third relates to excess pore pressure [15].

Criterion (a): Okusa [7] gauged a *liquefied state* for a non-cohesive soil when its vertical effective normal stress becomes zero, i.e.,

$$-(\gamma_s - \gamma_w)z \le \sigma_z', \tag{4.23}$$

in which the left-hand side of (4.23) represents effective geostatic stress, while the right-hand side is for the vertical effective normal stress.

Criterion (b): Tsai [12] extended Okusa's [7] second measure of a liquefied state, which requires the mean effective normal stresses to become zero in a three-dimensional condition, such that,

$$\frac{1}{3}\left[-(\gamma_s - \gamma_w)(1 + 2K_o)z - \left(\sigma_x' + \sigma_y' + \sigma_z'\right)\right] \le 0. \tag{4.24}$$

In this equation, the first term represents the mean of effective geostatic stress, while the second term represents the mean of wave-induced effective normal stresses.

Criterion (c): Zen and Yamazaki [15] suggested that liquefaction occurs for a two-dimensional progressive wave when

$$-(\gamma_s - \gamma_w)z + (P_b - p) \le 0. \tag{4.25}$$

This equation implies that liquefaction in a seabed may occur when the excess pore pressure becomes greater than the overburden soil pressure.

All the aforementioned criteria utilize the oscillatory effective normal stress and pore pressure, thus neglecting the residual effect, because knowledge of this kind is still in its infancy. Since the shear modulus (G) of a soil, as used in (3.13)–(3.15) is profoundly affected by the effective stress, thus the value of σ_z' computed from (3.57) for finite thickness may be grossly in error when it approaches zero. This implies that these equations, (3.13)–(3.15), may not be valid at low effective stresses. Consequently, the liquefied criteria based on the effective normal stress would not be

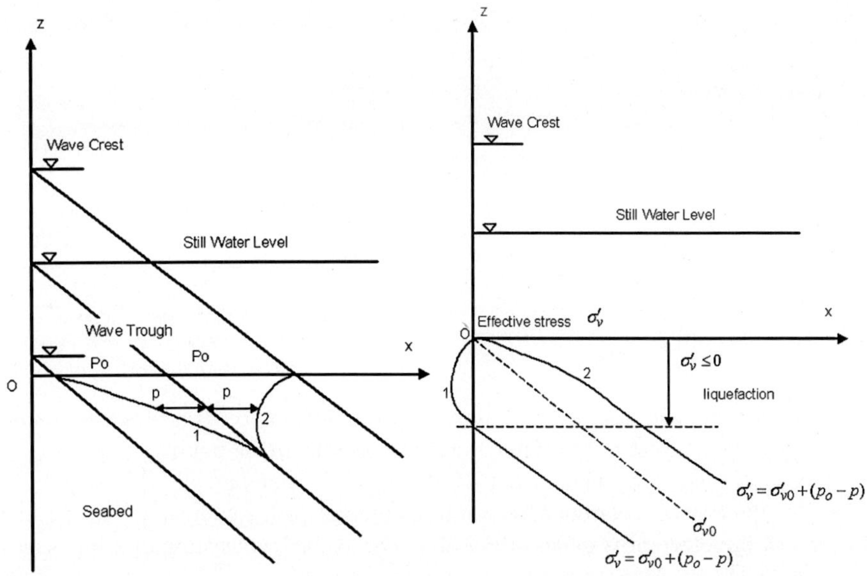

Fig. 4.3 Schematic sketch showing definition of liquefaction based on excess pore pressure

applicable. Thus, the liquefaction criterion (c), employing the excess pore pressure, may probably be the only meaningful condition [4].

Criterion (d): Now, criterion (c) would be extended to a three-dimensional wave condition. To apply the concept of excess pore pressure to a seabed, a schematic drawing of the pore pressure and effective stress distributions is illustrated in Fig. 4.3. The solid curves in the figure indicate the pore pressure beneath a wave trough and a wave crest. The excess pore pressure expressed by (4.22) is transient in nature, because both P_b and p are oscillatory and periodical in real ocean environments. Consequently, the effective stress varies periodically in accordance with the change of the excess pore pressure. If it attains zero or a negative value at certain depths below the seabed surface, the soil skelcton will reach a liquefied state (Fig. 4.3). Thus, the liquefaction criterion can be expressed as

$$-\frac{1}{3}(\gamma_s - \gamma_w)(1 + 2K_o)z \leq u_e. \tag{4.26}$$

The left-hand side of (4.26) represents the average effective geostatic stress, similar to that of (4.24) proposed by Tsai [12].

Two examples comparing the above criteria of liquefaction are illustrated in Fig. 4.4 for a seabed of infinite thickness. The wave and soil conditions used by Okusa [7] are employed in these examples. The regions between the seabed surface and the curves denote the liquefied areas. Criteria (a) and (c) are found to be identical to each other in these examples. The results obtained from criterion (b) are slightly greater than that from criteria (a) and (c), but less than that from criterion

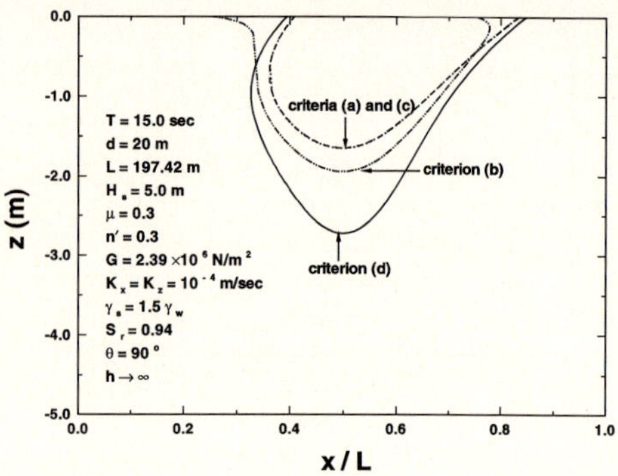

Fig. 4.4 Calculated liquefaction depth in a seabed of infinite thickness for the four liquefaction criteria cited

(d) in both examples. It is worth noting that liquefaction occurs near wave troughs for all criteria in a seabed of infinite thickness, as seen in the figures.

Using field data [16], obtained at HORF (Hazaki Oceanographical Research Facility) of the Port and Harbour Research Institute, Japan, a comparison of the four criteria of liquefaction is given in Fig. 4.5. The seabed, consisting of a fine sandy layer in relatively shallow water, was reported to be isotropic in an almost fully saturated condition, where dynamic wave pressure was significant [16].

Based on the soil and wave conditions reported [16], the liquefaction zone between the seabed surface and their lower boundaries for the four criteria mentioned above can be drawn. From the results presented in Fig. 4.5a, it can be found that all four criteria have predicted seabed liquefaction, but with different areas and maximum liquefied depths. It appears that only the curve of criterion (d), utilizing the excess pore pressure in a three-dimensional condition, encompasses the field point at $z = -0.2$ m (in the figure), whereas the curves for criterion (a)–(c) do not. At least from the case examined, criteria (a) and (b), based on the concept of effective normal stresses, may not be adequate. It is not expected that the predicted liquefied area of criterion (b) is located near the wave crests.

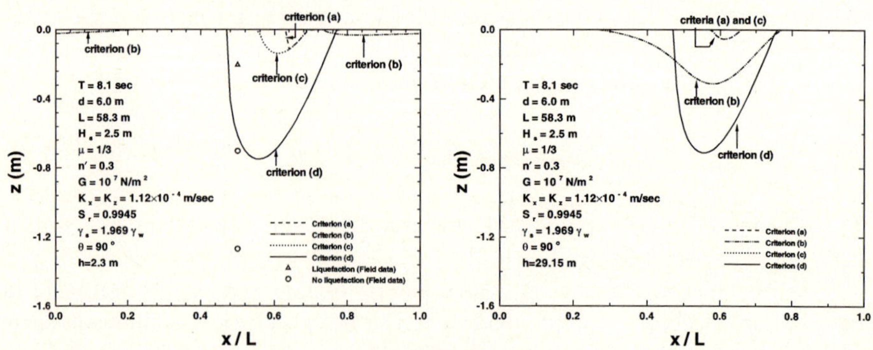

(a) (b)

Fig. 4.5 Calculated liquefaction depth in a sandy seabed of finite thickness for the four liquefaction criteria cited. Field data [16] are also included

A further comparison is presented in Fig. 4.5b, employing the same input data used in Fig. 4.5a, except the seabed thickness (now with $h/L = 0.5$, $h = 29.15$ m). The liquefied areas obtained from all criteria are located near the wave trough. Based on this preliminary assessment, it may be accepted that criterion (d) gives a more realistic result than the others. In fact, criterion (d) can also be derived by using the concept of seepage flow, to be discussed in next section.

In addition to the above criteria, another criterion for 3D liquefaction considering cohesion and friction of soil was proposed recently [14]. In the criterion, the initial consolidation was also included and the examples of wave-current induced liquefaction were given.

4.3.3 Seepage Force

Where a different water pressure gradient exists in the seabed, a seepage flow is generated. This is accompanied by the seepage forces on the soil skeleton in the direction of the flow due to the frictional effects. The seepage forces in the x-, y- and z-directions are given by

$$\begin{aligned} j_x &= \frac{\partial p}{\partial x} \\ &= \frac{imkp_o}{1-2\mu}\{(1-\lambda-2\mu)\big(C_2 e^{kz} - C_4 e^{-kz}\big) \\ &\quad + (1-\mu)\big(\delta^2 - k^2\big)\big(C_5 e^{\delta z} + C_6 e^{-\delta z}\big)\} \cos nky\, e^{i(mkx-\omega t)}, \end{aligned} \tag{4.27}$$

$$\begin{aligned} j_y &= \frac{\partial p}{\partial y} \\ &= \frac{-nkp_o}{1-2\mu}\{(1-\lambda-2\mu)\big(C_2 e^{kz} - C_4 e^{-kz}\big) \\ &\quad + (1-\mu)\big(\delta^2 - k^2\big)\big(C_5 e^{\delta z} + C_6 e^{-\delta z}\big)\} \sin nky\, e^{i(mkx-\omega t)}, \end{aligned} \tag{4.28}$$

$$\begin{aligned} j_z &= \frac{\partial p}{\partial z} \\ &= \frac{p_o}{1-2\mu}\{k(1-\lambda-2\mu)\big(C_2 e^{kz} + C_4 e^{-kz}\big) \\ &\quad + \delta(1-\mu)\big(\delta^2 - k^2\big)\big(C_5 e^{\delta z} - C_6 e^{-\delta z}\big)\} \cos nky\, e^{i(mkx-\omega t)}. \end{aligned} \tag{4.29}$$

The resultant total seepage force (j) in the x–z domain and its direction (θ_s) can be calculated by

$$j = \sqrt{j_x^2 + j_z^2} \quad \text{and} \quad \theta_s = \tan^{-1}\left(\frac{j_z}{j_x}\right), \tag{4.30}$$

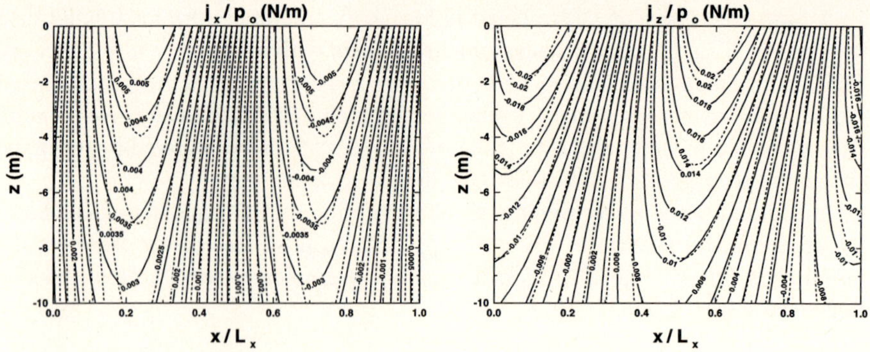

(a) j_x/p_o (b) j_z/p_o

Fig. 4.6 Contour of seepage force (**a**) j_x/p_o in the x-direction and (**b**) j_z/p_o in the z-direction. *Dashed lines* denote the results in a saturated condition, and *solid lines* for an unsaturated condition ($S_r = 0.975$). Remaining input data are the same as that in Fig. 4.2

where θ_s is the angle measured counter-clock-wise from the positive x-direction. Similarly, the total seepage force and the direction in the y–z and the x–y planes can also be derived.

The phenomenon of soil liquefaction resembles the action of boiling, because it is caused by the seepage flow arising through the void of the soil skeleton. Since the body force on the soil skeleton is the seepage forces and the self-weight of the soil particle, the vertical body force can be expressed by

$$-\frac{1+2K_o}{3}(\gamma_s - \gamma_w) - j_z. \tag{4.31}$$

Substituting $j_z = \partial p/\partial z$ into (4.31) and integrating it from 0 to z, renders the vertical effective stress

$$-\frac{1+2K_o}{3}(\gamma_s - \gamma_w)z + (P_b - p). \tag{4.32}$$

It has been well documented that liquefaction occurs when the vertical effective stress is equal to zero or a negative value, i.e.,

$$-\frac{1+2K_o}{3}(\gamma_s - \gamma_w)z + (P_b - p) \leq 0. \tag{4.33}$$

This expression is identical to criterion (d) presented in Sect. 4.3.2. It is beneficial to relate the liquefaction criterion to pore pressure, because it is readily measurable in the laboratory and in the field.

Substituting the differential pore pressure gradients in the x- and z-directions into (4.27) and (4.29) gives the seepage force j_x/p_o and j_z/p_o in each direction (Fig. 4.6). Two values of the degree of saturation ($S_r = 0.975$ and 1.0) are considered here. The other soil and wave characteristics are the same as those in Fig. 4.2b.

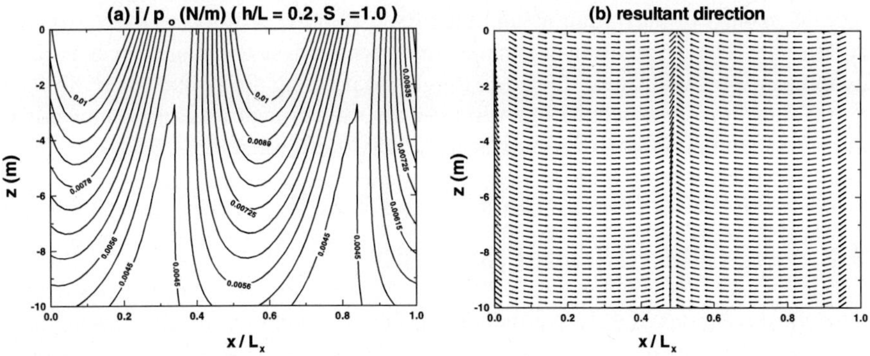

Fig. 4.7 Spatial distribution of (**a**) total seepage force j/p_o and (**b**) its resultant direction in a saturated seabed

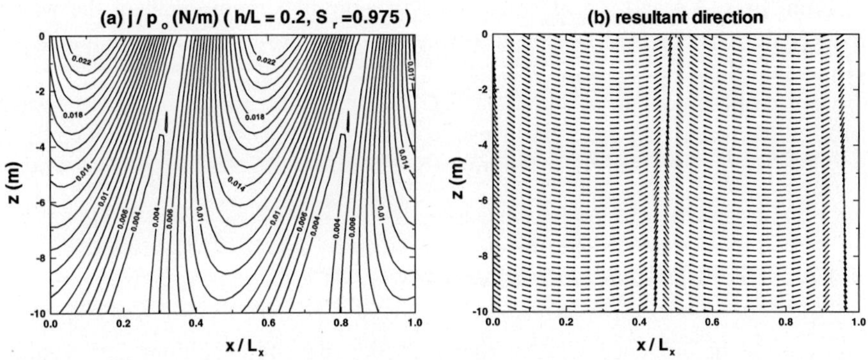

Fig. 4.8 Spatial distribution of (**a**) total seepage force j/p_o and (**b**) its resultant direction in an unsaturated seabed

It is observed that the greatest horizontal seepage forces are at one-quarter and three-quarters of the wave cycle, and become negligible under the wave crest and trough (Fig. 4.6a). The magnitude of horizontal seepage force (j_x/p_o) in a fully saturated seabed (dashed lines) is larger than that in an unsaturated seabed (solid lines), as seen in the figure.

For the vertical seepage forces (j_z/p_o) given in Fig. 4.6b, the trend is similar to that of the horizontal seepage force (in Fig. 4.6a), but with its maximum shifting toward the crest and the trough. The maximum seepage force in the vertical direction for soil of finite thickness is found to be four times larger than that in the horizontal direction (about 0.005 in Fig. 4.6a and 0.02 in Fig. 4.6b).

Figures 4.7 and 4.8 show the magnitude and resultant direction of the total seepage force for a saturated and unsaturated seabed, respectively. Since seabed instability occurs within a relatively thin layer beneath the seabed surface, only a small portion of the soil skeleton is shown (-10 m $\leq z \leq 0$, i.e., $-0.048 \leq z/L \leq 0$). The contours of the resultant seepage force exhibit identical pattern of skewed cells, with its centers of maximum magnitude offset from the position of the wave crest and trough. These figures also depict a strong tendency for seepage forces pushing up the soil at locations near the wave trough and pressing downwards near the wave crest. Thus, it is expected that the soil near the wave trough may move upwards in the vertical direction, thus resulting in instability of the seabed.

4.4 Wave-Induced Seabed Instability

Analytical solutions of the wave-induced soil response in front of a breakwater have been systematically developed in Chap. 3. Based on these general solutions, together with the criteria of shear failure and liquefaction presented in Sects. 4.2 and 4.3, it is possible to examine the influences of wave and soil characteristics on the wave-induced seabed instability.

From Figs. 4.4 and 4.5, it is expected that the maximum depth of the wave-induced liquefaction (z_{Lm}) occurs near wave troughs. However, the maximum depth of possible shear failure (z_{sm}) also occurs near wave crests, although it may cover the entire phase domain as shown in Fig. 4.2. In this section, the effects of wave and soil characteristics are fully investigated in relation to the maximum depths of possible liquefaction and shear failure, z_{Lm} and z_{sm}.

4.4.1 Effect of Wave Characteristics

To examine the influence of wave characteristics, the soil conditions are taken as: seabed thickness $h = 20$ m, soil shear modulus $G = 10^7$ N/m^2, Poisson's ratio $\mu = 1/3$ and porosity $n' = 0.3$. The seabed is considered as hydraulically isotropic with uniform permeability of $K_x = K_y = K_z = 10^{-4}$ m/s, and the unit weight of soil $\gamma_s = 2.0\gamma_w$. The degree of saturation is assumed to vary from 0.95 to 1.0.

The wave characteristics to be examined here include the wave period T, water depth d (in terms of relative water depth, d/L), wave height H_s (in term of wave steepness, H_s/L) and wave obliquity θ.

4.4.1.1 Wave Period

The distribution of the maximum depth of liquefaction (z_{Lm}) versus the degree of saturation (S_r) is presented in Fig. 4.9a for wave period (T) ranging from 10 s to 15 s and wave height of $H_s = 5.0$ m. Generally speaking, the maximum liquefied depth z_{Lm} decreases as wave period (T) decreases. This implies that a longer period wave has a greater liquefaction potential than a shorter period wave under the same wave height and water depth. This may be explained by the fact that the relative water depth d/L becomes smaller (i.e., in shallower water) for a longer wave period in the same water depth (see Fig. 4.10 later).

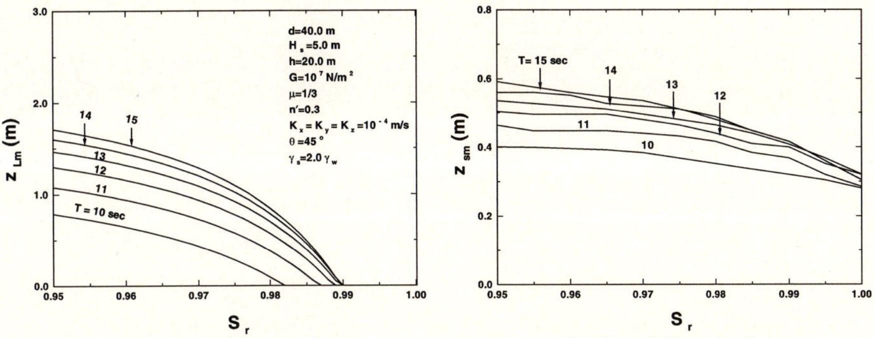

(a) liquefaction depth z_{Lm} (b) shear failure depth z_{sm}

Fig. 4.9 Distribution of the maximum (**a**) liquefaction depth z_{Lm} and (**b**) shear failure depth z_{sm} versus the degree of saturation S_r for various wave period T

Figure 4.9b shows the distribution of the maximum depth of shear failure (z_{sm}) versus the degree of saturation for various values of wave period. Again, the values of z_{sm} increase with the wave period T. However, the influence of wave period (T) on z_{sm} may be less significant than that on z_{Lm}. For example, the maximum liquefied depths z_{Lm} are 0.79 m and 1.7 m for $T = 10$ and 15 s, respectively at $S_r = 0.95$ (i.e., $z_{Lm(T=15)}/z_{Lm(T=10)} = 2.17$). However, the maximum depths of shear failure z_{sm} are 0.4 m and 0.58 m for $T = 10$ and 15 s, respectively at $S_r = 0.95$ (i.e., $z_{sm(T=15)}/z_{sm(T=10)} = 1.45$). This implies that wave period T affects the liquefaction potential more significantly than shear failure.

4.4.1.2 Water Depth

Relative water depth d/L has been recognized by coastal engineers as a dominant factor of wave characteristics. The wave field is considered as in shallow water, when $d/L \leq 0.1$, and as deep water for $d/L \geq 0.5$. In this example, the relative water depth considered is from 0.1 for shallow to 0.3 for intermediate water. The maximum liquefied depth z_{Lm} increases as relative depth d/L decreases (Fig. 4.10a). However, no liquefaction occurs when $d/L > 0.25$ with a wave height of 5 m. This implies that waves in a shallow water may cause a larger liquefaction depth than that in a deep water under a same wave height. However, it is expected that liquefaction may occur in deeper water under a larger wave (not shown in graphs).

The distribution of the maximum shear failure depth (z_{sm}) versus the degree of saturation for various values of relative water depth (d/L) is illustrated in Fig. 4.10b. It is observed that z_{sm} increases as relative water depth d/L decreases. The figure also shows that values of z_{sm} for different values of d/L become more significant when the degree of saturation S_r approaches unity (i.e., the fully saturated condition). It is worth noting that shear failure occurs at $d/L = 0.30$, at which no liquefaction is expected (Fig. 4.10).

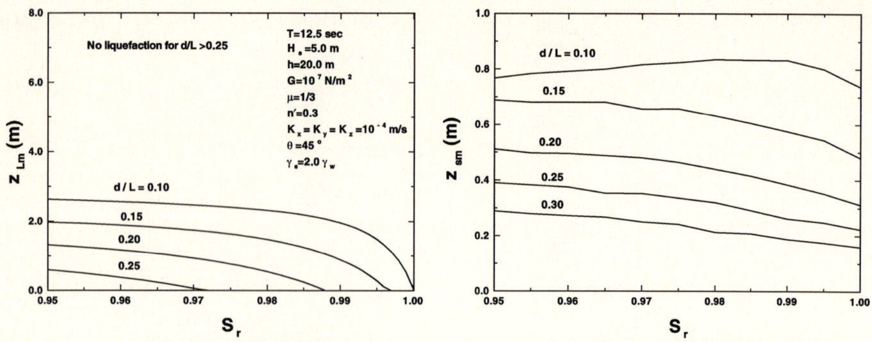

(a) liquefaction depth z_{Lm} (b) shear failure depth z_{sm}

Fig. 4.10 Distribution of the maximum (**a**) liquefaction depth z_{Lm} and (**b**) shear failure depth z_{sm} versus the degree of saturation S_r for various relative water depth d/L

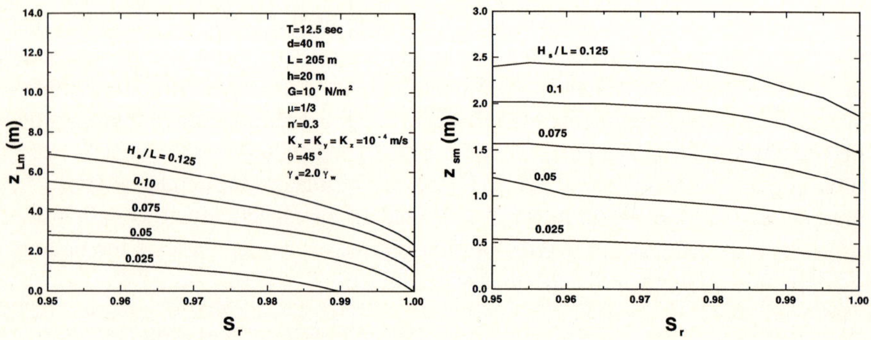

(a) liquefaction depth z_{Lm} (b) shear failure depth z_{sm}

Fig. 4.11 Distribution of the maximum (**a**) liquefaction depth z_{Lm} and (**b**) shear failure depth z_{sm} versus the degree of saturation S_r for various wave steepness H_s/L

4.4.1.3 Wave Height

In engineering practice, wave height has been commonly expressed in term of wave steepness, H_s/L, which is the ratio of wave height to wave length. In the examples presented in Fig. 4.11, wave steepnesses varying from 0.025 to 0.125 are considered. The maximum liquefied depth, z_{Lm}, increases as wave steepness H_s/L increases, as shown in Fig. 4.11a. For example, the maximum liquefied depth z_{Lm} is about 1.47 m for $H_s/L = 0.025$, and reaches 6.93 m for $H_s/L = 0.125$, at $S_r = 0.95$. However, this difference is reduced as S_r increases. It is worth noting that liquefaction does not occur in a saturated seabed under a small wave (e.g. $H_s/L = 0.025$, 0.05 in Fig. 4.11), but it occurs in such a seabed under a large wave (e.g. $H_s/L = 0.075$, 0.10, 0.125 in the same figure).

Similar trends can also be observed in Fig. 4.11b for the maximum depth of shear failure, z_{sm}. The figure clearly shows that z_{sm} increases with wave steepness

H_s/L. However, the differential increase in z_{sm} is significant for a larger wave. For example, the value of z_{sm} for a wave with $H_s/L = 0.125$ is almost five times of that with $H_s = 0.025$ at $S_r = 0.95$.

Upon comparing the results presented in Fig. 4.9, it seems that the influences of wave height on the magnitude of the wave-induced liquefaction and shear failure are more significant than that of wave period and water depth.

4.4.1.4 Wave Obliquity

Unlike other wave characteristics, wave obliquity does not directly affect the magnitude of the wave-induced liquefaction and shear failure, but it affects the location where they may occur. Based on the same wave and soil conditions in Fig. 4.11 (with $H_s = 5.0$ m), except for the wave obliquity θ to vary from 30° to 60°, the contours of liquefaction depth are plotted in Figs. 4.12–4.15, together with the water surface-profiles.

A three-dimensional short-crested wave system produced by oblique reflection in front of a vertical wall has a wave length (L_x) in the resultant progressive direction, and a crest length (L_y) normal to the former, as depicted in Fig. 3.1. For $\theta = 30°$, $L_x = L/\sin\theta = 410.18$ m and $L_y = L/\cos\theta = 236.82$ m, since the wave length $L = 205.09$ m for the wave period and water depth specified ($T = 12.5$ s, $d = 40$ m). At $\theta = 30°$, the three-dimensional perspective sketch and contour of water surface elevations are shown in Fig. 4.12, where intersections of the wave crests are also depicted.

For a porous seabed of fine sand ($h = 20$ m), the contours at 0.05 m intervals of the instantaneous liquefaction depth for $\theta = 30°$ are given in Fig. 4.13. From Figs. 4.12 and 4.13, it is obvious that transient liquefaction occurs in the vicinity of the wave trough. The calculated maximum liquefaction depth is about 0.8 m. The instantaneous liquefaction depth for $\theta = 45°$ and $\theta = 60°$ are shown in Figs. 4.14 and 4.15, respectively.

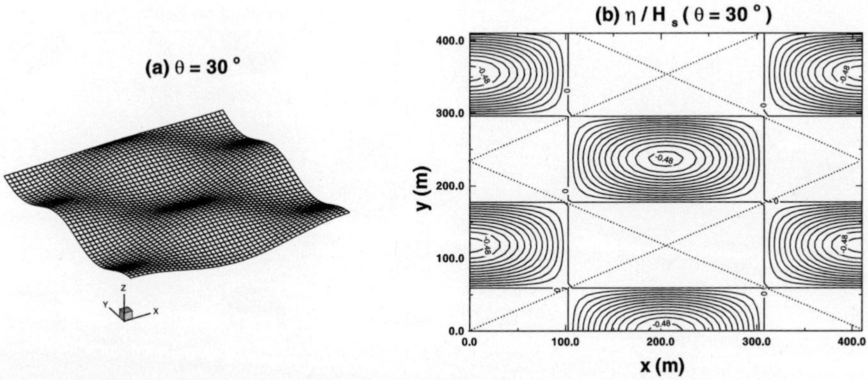

Fig. 4.12 (**a**) Three-dimensional perspective sketch and (**b**) contours of the water surface elevation for wave incident angle $\theta = 30°$

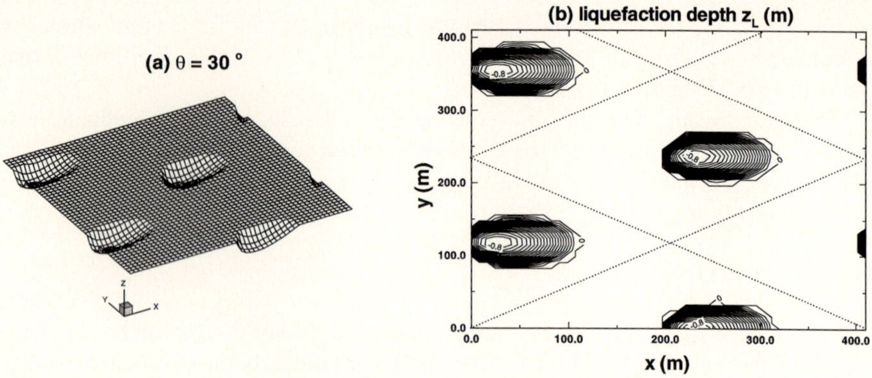

Fig. 4.13 (a) Three-dimensional perspective sketch and (b) contours of the wave-induced liquefaction for wave incident angle $\theta = 30°$

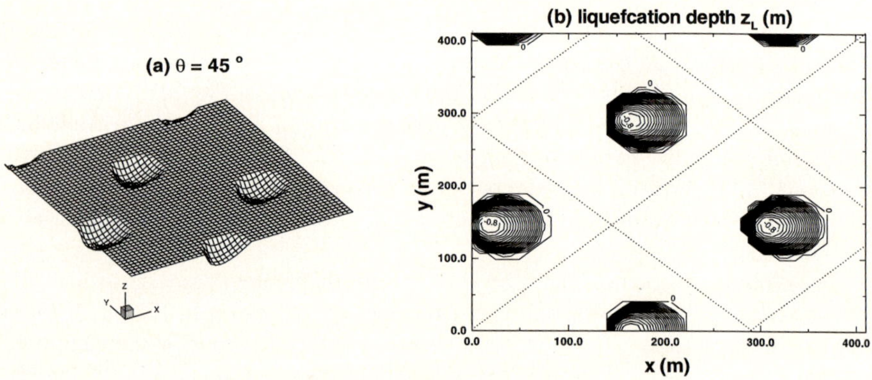

Fig. 4.14 (a) Three-dimensional perspective sketch and (b) contours of the wave-induced liquefaction for wave incident angle $\theta = 45°$

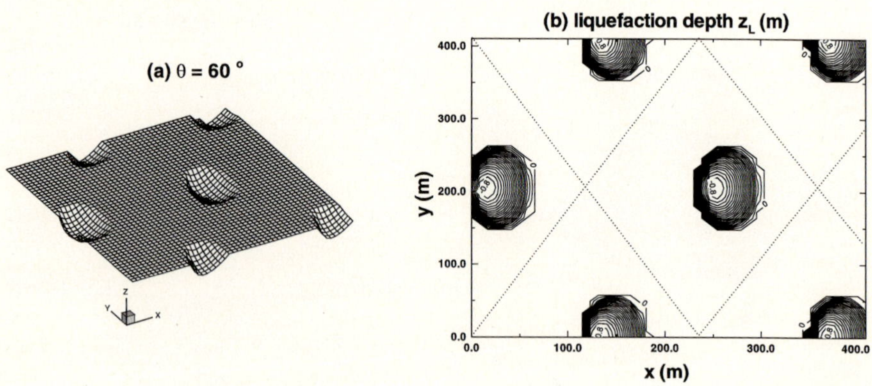

Fig. 4.15 (a) Three-dimensional perspective sketch and (b) contours of the wave-induced liquefaction for wave incident angle $\theta = 60°$

It is found that all liquefaction contours are skew, with their elliptic shapes and orientations varying with the angle of wave incident θ. Since the resultant wave system propagates in the x-direction (see Fig. 3.1), the transient liquefaction patterns have the potential to produce a scour trench along each of these alignments parallel to the wall where liquefaction is predicted. The actual intervals between any two consecutive trench alignments depend upon the incident wave angle θ.

In summary, the maximum depth of the wave-induced seabed instability (including liquefaction and shear failure) increases with wave periods (T) and wave heights (in terms of wave steepness H_s/L). However, it increases as water depth (in terms of d/L) decreases. Furthermore, the wave obliquity θ affects the occurring location of the maximum liquefaction depth. It is worth noting that consistent trends exist for the maximum liquefaction depth z_{Lm} and shear failure z_{sm} versus wave characteristics (Figs. 4.9–4.11).

4.4.2 Effect of Soil Characteristics

To investigate the influence of soil characteristics on the wave-induced seabed instability, a wave with period $T = 12.5$ s, height $H_s = 5$ m and water depth $d = 40$ m are used in the following examples. The resultant wave length L is 205 m. The soil characteristics examined here include seabed thickness, soil type (in terms of soil permeability) and the degree of saturation.

4.4.2.1 Seabed Thickness

The distribution of maximum liquefaction depth, z_{Lm}, versus the degree of saturation, S_r, for the range of relative seabed thickness h/L from 0.05 to 1.0 is presented in Fig. 4.16a. It covers the likely ranges of seabed thickness in practical applications. The figure clearly shows that the maximum liquefaction depth z_{Lm} decreases as h/L increases from 0.05 to 0.75, with similar results for $h/L = 0.75$ to 1.0. Conversely, the maximum shear failure depth, z_{Lm}, increases as h/L increases in an unsaturated seabed (Fig. 4.16b). However, the maximum value of z_{sm} occurs for $h/L = 0.2$ in a nearly saturated seabed (for example, $S_r \geq 0.982$ as in the figure).

Combining Figs. 4.16a and 4.16b, it may be stated that a thinner seabed is likely to be affected more by liquefaction than by shear failure. Thus, there may exist a critical seabed thickness which reduces the instability of the seabed to a minimum. This concept will be applied to an optimal design for seabed protection in Sect. 4.5.

Three types of non-cohesive soils have been examined by previous workers, each being associated with a specific value of permeability K_z. For example, $K_z = 10^{-4}$ m/s has been commonly assumed for fine sand and $K_z = 10^{-2}$ m/s for coarse sand [6], while $K_z = 10^{-1}$ m/s for gravel. With the wave and soil conditions used previously, except for seabed thickness $h = 20$ m and wave height $H_s = 5$ m, the effects of soil permeability K_z on the wave-induced seabed instability are now examined.

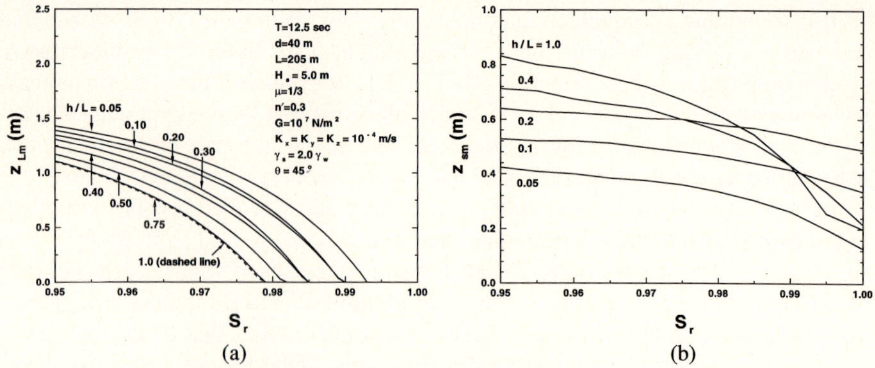

Fig. 4.16 Distribution of the maximum (**a**) liquefaction depth z_{Lm} and (**b**) shear failure depth z_{sm} versus the degree of saturation S_r for various relative seabed thickness h/L

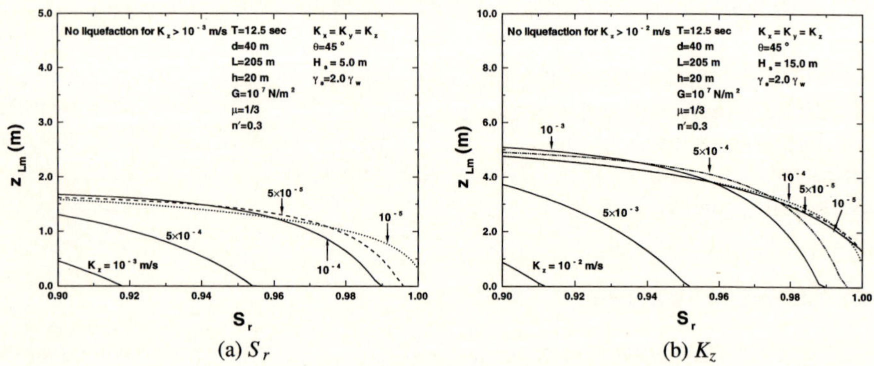

Fig. 4.17 Distribution of the maximum liquefaction depth z_{Lm} versus the degree of saturation S_r for various permeability K_z

Figure 4.17a illustrates the distribution of the maximum liquefaction depth z_{Lm} versus the degree of saturation for various values of K_z. The maximum liquefaction depth z_{Lm} increases as soil permeability K_z decreases, implying that a seabed with lower permeability may be liquefied much easier than that with higher permeability. For example, liquefaction always occurs in a seabed with $K_z = 10^{-5}$ m/s, but it only occurs in a seabed with $K_z = 10^{-3}$ m/s when the degree of saturation $S_r < 0.918$ (Fig. 4.17). It is worth noting that no liquefaction occurs in a seabed with permeability $K_z > 10^{-3}$ m/s in the example presented in Fig. 4.17a.

To further investigate the effect of permeability on the seabed instability, a case with a larger wave height ($H_s = 15$ m) is considered. It is not surprising that the maximum liquefaction depth due to this larger wave is about three times of that due to a smaller wave ($H_s = 5$ m), as shown in Figs. 4.17a and 4.17b. It is also found that liquefaction occurs in a seabed with permeability $K_z > 10^{-3}$ m/s (Fig. 4.17b),

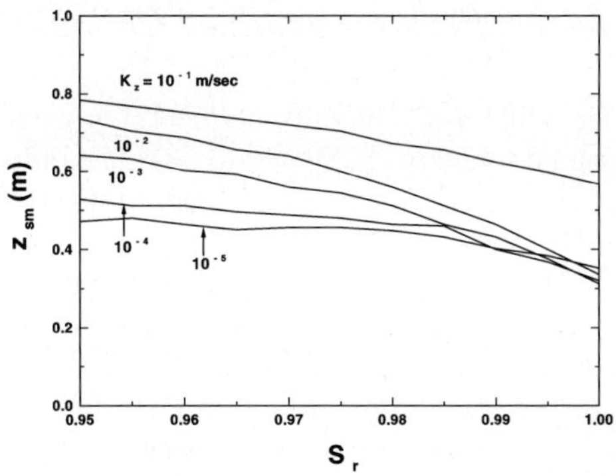

Fig. 4.18 Distribution of the maximum shear failure depth z_{sm} versus the degree of saturation S_r for various permeability K_z

which liquefaction does not occur under a small wave ($H_s = 5$ m). As we expected, this occurs at a lower degree of saturation.

The distribution of the maximum shear failure depth (z_{sm}) for several different values of permeability is presented in Fig. 4.18. Generally speaking, z_{sm} increases as the permeability K_z increases. The values of z_{sm} varies between 0.48 m (for $K_z = 10^{-5}$ m/s) and 0.78 m (for $K_z = 10^{-1}$ m/s) at $S_r = 0.95$, and this difference becomes smaller as S_r approaches full saturation.

4.4.2.2 Degree of Saturation

It was stated that wave-induced soil response is affected significantly by the degree of saturation, as discussed in Chap. 3. Therefore, it is necessary to further examine its influence on the wave-induced shear failure and liquefaction.

In the examples presented in Figs. 4.9–4.11 and 4.16–4.18 with various degrees of saturation (S_r), it is observed that the maximum liquefaction depth (z_{Lm}) decreases as S_r increases, and z_{Lm} reduces swiftly as S_r approaches unity. It is worth noting that no liquefaction occurs in a *saturated* seabed, except under the following conditions: (1) very shallow water, for example, $d/L = 0.1$ in Fig. 4.10a; (2) large wave, for instance, $H_s/L > 0.05$ in Fig. 4.11a; (3) a seabed with very low permeability, such as $K_z \leq 10^{-5}$ m/s in Fig. 4.17.

Similarly, the maximum shear failure depth (z_{sm}) decreases as S_r increases. However, the gradients of the reduction in z_{sm} versus S_r is rather small, compared with that of z_{Lm}. It is also observed that shear failure still occurs in the cases where liquefaction does not.

In summary, at least from the examples presented in Figs. 4.16–4.18, the maximum liquefaction depth z_{Lm} increases and the maximum shear failure depth z_{sm} decreases as seabed thickness and permeability decrease. However, both z_{Lm} and z_{sm} decrease as S_r increases. This implies that the influence of soil characteristics on the wave-induced liquefaction and shear failure may not be consistent, unlike the wave characteristics.

4.4.3 Effect of Combined Obliquity-Permeability Parameter

The combined obliquity-permeability parameter *KMN* in the δ and λ coefficients can be found in (3.27) and (3.33). The influence of *KMN* on the wave-induced soil response was examined in Chap. 3. Its effect on the seabed instability is discussed in this section.

The maximum liquefaction depth z_{Lm} versus the degree of saturation for various values of *KMN* is examined (graphs not shown here). For the range of *KMN* from 0.5 to 100, the liquefaction depth is found to be independent of *KMN*. Similar results are also true for the maximum shear failure depth z_{sm}.

Based on the examples presented in Sect. 4.4, it may be concluded that the maximum liquefaction depth (z_{Lm}) is about two or three times the maximum failure depth (z_{sm}). However, liquefaction occurs only near the wave trough, while shear failure occurs over a thinner layer near the seabed surface throughout the whole wave cycle.

4.4.4 Temporal Variation in Wave-Induced Liquefaction

It is well known that a standing wave profile fluctuates not only spatially, but also temporally. Therefore, its wave phase in the temporal domain may be different from that in a spatial domain. This differs from a progressive wave , in which the phase in a temporal domain is consistent with that in a spatial domain.

Figures 4.19 and 4.20 illustrate the water surface undulations and the wave-induced liquefaction in spatial domain with various temporal phases for progressive and standing waves. Input data for the analysis are specified in the figures, while the wave height of an incident progressive wave is half that of the standing waves resulting from full reflection.

As shown in Fig. 4.19a, the water surface profile of a progressive wave oscillates progressively in spatial domain and propagates in the x-direction. Thus, wave pres-

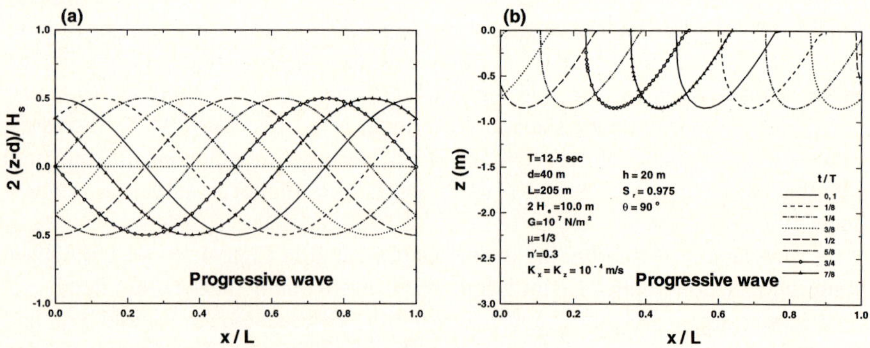

Fig. 4.19 (**a**) Water surface undulations and (**b**) wave-induced liquefaction due to progressive waves with different temporal intervals

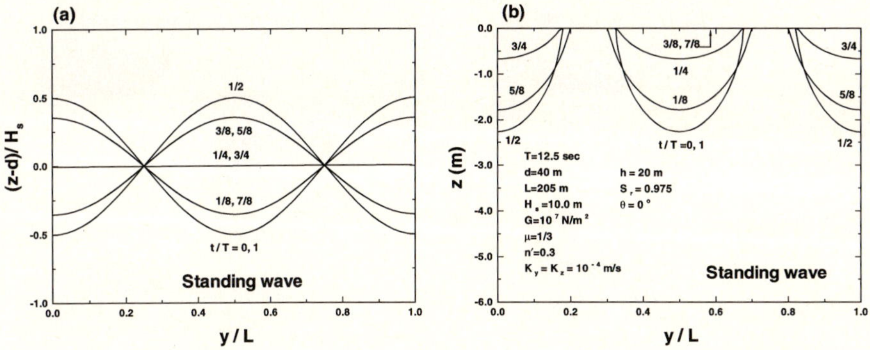

Fig. 4.20 (**a**) Water surface undulations and (**b**) wave-induced liquefaction due to standing waves with different temporal intervals

sure at the seabed surface also progresses in the spatial domain. This results in the liquefaction regions moving progressively in the x-direction with time, as depicted in Fig. 4.19b. On the other hand, the water surface oscillation of a standing wave contains crests at anti-nodes ($x/L = 0, 1/2, 1, \ldots,$ etc.), as shown in Fig. 4.20a. The water level remains constant at points mid-way between the anti-nodes, forming nodal points ($y/L = 1/4, 3/4, \ldots,$ etc.). Thus, the water particles move up and down at the anti-nodes in a standing wave system. As seen in Fig. 4.20b, the liquefied regions due to a standing wave do not move progressively in the spatial domain, but occur near the anti-nodes. This phenomenon arises because the water particles in a standing wave system only move up and down, not progressively. It is worth to note that the maximum liquefaction depth due to standing wave at anti-nodes is about three times that in a progressive wave, as shown in Figs. 4.19b and 4.20b.

Since progressive and standing waves are the two limiting conditions of a short-crested wave system, it is useful to find out the temporal variation in the wave-induced seabed instability in a three-dimensional wave system.

Figure 4.21 illustrates the contours of liquefaction depth due to a short-crested wave ($\theta = 45°$) for various phases (t/T). The liquefied zones advance progressively in the x-direction at the same y/L_y at different time intervals (comparing Fig. 4.21a for $t/T = 0$ and Fig. 4.21b for $t/T = 1/8, \ldots$, and Fig. 4.21g for $t/T = 7/8$). The result is similar to that for a progressive wave presented in Fig. 4.19b. However, the liquefied zones do not move laterally in the y-direction, similar to that in a standing wave shown in Fig. 4.20b. This is because that the combined wave crests of a short-crested wave system are propagating only in the positive x-direction (Fig. 3.1).

4.5 Seabed Protection

The subject of seabed protection has attracted great attention from coastal engineers and marine geotechnical engineers in recent years. In general, it involves placement of some suitable material on an existing seabed to minimize or prevent the wave-

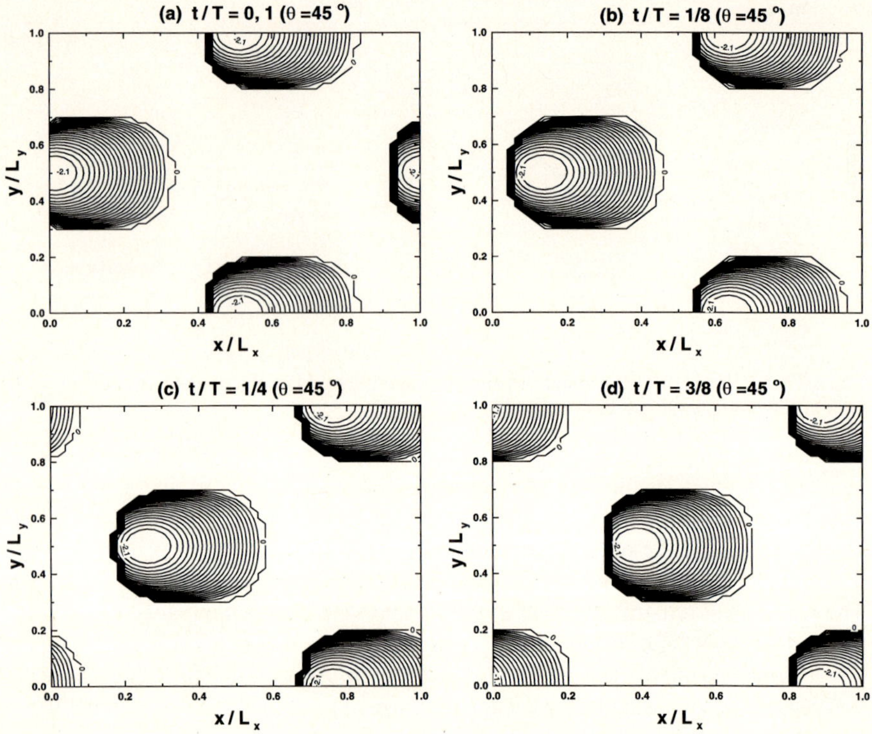

Fig. 4.21 Wave-induced liquefaction due to short-crested waves with incident wave angle $\theta = 45°$ for (**a**) $t/T = 0, 1$, (**b**) $t/T = 1/8$, (**c**) $t/T = 1/4$, (**d**) $t/T = 3/8$, (**e**) $t/T = 1/2$, (**f**) $t/T = 5/8$, (**g**) $t/T = 3/4$ and (**h**) $t/T = 7/8$

induced seabed instability. On the basis of the effects of wave and soil characteristics on the two different types of seabed instability (liquefaction and shear failure) discussed previously, it is beneficial to further propose a suitable method for seabed protection.

Generally speaking, a natural seabed consists of multiple layers with different soil properties, for example, the seabed in the Ekofisk oil field [1] mentioned previously. Thus, it is necessary to examine the various influences of a top layer on the wave-induced seabed instability before any suitable method of seabed protection can be discussed.

4.5.1 Effects of a Top Layer

Four possible layer combinations are considered in the analysis of a two-layered seabed consisting of fine sand (Fs), coarse sand (Cs) and gravel (Gv), with their soil properties given in Table 4.1. The seabed of interest has a total thickness of 50 m for all cases studied:

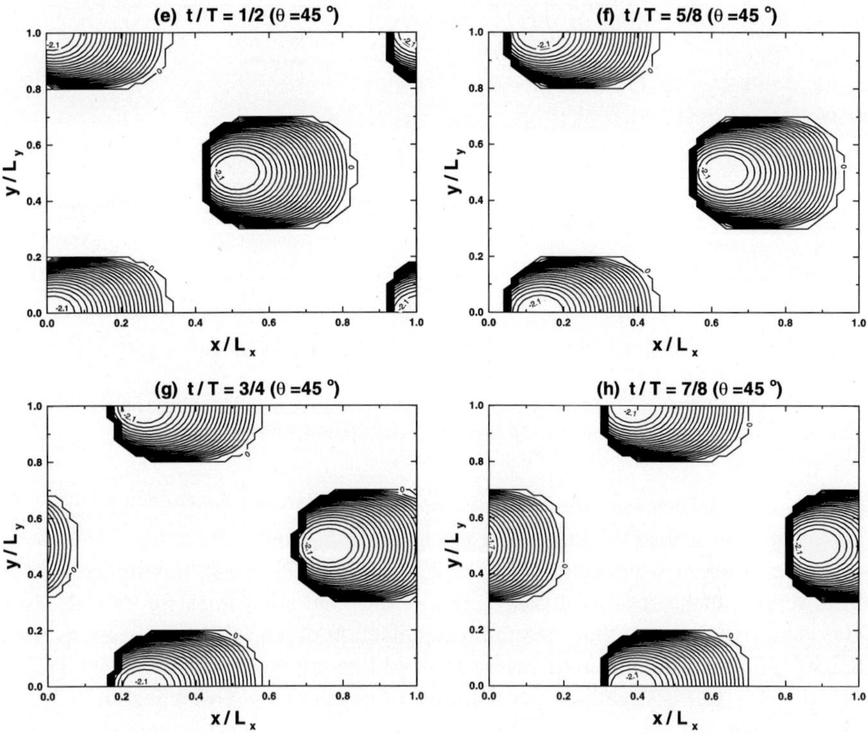

Fig. 4.21 (Continued)

Table 4.1 Soil characteristics of three different soil types

Soil characteristics	Gravel	Coarse sand	Fine sand
Permeability K_x, K_y, K_z:	10^{-1} m/s	10^{-2} m/s	10^{-4} m/s
Poisson's ratio μ:	0.4	0.3333	0.3333
Porosity n':	0.4	0.3	0.3
Shear modulus G:	5×10^7 N/m^2	10^7 N/m^2	10^7 N/m^2
Unit weight of soil γ_s:	$2.0\gamma_w$	$2.0\gamma_w$	$2.0\gamma_w$

- Case 1: fine sand on top of coarse sand (denoted as "Fs + Cs"),
- Case 2: fine sand on top of gravel (denoted as "Fs + Gv"),
- Case 3: coarse sand on top of fine sand (denoted as "Cs + Fs"),
- Case 4: gravel on top of fine sand (denoted as "Gv + Fs").

Based on the failure criteria presented in Sects. 4.2 and 4.3 and employing the layered solution presented in Sect. 3.4.4, the potential failure zone for all cases are calculated under an unsaturated condition ($S_r = 0.975$).

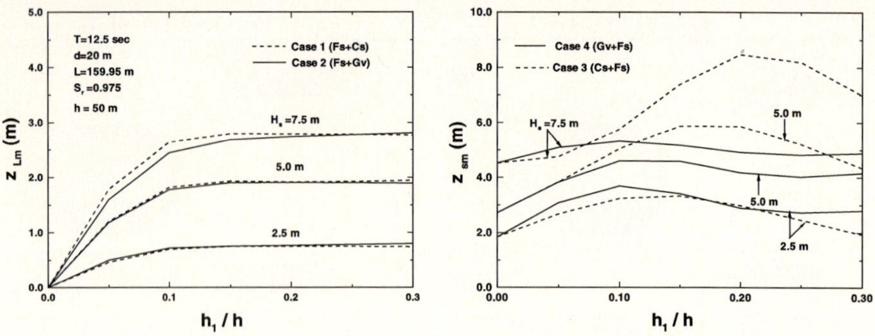

(a) S_r=1 (b) shear failure depth z_{sm} (Cases 3 and 4)

Fig. 4.22 Distribution of the maximum (**a**) liquefaction depth z_{Lm} and (**b**) shear failure depth z_{sm} versus relative seabed thickness of a top layer h_1/h for various values of H_s

Figure 4.22a illustrates the distribution of the maximum liquefaction depth (z_{Lm}) versus relative seabed thickness of a top layer to the total seabed thickness (h_1/h) for three different wave heights ($H_s = 2.5$, 5.0 and 7.5 m). In this figure, dashed lines represent the results of case 1 (Fs + Cs), and solid lines for case 2 (Fs + Gv). The figure shows that maximum liquefaction depth (z_{Lm}) increases as wave height (H_s) increases in both cases, as would be expected. The resultant z_{Lm} of case 1 (Fs + Cs) is slightly larger than that of case 2 (Fs + Gv) when $H_s \geq 5.0$ m. This indicates that fine sand has slightly worse effect, whereas the underlying gravel provides better flow conditions.

It was not expected that liquefaction would occur in cases 3 (Cs + Fs) and 4 (Gv + Fs) under a small wave height, as discussed in Sect. 3.4.4. This implies that placing a coarser material as the top layer may help in protecting a seabed from the wave-induced liquefaction. However, these cases do not prevent shear failure from occurring, as shown in Fig. 4.22b. As seen in this figure, the maximum shear failure depth z_{sm} of case 3 (Cs + Fs) is larger than that of case 4 (Gv + Fs). The difference becomes much greater as wave height increases (Fig. 4.22b).

Figure 4.22 also demonstrate the effect of top layer thickness (h_1/h) on the wave-induced seabed instability for the four cases examined. Generally speaking, maximum liquefaction depth z_{Lm} increases as h_1/h increases, when fine sand is on top of a coarse material (cases 1 and 2 in Fig. 4.22a). The coarser material (coarse sand or gravel), serving as a lower layer, will provide better flow conditions and thus reduce liquefaction. As seen in Fig. 4.22b, the maximum shear failure depth (z_{sm}) increases as h_1/h increases until a critical h_1/h value is reached, and it then decreases as h_1/h increases.

The results presented in Fig. 4.22 are derived from a moderate wave condition ($H_s = 7.5$ m, $H_s/L = 0.047$) which is in the limit of the linear wave theory. The effect of a large wave amplitude on seabed instability is now examined. According to Tsai *et al.* [13], the maximum wave steepness of a standing wave can be expressed as

$$(H_s/L)_{\max} = 0.204 \tanh kd, \tag{4.34}$$

Table 4.2 Maximum depths of liquefaction and shear failure under a large wave

Maximum depth of liquefaction z_{Lm}					Maximum depth of shear failure z_{sm}				
$h_1/h(h_1)$	Case 1	Case 2	Case 3	Case 4	$h_1/h(h_1)$	Case 1	Case 2	Case 3	Case 4
0.00 (0 m)	5.23	0	9.53	9.53	0.00 (0 m)	9.33	10.40	7.16	7.16
0.05 (2.5 m)	8.49	6.29	2.98	2.79	0.05 (2.5 m)	6.53	8.02	6.77	6.69
0.10 (5.0 m)	8.45	7.41	0	0	0.10 (5.0 m)	6.32	7.79	6.34	6.21
0.15 (7.5 m)	8.45	7.92	0	0	0.15 (7.5 m)	6.11	5.68	5.70	5.58
0.20 (10.0 m)	8.47	8.05	0	0	0.20 (10.0 m)	6.00	5.68	6.00	5.01
0.25 (12.5 m)	8.64	8.95	0	0	0.25 (12.5 m)	5.89	5.68	7.81	7.53
0.30 (15.0 m)	8.85	9.47	0	0	0.30 (15.0 m)	5.89	5.47	8.50	8.98

from which the $(H_s)_{\max}$ is about 21.4 m for a standing wave using input data cited in Fig. 4.22. In fact, the maximum wave steepness of a short-crested wave system has been reported to be larger than that in a standing wave in deep water [13]. Unlike the two-dimensional progressive and standing waves for which breaking index has been well established, the breaking criterion for a short-crested waves in a shallow water is not available. Consequently, the maximum wave height $H_s = 21.4$ m is conservative for the present short-crested waves.

Table 4.2 illustrates the maximum depths of liquefaction and shear failure (z_{Lm} and z_{sm}) for the four cases examined with various values of the relative seabed thickness (h_1/h), based on a total thickness $h = 50$ m but with variable top layer thickness h_1. Two conclusions may be drawn from the cases examined above:

- Putting coarse sand or gravel on top of fine sand (i.e., cases 3 and 4) could completely protect seabed from liquefaction, when the thickness of the top layer is larger than a critical value (e.g., $h_1/h \geq 0.1$ in this example). However, in cases 1 and 2 (putting fine sand on coarse sand or gravel), the depth of liquefaction is greater.
- In cases 1 and 2, the maximum shear failure depth (z_{sm}) decreases as thickness of the top layer (h_1/h) increases. However, in cases 3 and 4, it decreases first as h_1/h increases from zero to 0.15, then increases for $h_1/h > 0.15$. This implies that there is a critical value of h_1/h which produces a minimum depth of shear failure.

4.5.2 Methodology of Seabed Protection

Among the wave characteristics discussed previously, wave period (T), water depth (d), wave height (H_s) and wave obliquity (θ), it may be difficult to change these once a marine structure is constructed. Also, the combined obliquity-permeability parameter (KMN) has insignificant effect on seabed instability. In the field, engineers may have difficulty in modifying the degree of saturation of the soil, although

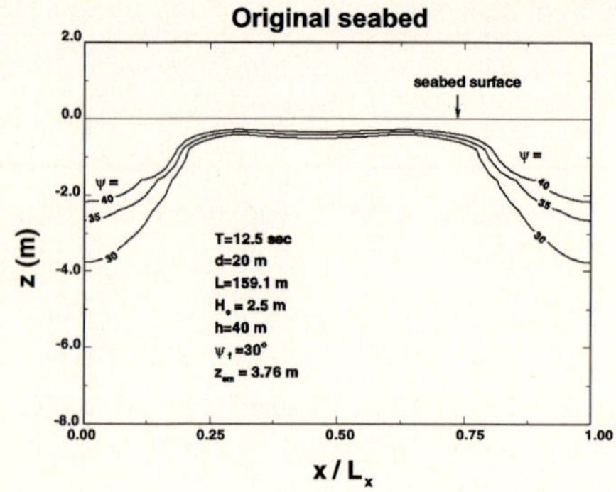

(a) original seabed

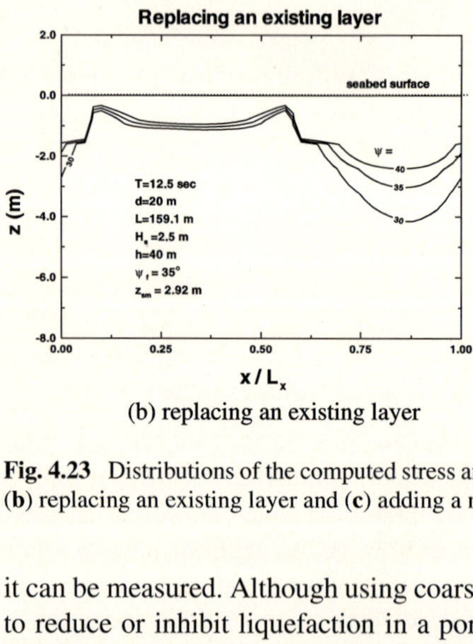

(b) replacing an existing layer

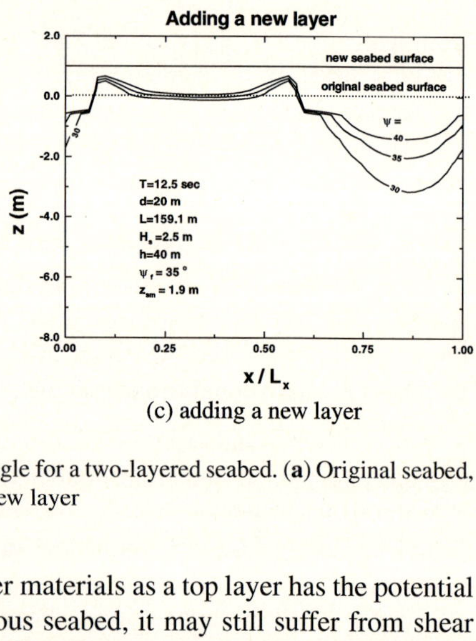

(c) adding a new layer

Fig. 4.23 Distributions of the computed stress angle for a two-layered seabed. (**a**) Original seabed, (**b**) replacing an existing layer and (**c**) adding a new layer

it can be measured. Although using coarser materials as a top layer has the potential to reduce or inhibit liquefaction in a porous seabed, it may still suffer from shear failure (as discussed in Sect. 4.5.1). Therefore, the control of relative thickness of the top layer (h_1/h) may be taken as an effective means to combat the problem of seabed liquefaction.

In general, two different categories of placement are available, namely, replacing an existing layer or adding a new layer to the original seabed. Concrete blocks and large gravel are commonly used for seabed protection, but their properties are quite different from that of a sandy material. Within the limitation of the present solution, only small gravel is considered as the covering material on a sandy seabed.

As an example, the original seabed is considered as a two-layered medium consisting of coarse sand (10 m) on top of fine sand (30 m). Wave height of $H_s = 2.5$ m is used in this example. The calculated maximum shear failure depth is 3.76 m, occurring near the wave crest (Fig. 4.23a). Upon placing a layer of gravel, it becomes a three-layered seabed. The influence of the thickness of the top layer on the wave-induced seabed instability for two different placement methods is now investigated.

4.5.2.1 Replacing an Existing Layer

Generally speaking, engineering practice under this category includes the placement of other suitable materials to fill the trench formed by excavating sediments from the original seabed. This has been considered for seabed protection in engineering practice.

Figure 4.23b shows the distribution of the stress angles resulting from replacing an existing layer of the original seabed under saturated conditions. The thickness of a top layer with gravel is taken as $h_1 = 1$ m (i.e., $h_1/h = 0.025$). It is observed that the maximum shear failure depth is about 2.92 m occurring at $x = 0.85L_x$ (Fig. 4.23b), which is about 77 % of the resultant depth z_{sm} without any protection (Fig. 4.23a).

4.5.2.2 Adding a New Layer

Adding a new layer by dumping coarse material directly on an existing seabed is a popular method of seabed protection. Unlike replacing an existing layer of the original seabed, the total seabed thickness is the sum of the thickness of the original seabed and the extra top layer.

Now, consider a top layer of 1 m thick on an original seabed consisting of 10 m coarse sand on top of fine sand of 30 m (Fig. 4.23a). The distribution of the computed stress angles after adding a new layer on the original seabed is presented in Fig. 4.23c. The maximum shear failure depth z_{sm} is reduced to 1.9 m below the original seabed surface. It is about 51 % of the resultant depth z_{sm} without any protection. Compared with the method of replacing an existing layer of the original seabed, adding a new layer on the original seabed may cause a smaller shear failure depth (see Figs. 4.23b and 4.23c).

4.5.2.3 Optimal Design Consideration of Seabed Protection

At least from the two examples presented in Fig. 4.23, it has become possible to reduce the wave-induced shear failure potential by replacing an existing layer or adding a new layer onto the original seabed. However, a practical question remains: *What is the optimal relative thickness of a top layer h_1/h for seabed protection?*

Figure 4.24 illustrates the maximum shear failure depths z_{sm} versus relative seabed thickness of a top layer h_1/h under saturated and unsaturated conditions, respectively. Two different wave heights, $H_s = 2.5$ m and 5.0 m, are considered. These figures reveal the maximum shear failure depth z_{sm} caused by replacing part of an original seabed is larger than that by adding a new layer.

For adding a new layer to the original seabed, there is a critical thickness of the top layer h_1 (or h_1/h) which may reduce the wave-induced shear failure to a minimum (Fig. 4.24). For example, a minimum shear failure depth occurs at $h_1/h = 0.0375$ ($h_1 = 1.5$ m, $h = 40$ m) under saturated condition, while it only

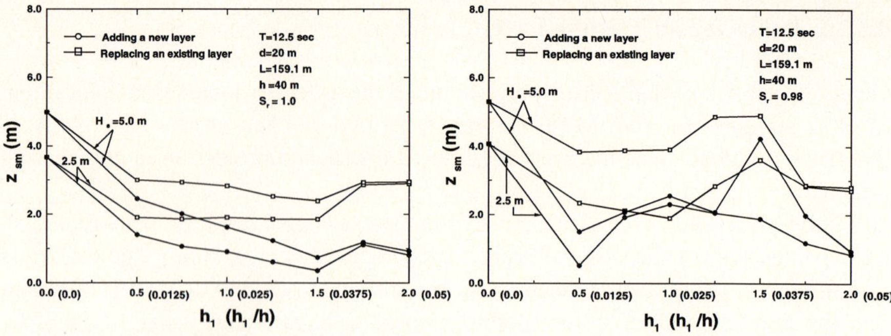

(a) S_r=1 (b) S_r=0.98

Fig. 4.24 Distribution of the maximum shear failure depth z_{sm} versus relative seabed thickness of a top layer h_1/h for replacing an existing layer and adding a new layer on the original seabed. (**a**) $S_r = 1.0$ and (**b**) $S_r = 0.98$

Table 4.3 Maximum depths of liquefaction and shear failure by replacing an existing layer and adding a new layer on the original seabed under a large wave

Saturated seabed ($S_r = 1.0$)					Unsaturated seabed ($S_r = 0.98$)				
h_1	Liquefaction z_{Lm}		Shear failure z_{sm}		h_1	Liquefaction z_{Lm}		Shear failure z_{sm}	
	add	replace	add	replace		add	replace	add	replace
0	0	0	6.82	6.82	0	5.57	5.57	8.00	8.00
0.25	0	0	4.5	4.92	0.25	1.04	1.24	5.50	5.85
0.50	0	0	4.03	4.61	0.50	0.17	0.79	5.05	5.65
0.75	0	0	4.95	5.75	0.75	3.64	1.51	6.15	6.89
1.00	0	0	3.6	4.45	1.00	4.02	5.21	4.80	5.70
1.25	0	0	2.70	3.97	1.25	3.76	5.09	4.20	5.65
1.50	0	0	2.55	4.01	1.50	2.40	4.29	3.60	5.60
1.75	0	0	2.14	3.98	1.75	2.45	4.15	3.45	4.85
2.0	0	0	1.95	3.95	2.0	0.0	2.30	3.05	4.80

requires $h_1/h = 0.0125$ (or $h_1 = 0.5$ m, $h = 40$ m) under the unsaturated condition of $S_r = 0.98$. This implies that adding a thicker layer of gravel, such as $h_1 > 1.5$ m (Fig. 4.24a), may not be economical in relation to the minimum shear failure depth (z_{sm}).

With a large wave of $H_s = 21.4$ m explained in Sect. 4.5.1 and Table 4.2, the calculated z_{Lm} and z_{sm} values are presented in Table 4.3 for a seabed under fully saturated ($S_r = 1.0$) and unsaturated ($S_r = 0.975$) conditions, respectively. As seen in the table, both methods can be used to protect the seabed from liquefaction at $S_r = 1$ (i.e., saturated condition). However, liquefaction still occurs under the unsaturated condition at $S_r = 0.98$, which is very close to unity. For the maximum depth of shear failure, it seems that z_{sm} decreases as the thickness of the top layer (h_1) increases, except at $h_1 = 0.5$ m, which results in a minimum value of z_{sm}

for $0.0 \leq h_1 \leq 1.0$ m. It may be concluded that adding a new layer on an original seabed provides better seabed protection than replacing an existing layer. This trend is similar to the results under a smaller wave (Fig. 4.24). However, the values of the critical thickness of a top layer (h_1/h) are also affected by other wave and soil characteristics at different sites.

4.6 Summary

In this chapter, a new criterion is proposed for the wave-induced liquefaction. Based on the general solution presented in Chap. 3, this criterion is shown to be applicable for a seabed of infinite and finite thickness, supported by field data available.

Employing the general solution of a layered seabed, four different combinations of soil type are examined. At least from the numerical examples presented, gravel and coarse sand may serve to reduce liquefaction potential. Two common methods of seabed protection are also examined. It is found that adding a new layer of gravel to the original seabed produces a smaller shear failure depth than replacing an existing layer.

From the wave and soil characteristics considered here, it appears shear failure depth can be controlled by the relative thickness of a top layer. As numerical results presented, an optimal value of h_1/h exists which generates a minimum shear failure depth.

References

1. Bjerrum, J.: Geotechnical problem involved in foundations of structures in the North Sea. Géotechnique **23**(3), 319–358 (1973)
2. Griffiths, D.V.: Some theoretical observations on conical failure criteria in principal stress space. Int. J. Solids Struct. **22**, 553–565 (1986)
3. Griffiths, D.V.: Failure criteria interpretation based on Mohr-Coulomb friction. J. Geotech. Eng. **116**, 986–999 (1990)
4. Hsu, J.R.C., Jeng, D.S., Lee, C.P.: Oscillatory soil response and liquefaction in an unsaturated layered seabed. Int. J. Numer. Anal. Methods Geomech. **19**(12), 825–849 (1995)
5. Lundgren, H., Lindhardt, J.H.C., Romold, C.J.: Stability of breakwaters on porous foundation. In: Proceeding of 12th International Conference on Soil Mechanics and Foundation Engineering, vol. 1, pp. 451–454 (1989)
6. Mei, C.C., Foda, M.A.: Wave-induced response in a fluid-filled poro-elastic solid with a free surface-a boundary layer theory. Geophys. J. R. Astron. Soc. **66**, 597–631 (1981)
7. Okusa, S.: Wave-induced stress in unsaturated submarine sediments. Géotechnique **35**(4), 517–532 (1985)
8. Rahman, M.S.: Wave-induced instability of seabed: Mechanism and conditions. Mar. Geotechnol. **10**, 277–299 (1991)
9. Scott, R.F.: Principle of Soil Mechanics. Addison-Publishing, Massachusetts (1963)
10. Silvester, R., Hsu, J.R.C.: Coastal Stabilization. PTR Prentice-Hall Inc., New Jersey (1993)
11. Smith, A.W., Gordon, A.D.: Large breakwater toe failures. J. Waterways Harbor Coast. Eng. Div. **109**(2), 253–255 (1983)

12. Tsai, C.P.: Wave-induced liquefaction potential in a porous seabed in front of a breakwater. Ocean Eng. **22**(1), 1–18 (1995)
13. Tsai, C.P., Jeng, D.S., Hsu, J.R.C.: Computations of the almost highest short-crested waves in deep water. Appl. Ocean Res. **16**(6), 317–326 (1994)
14. Ye, J.: 3d liquefaction criteria for seabed considering the cohesion and friction of soil. Appl. Ocean Res. **37**, 111–119 (2012)
15. Zen, K., Yamazaki, H.: Mechanism of wave-induced liquefaction and densification in seabed. Soil Found. **30**(4), 90–104 (1990)
16. Zen, K., Yamazaki, H.: Field observation and analysis of wave-induced liquefaction in seabed. Soil Found. **31**(4), 161–179 (1991)

Chapter 5
Wave-Induced Seabed Response in Non-homogeneous Anisotropic Seabed

Abstract In a natural seabed, the soil properties is complicated, which normally vary with soil depth, change of temperature, and geographic characteristics. Furthermore, anisotropic soil behavior is commonly observed in marine sediments. However, most previous studies have been limited to the case of isotropic seabed with uniform soil characteristics. This chapter consists of three major components: (i) analytical solutions for a seabed with variable permeability; (ii) analytical solution for a cross-isotropic seabed; and (iii) numerical model for a non-homogeneous seabed with cross-anisotropic soil behavior. With the newly analytical solutions and numerical model, effects of variable permeability and cross-isotropic soil behavior on the wave-induced seabed response were examined.

Keywords Variable soil characteristics · Cross-anisotropic soil behavior · Pore pressure · Liquefaction · Effective stresses

5.1 Introduction

Soil permeability is a measure of how fast the pore fluid is transmitted through the voids between grains. Marine sediments below the seabed surface undergo consolidation due to both the overburden soil pressure and the water pressure above it, resulting in a decrease in their permeability. An example of the vertical permeability varying with burial depth (z) for marine sediments in the Gulf of Mexico was reported in [6]. Similar evidence for soil consolidation versus depth was also reported [3, 4, 38].

The major difficulty in analyzing the wave-seabed interaction problem with variable permeability arises from the governing partial differential equation containing variable coefficients. In general, it is not possible to find exact solutions to even simple differential equations with rapidly varying coefficient (permeability). However, it has been shown [46] that if the governing partial differential equations can be written in a certain canonical form and a depth function $a(z)$ is described in terms of an Nth-order VS function (VS_N), then an analytical solution would become possible. Moreover, since VS_N contains $2N+1$ arbitrary parameters, it is possible to select an appropriate VS_N function to closely approximate a realistic depth function.

D.-S. Jeng, *Porous Models for Wave-seabed Interactions*,
DOI 10.1007/978-3-642-33593-8_5,

Recently, Jeng and Seymour [24–26, 39] have developed a series of analytical solutions for the wave-induced seabed response with variable permeabilities, applying the VS_1 function. Based on these solutions, the effects of variable permeability on the wave-induced pore pressure and effective stresses are discussed here, together with the wave-induced seabed instability.

In nature, most marine sediments display some degree of anisotropy, having different elastic properties in different directions, due to the manner of their deposition, particle shape and stress history. However, many materials show more limited forms of anisotropy. For example, a *cross-anisotropic* material possesses an axis of symmetry in the sense that its properties are independent of the rotation of a sample about its axis of symmetry. This implies that a cross-anisotropic material has the same properties in any horizontal directions, but different in the vertical direction. When a soil is deposited vertically and then subjected to equal horizontal stresses, it is expected to exhibit a vertical axis of symmetry and be transversely isotropic. This assumption is valid for most soils [11] .

Unlike an isotropic material, the elastic behavior of a cross-anisotropic material is described by five independent elastic parameters [2, 28, 32]. Among these parameters, only three can be obtained from standard triaxial tests, the others require more advanced testing using a cubical triaxial apparatus and a torsional shear device [1, 11, 27, 50].

Because the requirement of advanced experimental technique for anisotropic constants, geotechnical engineers have commonly used two-parameter isotropic models in practical applications, neglecting the effects of anisotropy. However, the effects of anisotropy may not always be negligible. Based on three rather than five parameters (and two additional assumptions), a simplified anisotropic model for lightly over-consolidated post-glacial soils was proposed [11]. The results indicate that the three-parameter anisotropic model reduced the error in the prediction of strains by 30 %–40 %, compared to the prediction of the isotropic model [11]. Other similar evidence for soil anisotropy has also been reported in the literature [10, 36, 37].

Gatmiri [9] might have been the first to investigate the wave-induced soil response in a cross-anisotropic seabed, using a finite element analysis. However, only soil response under a fully saturated condition was considered. Since the degree of saturation of marine sediment in the field normally varies in the range of 0.9 to 1.0 [7], it is believed that most marine sediments have a degree of saturation very close to unity, but not fully saturated [33]. On the other hand, it is rare that full saturation can be attained in the field or laboratory conditions, except for an ideal situation. Thus, it is beneficial to develop analytical solutions for an unsaturated condition for engineering applications.

Recently, Jeng [14, 15, 18, 19] has developed a series of analytical solutions for wave-induced soil response in a cross-anisotropic seabed. Based on the general solutions, together with the criteria of liquefaction and shear failure, the effect of cross-anisotropic parameters on the wave-induced seabed response is now discussed.

In addition to analytical solutions for either variable permeability or anisotropic seabed, a one-dimensional Finite Element Model for the wave-induced seabed response with combined non-homogeneous (variable permeability and modulus) and cross-anisotropic soil behavior was outlined in Sect. 5.4. Based on the numerical model, the combined effects on the wave-induced soil response will be discussed in detail.

5.2 Analytical Solution for a Seabed with Variable Permeability

Consider the wave-seabed interaction subject to a three-dimensional short-crested wave system in front of a vertical wall, as shown in Fig. 3.1. The basic assumptions for this problem are the same as that mentioned in Chap. 3, except for the variable soil permeability.

5.2.1 Boundary Value Problem

The change in the vertical permeability, K_z, with depth below seabed surface is represented by a *depth function* $a(z)$

$$K_z = a^2(z) K_{zo}, \tag{5.1}$$

with respect to a referenced value of permeability K_{zo} taken at the seabed surface.

Substituting (5.1) into (3.7), the consolidation equation renders

$$\frac{K_x}{K_{zo}} \frac{\partial^2 p}{\partial x^2} + \frac{K_y}{K_{zo}} \frac{\partial^2 p}{\partial y^2} + \frac{\partial}{\partial z}\left(a^2(z) \frac{\partial p}{\partial z}\right) - \frac{\gamma_w n' \beta}{K_{zo}} \frac{\partial p}{\partial t} = \frac{\gamma_w}{K_{zo}} \frac{\partial \varepsilon}{\partial t}, \tag{5.2}$$

where p is wave-induced pore pressure, K_x and K_y are the soil permeabilities in the x- and y-directions, respectively, n' is the soil porosity, β is the compressibility of the pore fluid defined by (3.9), ε is the volume strain defined by (3.8), and t is time.

From the effective stress concept, the equations for overall equilibrium of the poro-elastic medium are

$$G\nabla^2 u + \frac{G}{1-2\mu} \frac{\partial \varepsilon}{\partial x} = \frac{\partial p}{\partial x}, \tag{5.3}$$

$$G\nabla^2 v + \frac{G}{1-2\mu} \frac{\partial \varepsilon}{\partial y} = \frac{\partial p}{\partial y}, \tag{5.4}$$

$$G\nabla^2 w + \frac{G}{1-2\mu} \frac{\partial \varepsilon}{\partial z} = \frac{\partial p}{\partial z}, \tag{5.5}$$

in the x-, y- and z-directions, respectively. These are the same as (3.16)–(3.18) presented previously.

Based on Hooke's law, the relationship between stress and soil displacements given by (3.13)–(3.15) as mentioned in Chap. 3. For a soil matrix, mathematical expressions for the wave-induced seabed response can be derived subject to some appropriate boundary conditions at the rigid impermeable bottom and at the seabed surface. The boundary conditions given in Sect. 3.3.2 are summarized as follows:

(a) SBC: Boundary Conditions at the Seabed Surface ($z = 0$)

$$\sigma_z' = \tau_{xz} = \tau_{yz} = 0 \quad \text{at } z = 0. \tag{5.6}$$

and

$$\begin{aligned} p &= \frac{\gamma_w H_s}{2\cosh kd} \cos nky \cos(mkx - \omega t) \\ &= p_o \cos nky \cos(mkx - \omega t) \quad \text{at } z = 0, \end{aligned} \tag{5.7}$$

(b) BBC: Boundary Conditions at the Rigid Bottom ($z \to -\infty$ or $z = -h$) For a seabed of infinite thickness,

$$u = v = w = p = 0 \quad \text{as } z \to -\infty. \tag{5.8}$$

For a seabed of finite thickness and a layered seabed,

$$u = v = w = \frac{\partial p}{\partial z} = 0 \quad \text{at } z = -h. \tag{5.9}$$

(c) MBC: Matching Boundary Conditions Between Two Consecutive Sub-layers in a Layered Seabed ($z = -h_j$)

$$\begin{aligned} &p_j = p_{j+1}, \quad u_j = u_{j+1}, \quad v_j = v_{j+1}, \quad w_j = w_{j+1}, \\ &\sigma'_{z,j} = \sigma'_{z,j+1}, \quad \tau_{xz,j} = \tau_{xz,j+1}, \quad \tau_{yz} = \tau_{yz,j+1}, \\ &K_{z,j}\frac{\partial p_j}{\partial z} = K_{z,j+1}\frac{\partial p_{j+1}}{\partial z}, \quad \text{at } z = -h_j,\ j = 1, 2, 3, \ldots, N-1, \end{aligned} \tag{5.10}$$

in which subscript "j" denotes the jth sub-layer in a N-layered seabed.

5.2.2 General Solutions

The boundary value problem, described by governing equations (5.2)–(5.5) and the boundary conditions (5.6)–(5.10), represents the water wave-seabed interaction in a porous seabed with variable permeability. In this section, analytical solutions for such problems are derived for a seabed of infinite and finite thickness, as well as for a layered seabed.

5.2.2.1 Basic Framework

Similar to the analytical solution for the soil response of a porous seabed with uniform permeability developed in Sect. 3.4, the soil displacements and pore pressure for the case of variable permeability can be expressed as:

$$\begin{Bmatrix} u \\ v \\ w \\ p \end{Bmatrix} = p_o \begin{Bmatrix} m\mathcal{U}(z)\cos nky \\ in\mathcal{U}(z)\sin nky \\ \mathcal{W}(z)\cos nky \\ \mathcal{P}(z)\cos nky \end{Bmatrix} e^{i(mkx-\omega t)}. \tag{5.11}$$

In (5.11), only the real parts of the complex functions are used. With this simplified formulation there are only three independent unknown functions [25]. Substituting expressions (5.11) into (5.2)–(5.5) produces three Ordinary Differential Equations (**ODE**):

$$\mathcal{U}' - ik\mathcal{W} = B_1 e^{kz} + B_2 e^{-kz}, \tag{5.12}$$

$$\mathcal{U}'' - k^2\mathcal{U} = \frac{ik(1-2\mu)}{2G(1-\mu)}\mathcal{P} + \frac{k}{2(1-\mu)}\left(B_1 e^{kz} - B_2 e^{-kz}\right), \tag{5.13}$$

and

$$\left(a^2(z)\mathcal{P}'\right)' - \delta^2\mathcal{P} = \Phi_o\left(B_1 e^{kz} - B_2 e^{-kz}\right), \tag{5.14}$$

where B_1 and B_2 are undetermined variables. The complex parameter δ was given in (3.27) and Φ_o is given by

$$\Phi_o = \frac{\omega\gamma_w(1-2\mu)}{2K_{zo}(1-\mu)}. \tag{5.15}$$

Equation (5.14) contains the *depth function*, $a(z)$, which describes the variation in the soil permeability with depth. Generally, it is difficult to obtain analytical solutions for the differential equation in the form of (5.14). However, an exact solution of (5.14) is possible to obtained by following the approach of Varley and Seymour [46] (see Sect. 5.5).

The independent solution to the homogeneous version of (5.14),

$$\left(a^2(z)\mathcal{P}'\right)' - \delta^2\mathcal{P} = 0, \tag{5.16}$$

can be written in the form

$$E_1(z) = \frac{e^{\delta R(z)}[\lambda_1 + \ell_1\delta/\sqrt{a(R(z))}]}{[\lambda_1 + \ell_1\delta/\sqrt{a(0)}]} \tag{5.17}$$

$$E_2(z) = \frac{e^{-\delta R(z)}[\lambda_1 - \ell_1\delta/\sqrt{a(R(z))}]}{[\lambda_1 - \ell_1\delta/\sqrt{a(0)}]}. \tag{5.18}$$

The *VS* function, $a(R(z))$, in (5.17)–(5.18) are defined as follows:

(i) if $a(R(z))$ is a decreasing function with z,

$$\sqrt{a\big(R(z)\big)} = \frac{\ell_1}{\sqrt{\lambda_1}}\coth\left[\sqrt{\lambda_1}\big(R(z)+r_1\big)\right], \quad \lambda_1 \ge 0, \tag{5.19}$$

or

$$\sqrt{a\big(R(z)\big)} = -\frac{\ell_1}{\sqrt{-\lambda_1}} \cot\big[\sqrt{-\lambda_1}\big(R(z)+r_1\big)\big], \quad \lambda_1 \le 0, \tag{5.20}$$

(ii) if $a(R(z))$ is an increasing function with z,

$$\sqrt{a\big(R(z)\big)} = \frac{\ell_1}{\sqrt{\lambda_1}} \tanh\big[\sqrt{\lambda_1}\big(R(z)+r_1\big)\big], \quad \lambda_1 \ge 0, \tag{5.21}$$

or

$$\sqrt{a\big(R(z)\big)} = -\frac{\ell_1}{\sqrt{-\lambda_1}} \tan\big[\sqrt{-\lambda_1}\big(R(z)+r_1\big)\big], \quad \lambda_1 \le 0, \tag{5.22}$$

where

$$R(z) = \int^{z} \frac{ds}{a(s)}. \tag{5.23}$$

In (5.17) and (5.18), $E_1(z)$ and $E_2(z)$ satisfy the boundary conditions:

$$E_1(0) = E_2(0) = 1, \qquad E_1(-\infty) = 0, \quad \text{and} \quad E_2(-\infty) = \infty. \tag{5.24}$$

In (5.19)–(5.22), the depth function $a(R(z))$ contains three arbitrary constants which can be used to curve-fit many different variations in material properties with depth. Herein, we only consider several monotonic depth functions.

The general solution of (5.14) can be expressed as

$$\mathcal{P}(z) = B_3 E_1(z) + B_4 E_2(z) + \Phi_o \int_{-h}^{z} F(z,s)\big(B_1 e^{ks} - B_2 e^{-ks}\big)\, ds, \tag{5.25}$$

where $F(z,s)$ is given by

$$\begin{aligned} F(z,s) &= \frac{E_1(z)E_2(s) - E_1(s)E_2(z)}{a^2(s)[E_1'(s)E_2(s) - E_1(s)E_2'(s)]} \\ &= \frac{E_1(z)E_2(s) - E_1(s)E_2(z)}{D_o a(s)} \\ &= E_1(z)\kappa_2(s) - E_2(z)\kappa_1(s), \end{aligned} \tag{5.26}$$

in which

$$\kappa_1(s) = \frac{E_1(s)}{D_o a(s)}, \qquad \kappa_2(s) = \frac{E_2(s)}{D_o a(s)}, \qquad D_o = \frac{2\delta\lambda_1^2(\ell_1 - \delta^2)}{\ell_1^2 - \lambda_1^2\delta^2/a(0)}. \tag{5.27}$$

The general solution for wave-induced seabed response with variable permeability can be obtained by substituting (5.25) into (5.12) and (5.13), together with appropriate boundary conditions, respectively. Analytical solutions for a seabed of infinite, finite thickness and a layered medium are developed systematically in the following subsections.

5.2.2.2 Analytical Solutions

For a porous seabed of infinite thickness, the bottom boundary condition, (5.8), requires $B_2 = B_4 = 0$. Thus, the pore pressure can be expressed as

$$\mathcal{P}(z) = B_3^\infty E_1(z) + \phi_o B_1^\infty \{S_1^\infty(z) - S_2^\infty(z)\}, \tag{5.28}$$

where $S_1^\infty(z)$ and $S_2^\infty(z)$ are defined as

$$S_1^\infty(z) = \int_0^z \kappa_2(s)\, e^{ks}\, ds, \qquad S_2^\infty(z) = \int_{-\infty}^z \kappa_1(s)\, e^{-ks}\, ds. \tag{5.29}$$

Again, from (5.12) and (5.13), the soil displacements $\mathcal{U}$ and $\mathcal{W}$ can be derived by:

$$\mathcal{U}(z) = \left(B_5^\infty + \frac{z}{4(1-\mu)} B_1^\infty\right) e^{kz} + M\left[e^{kz} \int_0^z e^{-ks}\, P^\infty(s)\, ds - e^{-kz} \int_{-\infty}^z e^{ks}\, P^\infty(s)\, ds\right], \tag{5.30}$$

and

$$\mathcal{W}(z) = i\left\{B_5^\infty e^{kz} + \left[\frac{(kz+1)}{4(1-\mu)} - \frac{1}{k}\right] B_1^\infty e^{kz} + M\left[e^{kz} \int_0^z e^{-ks}\, P(s)\, ds + e^{-kz} \int_{-\infty}^z e^{ks}\, P(s)\, ds\right]\right\}, \tag{5.31}$$

in which

$$M = \frac{i(1-2\mu)}{4G(1-\mu)}. \tag{5.32}$$

The three unknown coefficients B_1^∞, B_3^∞ and B_5^∞ in (5.28), (5.30) and (5.31) for pore pressure and soil displacements can be solved by applying boundary conditions at the seabed surface, (5.6) and (5.7). They can be expressed in closed-forms as

$$B_1^\infty = \frac{8M(1-\mu)^2}{(1-2\mu)(3-4\mu)}, \qquad B_3^\infty = 1 + \frac{8M\Phi_o(1-\mu)^2}{(1-2\mu)(3-4\mu)} S_3^\infty(0), \tag{5.33}$$

$$B_5^\infty = M S_3^\infty(0)\left[1 + \frac{8M\Phi_o(1-\mu)^2}{(1-2\mu)(3-4\mu)} S_3^\infty(0)\right] + \frac{2M(1-\mu)}{(1-2\mu)}(3-4\mu)\big[(1-2\mu) - 4kM\Phi_o(1-\mu) S_4^\infty(0)\big], \tag{5.34}$$

where

$$S_3^\infty(z) = \int_{-\infty}^z E_1(s)\, e^{ks}\, ds, \tag{5.35}$$

$$S_4^\infty(z) = \int_{-\infty}^{z} E_1(s)\, e^{ks} \int_0^s \kappa_2(\alpha)\, d\alpha\, ds \\ - \int_{-\infty}^{z} E_2(s)\, e^{-ks} \int_{-\infty}^{s} \kappa_1(\alpha)\, d\alpha\, ds. \tag{5.36}$$

From Hooke's law, (3.13)–(3.15) render the wave-induced effective normal stresses

$$\sigma'_x = 2iGp_o\left\{m^2 k\mathcal{U}(z) - \frac{2M\mu}{1-2\mu}\mathcal{P}(z) + \frac{\mu}{2(1-\mu)} B_1^\infty e^{kz}\right\} \cos nky\, e^{i(mkx-\omega t)}, \tag{5.37}$$

$$\sigma'_y = 2iGp_o\left\{n^2 k\mathcal{U}(z) - \frac{2M\mu}{1-2\mu}\mathcal{P}(z) + \frac{\mu}{2(1-\mu)} B_1^\infty e^{kz}\right\} \cos nky\, e^{i(mkx-\omega t)}, \tag{5.38}$$

$$\sigma'_z = \frac{2iGp_o}{1-2\mu}\left\{-k(1-2\mu)\mathcal{U}(z) - 2M(1-\mu)\mathcal{P}(z) \\ + \frac{1}{2}(1-2\mu) B_1^\infty e^{kz}\right\} \cos nky\, e^{i(mkx-\omega t)}, \tag{5.39}$$

and shear stresses are given by

$$\tau_{xy} = -2mnkGp_o\mathcal{U}(z) \sin nky\, e^{i(mkx-\omega t)}, \tag{5.40}$$

$$\tau_{xz} = mGp_o\left\{2ik\mathcal{W}(z) + B_1^\infty e^{kz}\right\} \cos nky\, e^{i(mkx-\omega t)}, \tag{5.41}$$

$$\tau_{yz} = inGp_o\left\{2ik\mathcal{W}(z) + B_1^\infty e^{kz}\right\} \sin nky\, e^{i(mkx-\omega t)}. \tag{5.42}$$

For a porous seabed of finite thickness, the wave-induced pore pressure, (5.25) can be rewritten as:

$$\mathcal{P}(z) = B_3 E_1(z) + B_4 E_2(z) + \Phi_o \int_{-h}^{z} F(z,s)\left(B_1 e^{ks} - B_2 e^{-ks}\right) ds \\ = B_3 E_1(z) + B_4 E_2(z) + \Phi_o\left\{B_1 S_1(z) - B_2 S_2(z)\right\} \tag{5.43}$$

where

$$S_1(z) = \int_{-h}^{z} F(z,s)\, e^{ks}\, ds, \qquad S_2(z) = \int_{-h}^{z} F(z,s)\, e^{-ks}\, ds, \tag{5.44}$$

in which $F(z,s)$ is defined by (5.26), and $E_1(z)$, $E_2(z)$ by (5.17) and (5.18).

Again, substituting (5.43) into (5.12) and (5.13), the amplitude of the soil displacements $\mathcal{U}$ and $\mathcal{W}$ are given as,

$$\mathcal{U}(z) = B_5 e^{kz} + B_6 e^{-kz} + \frac{z}{4(1-\mu)}\left(B_1 e^{kz} + B_2 e^{-kz}\right) \\ + 2M \int_{-h}^{z} \sinh k(z-s) P(s)\, ds, \tag{5.45}$$

$$\mathcal{W}(z) = -i\left\{B_5\, e^{kz} - B_6\, e^{-kz} + \left(\frac{kz+1}{4k(1-\mu)} - \frac{1}{k}\right) B_1\, e^{kz} \right.$$
$$\left. + \left(\frac{1-kz}{4k(1-\mu)} - \frac{1}{k}\right) B_2\, e^{-kz} + 2M \int_{-h}^{z} \cosh k(z-s) P(s)\, ds \right\}. \quad (5.46)$$

The effective normal stresses can be further derived as:

$$\sigma_x' = 2iGp_o\left\{m^2 k\mathcal{U}(z) - \frac{2M\mu}{1-2\mu}\mathcal{P}(z) \right.$$
$$\left. + \frac{\mu}{2(1-\mu)}\left(B_1\, e^{kz} - B_2\, e^{-kz}\right)\right\} \cos nky\; e^{i(mkx-\omega t)}, \quad (5.47)$$

$$\sigma_y' = 2G\left\{n^2 k\mathcal{U}(z) - \frac{2M\mu}{1-2\mu}\mathcal{P}(z) \right.$$
$$\left. + \frac{\mu}{2(1-\mu)}\left(B_1\, e^{kz} - B_2\, e^{-kz}\right)\right\} \cos nky\; e^{i(mkx-\omega t)}, \quad (5.48)$$

$$\sigma_z' = \frac{2iGp_o}{1-2\mu}\left\{-(1-2\mu)k\mathcal{U}(z) - 2M(1-\mu)\mathcal{P}(z) \right.$$
$$\left. + \frac{1}{2}(1-2\mu)\left(B_1\, e^{kz} - B_2\, e^{-kz}\right)\right\} \cos nky\; e^{i(mkx-\omega t)}, \quad (5.49)$$

and shear stresses are

$$\tau_{xy} = -2mnkGp_o\mathcal{U}(z) \sin nky\; e^{i(mkx-\omega t)}, \quad (5.50)$$

$$\tau_{xz} = Gp_o\left(2ik\mathcal{W}(z) + B_1\, e^{kz} + B_2\, e^{-kz}\right) \cos nky\; e^{i(mkx-\omega t)}, \quad (5.51)$$

$$\tau_{yz} = Gp_o\left(2ik\mathcal{W}(z) + B_1\, e^{kz} + B_2\, e^{-kz}\right) \sin nky\; e^{i(mkx-\omega t)}. \quad (5.52)$$

The six unknown coefficients B_1–B_6 in (5.43), and (5.45)–(5.52) for the wave-induced soil response can be solved by applying boundary conditions at the seabed surface and rigid bottom, (5.6), (5.7) and (5.9).

A set of symbolic linear equations can be written in a matrix form as

$$\mathbf{A} \bullet \mathbf{B} = \mathbf{C}, \quad (5.53)$$

where $\mathbf{B}$ is a column matrix containing the six unknown B_i coefficients. $\mathbf{C}$ is also a column matrix representing the resultant boundary values to be satisfied, such that

$$\mathbf{B} = \{B_1, B_2, B_3, B_4, B_5, B_6\}^{\mathbf{T}}, \quad (5.54)$$

$$\mathbf{C} = \left\{i, \frac{i}{2Gk}, 0, 0, 0, 0\right\}^{\mathbf{T}}, \quad (5.55)$$

and **A** is a 6×6 symbolic coefficient matrix given by

$$\mathbf{A} = \left\{ \begin{array}{cccccc} a_{11} & a_{12} & 1 & 1 & 0 & 0 \\ a_{21} & a_{22} & -2MS_5(0) & -2MS_6(0) & -1 & -1 \\ a_{31} & a_{32} & 4k & 4k & 2k & -2k \\ a_{41} & a_{42} & 0 & 0 & e^{-kh} & e^{kh} \\ a_{51} & a_{52} & 0 & 0 & e^{-kh} & e^{kh} \\ 0 & 0 & E_1'(-h) & E_2'(-h) & 0 & 0 \end{array} \right\}, \tag{5.56}$$

$$\begin{aligned} a_{11} &= \Phi_o S_1(0), & a_{12} &= -\Phi_o S_2(0), \\ a_{21} &= -2\Phi_o M S_3(0) + \frac{1}{2k}, & a_{22} &= 2\Phi_o M S_4(0) - \frac{1}{2k}, \\ a_{31} &= \frac{2\mu - 1}{2(1-\mu)} + 4Mk\Phi_o S_7(0), & a_{32} &= \frac{2\mu - 1}{2(1-\mu)} - 4Mk\Phi_o S_8(0), \\ a_{41} &= \frac{-h}{4k(1-\mu)} e^{-kh}, & a_{42} &= \frac{-h}{4k(1-\mu)} e^{kh}, \\ a_{51} &= \left(\frac{1-kh}{4k(1-\mu)} - \frac{1}{k} \right) e^{-kh}, & a_{52} &= \left(\frac{1+kh}{4k(1-\mu)} - \frac{1}{k} \right) e^{kh}, \end{aligned} \tag{5.57}$$

where $S_i(z)$ ($i = 3$–10) are given by

$$S_3(z) = \int_{-h}^{z} \sinh k(z-s) \int_{-h}^{s} F(s,\alpha) e^{k\alpha} \, d\alpha \, ds, \tag{5.58}$$

$$S_4(z) = \int_{-h}^{z} \sinh k(z-s) \int_{-h}^{s} F(s,\alpha) e^{-k\alpha} \, d\alpha \, ds, \tag{5.59}$$

$$S_5(z) = \int_{-h}^{z} \sinh k(z-s) E_1(s) \, ds, \tag{5.60}$$

$$S_6(z) = \int_{-h}^{z} \sinh k(z-s) E_2(s) \, ds, \tag{5.61}$$

$$S_7(z) = \int_{-h}^{z} \cosh k(z-s) \int_{-h}^{s} F(s,\alpha) e^{k\alpha} \, d\alpha \, ds, \tag{5.62}$$

$$S_8(z) = \int_{-h}^{z} \cosh k(z-s) \int_{-h}^{s} F(s,\alpha) e^{-k\alpha} \, d\alpha \, ds, \tag{5.63}$$

$$S_9(z) = \int_{-h}^{z} \cosh k(z-s) E_1(s) \, ds, \tag{5.64}$$

$$S_{10}(z) = \int_{-h}^{z} \cosh k(z-s) E_2(s) \, ds. \tag{5.65}$$

Once the six unknown coefficients are obtained, the wave-induced pore pressure and soil displacements can be calculated from (5.43) and (5.45)–(5.46), and the effective stresses can then be obtained from (5.47)–(5.52).

Following the procedure outlined previously, the wave-induced pore pressure and soil displacements in a layered seabed with variable permeability can be obtained, but not shown here.

5.2.2.3 A Special Case: A Seabed with Uniform Permeability

For a porous seabed of infinite thickness, when $a(z) = 1$, the solution given in Sect. 5.2.2.2 yields

$$E_1^u(z) = e^{\delta z}, \qquad E_2^u(z) = e^{-\delta z}, \tag{5.66}$$

$$S_{1,u}^{\infty}(z) = -\frac{1}{2\delta(\delta-k)}\left[e^{-(\delta-k)z} - 1\right], \qquad S_{2,u}^{\infty}(z) = -\frac{1}{2\delta(\delta-k)}\,e^{-(\delta-k)z}, \tag{5.67}$$

$$S_{3,u}^{\infty}(z) = \frac{1}{\delta+k}\,e^{(\delta+k)z}, \qquad S_{4,u}^{\infty}(z) = -\frac{1}{2k}\,e^{2kz}, \tag{5.68}$$

where subscript "u" denotes the solution for a seabed with uniform permeability.

Introducing (5.66)–(5.68) into (5.33)–(5.34) results in the expression for the B_i^{∞} coefficients, which in turn produces the solution for the soil response in a seabed of infinite thickness presented in Sect. 3.4.2. An example of the numerical comparison between the reduced form of the present solution and the conventional solution with uniform permeability is presented in Sect. 5.2.3.

Similarly, for a porous seabed of finite thickness, the solution given in Sect. 5.2.2.2 can be simplified to

$$S_1^u(z) = \frac{1}{\delta(1+\delta^2)}\left\{\delta e^{kz} - e^{-kh}\left[\sinh\delta(z+h) + \delta\cosh\delta(z+h)\right]\right\}, \tag{5.69}$$

$$S_2^u(z) = \frac{1}{\delta(1+\delta^2)}\left\{-\delta e^{-kz} - e^{-kh}\left[\sinh\delta(z+h) + \delta\cosh\delta(z+h)\right]\right\}. \tag{5.70}$$

The other $S_i^u(z)$ ($i = 3$–10) can also be derived, following their definitions as used previously.

Substituting all $S_i^u(z)$ ($i = 1$–10) into the symbolic coefficient matrix, (5.56), the six unknown coefficients for a seabed with uniform permeability can be obtained. The wave-induced soil response can then be calculated numerically.

5.2.3 Results and Discussion

It is well known that soil permeability is one of the most important parameters for categorizing marine sediments. Based on the numerical results presented in Chaps. 3

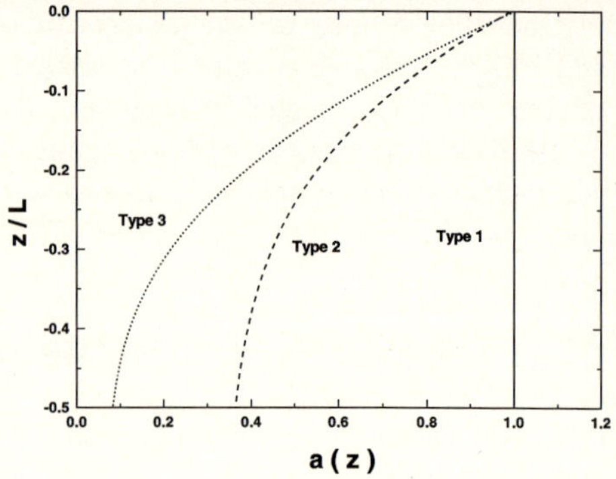

Fig. 5.1 Depth functions $a(z)$ considered for a soil matrix with variable permeability

Table 5.1 The depth functions $a(z)$ used in this study

Type 1	$a(R(z)) = 1$
Type 2	$a(R(z)) = \frac{(-0.4124)^2}{5} \coth^2[\sqrt{5}(R(z) - 0.3)]$
Type 3	$a(R(z)) = \frac{(-0.7413)^2}{0.8} \coth^2[\sqrt{0.8}(R(z) - 0.3)]$

and 4, the wave-induced soil response is affected significantly by the soil permeability in a porous seabed with uniform permeability. Since analytical solutions for soil response in a seabed with variable permeability are now available, the effects of variable permeability on the wave-induced soil response are investigated here.

Three different types of decreasing depth function $a(z)$ are considered here (Fig. 5.1 and Table 5.1). Type 1 (in solid line) is for $a(z) = 1$, which represents the conventional assumption of uniform permeability. Type 2 (in dashed line) and Type 3 (in dotted line) describe two different gradual reduction functions in permeability with depth.

As an example, we consider a short-crested system with period $T = 12.5$ seconds propagating over a uniform water depth $d = 20$ m. Three different types of soils are used, namely gravel, coarse and fine sand. Their soil characteristics are specified in graphs.

Since the results for soil response in a seabed of infinite and finite thickness plus a layered medium were discussed in previous chapters, only the numerical results for a porous seabed of infinite thickness are presented in the following sections.

5.2.3.1 Verification

Because experimental data for the wave-induced soil response with variable permeability are not yet available, verification of the present theory is restricted to the special case, $a(z) = 1$. Comparisons between the present solution for $a(z) = 1$ (Type 1 in solid line) and the conventional solution reported by [13] (in circles "o") for uniform permeability are given in Fig. 5.2. As shown in these figures. for pore

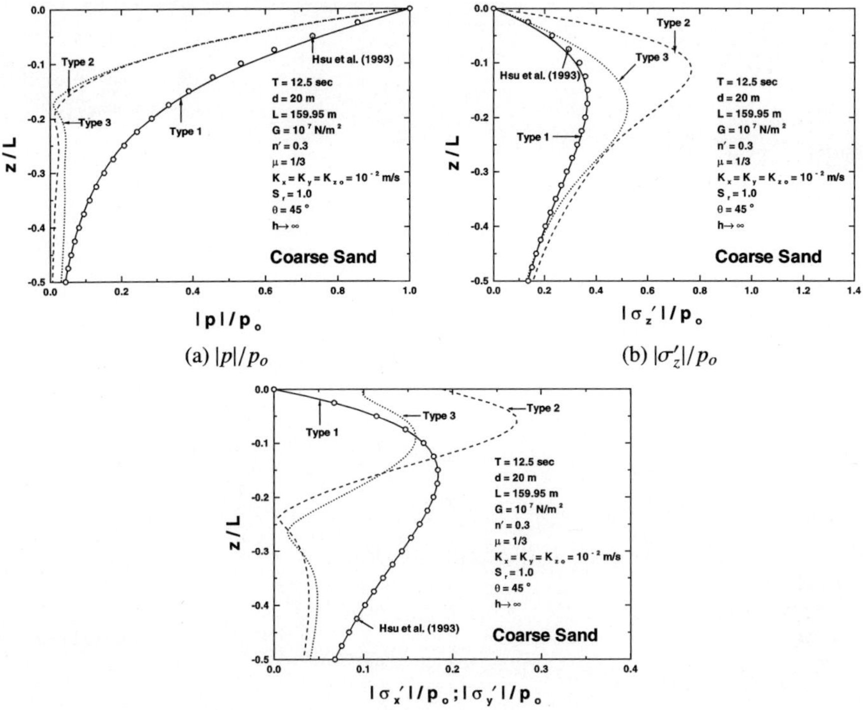

(a) $|p|/p_o$ (b) $|\sigma_z'|/p_o$

(c) $|\sigma_x'|/p_o$ & $|\sigma_y'|/p_o$

Fig. 5.2 Vertical distribution of the wave-induced soil response versus z/L in saturated coarse sand. **(a)** $|p|/p_o$; **(b)** $|\sigma_z'|/p_o$; and **(c)** $|\sigma_x'|/p_o$ and $|\sigma_y'|/p_o$; **(d)** $|\tau_{xz}|/p_o$; $|\tau_{yz}|/p_o$; **(e)** $|\tau_{xy}|/p_o$; **(f)** $|2Gku|/p_o$; and **(g)** $|2Gkw|/p_o$

pressure, effective stresses and soil displacements, the numerical results for these two different approaches are identical under the same input data.

5.2.3.2 Effect of Variable Permeabilities

For a fully saturated seabed of infinite thickness with uniform permeability, it has been reported that the wave-induced seabed response was independent of permeability [13, 51]. With the general solution for seabed response in soil of variable permeability, it is desirable to re-examine this statement.

For saturated coarse sand, the vertical distributions of the maximum soil response against the relative soil depth z/L are illustrated in Fig. 5.2, in which three different depth functions from Fig. 5.1 are used. The numerical values of the pore pressure and vertical effective normal stress are also tabulated in Table 5.2. The wave-induced pore pressure $|p|/p_o$ calculated from uniform permeability (Type 1) is greater than that from variable permeability (Types 2 and 3) (Fig. 5.2a). However, an opposite trend is observed for $|\sigma_z'|/p_o$ in Fig. 5.2b and for $|\tau_{xz}|/p_o$ in Fig. 5.2d, due to the

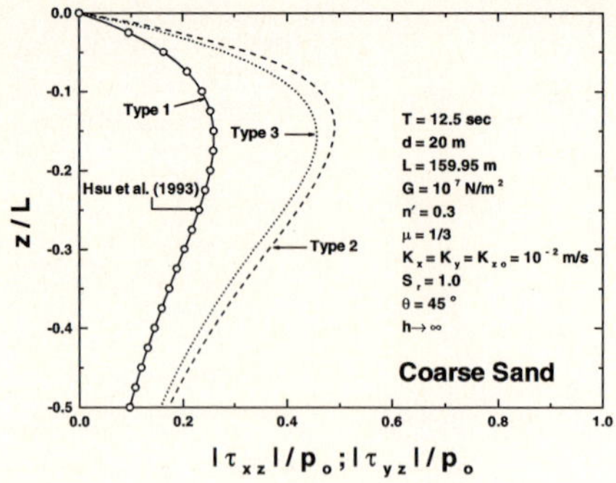

(d) $|\tau_{xz}|/p_o$; $|\tau_{yz}|/p_o$

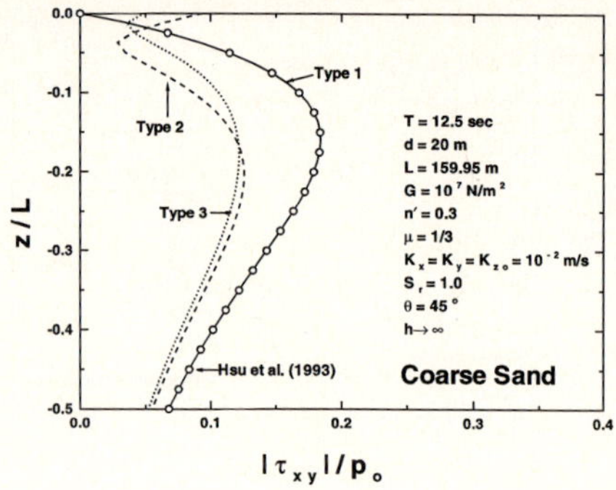

(e) $|\tau_{xy}|/p_o$

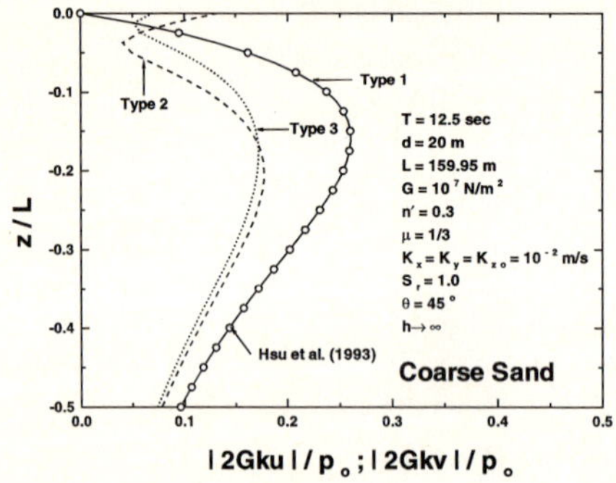

(f) $|2Gku|/p_o$

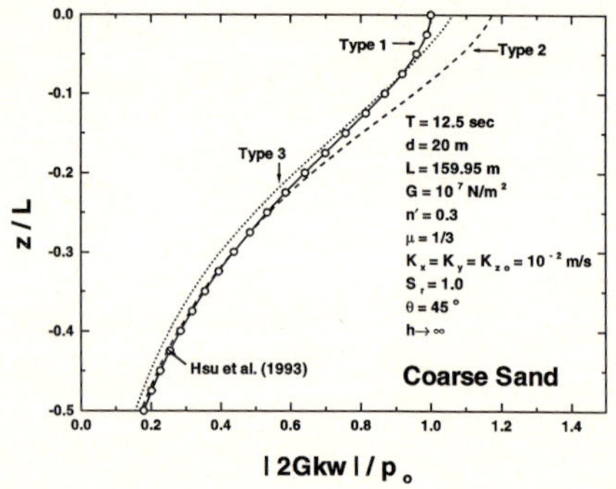

(g) $|2Gkw|/p_o$

Fig. 5.2 (Continued)

compensating nature between the values of $|p|$ and $|\sigma_z'|$. This implies that the conventional solutions with uniform permeability [13] may over-estimate the pore pressure $|p|/p_o$, but under-estimate the effective stresses, $|\sigma_z'|/p_o$ and $|\tau_{xz}|/p_o$.

For other soil response parameters, such as $|\sigma_x'|/p_o$, $|\sigma_y'|/p_o$, $|\tau_{xy}|/p_o$, $|2Gku|/p_o$ and $|2Gkv|/p_o$, their values for variable permeability are greater than those for uniform permeability near the seabed surface. For example, the maximum horizontal effective normal stress $|\sigma_x'|/p_o$ for Type 3 (variable permeability) is larger than that of Type 1 (uniform permeability) when $-0.1 \leq z/L \leq 0$, as shown in Fig. 5.2c. Similar trends can also be observed in Figs. 5.2e and 5.2f for $|\tau_{xy}|/p_o$ and $|2Gku|/p_o$, respectively.

At least from the examples presented above for saturated coarse sand, the influence of variable permeability on the vertical soil displacement $|2Gkw|/p_o$ is rather insignificant (see Fig. 5.2g), compared with other soil response parameters.

As shown in Fig. 5.2, the wave-induced soil response (except the vertical soil displacements) in a saturated seabed of infinite thickness is affected significantly by variable permeability. This implies that soil response in a field condition would be affected by soil permeability. This finding contradicts the conclusion based on the

Table 5.2 Maximum pore pressure and effective stress amplitudes in a saturated coarse sand at various depths for Type 1 and Type 2 soils+

z/L	$\lvert p_{type\,1}\rvert/p_o$	$\lvert p_{type\,2}\rvert/p_o$ $\Delta p/p_o{}^*$	$\lvert\sigma'_{z,type\,1}\rvert/p_o$	$\lvert\sigma'_{z,type\,2}\rvert/p_o$ $\Delta\sigma'_z/p_o{}^*$
0.0	1.0	1.0 (0.0)	0.0	0.0 (0.0)
−0.05	0.7572	0.5213 (−0.2359)	0.2067	0.5580 (0.3513)
−0.1	0.5545	0.2265 (−0.3280)	0.3199	0.7620 (0.4421)
−0.15	0.3986	0.0674 (−0.3312)	0.3629	0.7237 (0.3608)
−0.2	0.2850	0.0124 (−0.2726)	0.3597	0.5907 (0.2310)
−0.25	0.2046	0.0238 (−0.1808)	0.3304	0.4673 (0.1369)
−0.3	0.1081	0.0204 (−0.1276)	0.2893	0.3777 (0.0877)
−0.35	0.1481	0.0110 (−0.0934)	0.2457	0.3066 (0.0609)
−0.4	0.0792	0.0110 (−0.0682)	0.2043	0.2462 (0.0419)
−0.45	0.0581	0.0085 (−0.0486)	0.1673	0.1959 (0.0286)
−0.5	0.0426	0.0064 (−0.0362)	0.1355	0.1548 (0.0193)

+Input data is the same as the legend in Fig. 5.2

$^*\Delta p = (\lvert p_{type\,2}\rvert - \lvert p_{type\,1}\rvert)$, $\Delta\sigma'_z = (\lvert\sigma'_{z,type\,2}\rvert - \lvert\sigma'_{z,type\,1}\rvert)$

conventional solutions with uniform permeability [13, 51] in that seabed response is independent of permeability.

Since the wave-induced soil response strongly depends on soil characteristics as discussed above, the effects of soil types and the degree of saturation on the soil response with variable permeability are further examined.

5.2.3.3 Effect of Soil Types

It is well known that soil type is one of the dominant factors in analyzing the wave-seabed interaction problem. Thus, it is important to re-examine its influence on the wave-induced seabed response with variable permeability. The vertical distribution of the maximum pore pressure $|p|/p_o$ and vertical effective normal stress $|\sigma'_z|/p_o$ are presented in Fig. 5.2 for saturated coarse sand, Fig. 5.3 for unsaturated coarse sand (with $K_{zo} = 10^{-2}$ m/s), Figs. 5.4–5.5 for fine sand ($K_{zo} = 10^{-4}$ m/s) and Figs. 5.6–5.7 for gravel (with $K_{zo} = 10^{-1}$ m/s), with S_r varying as 1 and 0.975.

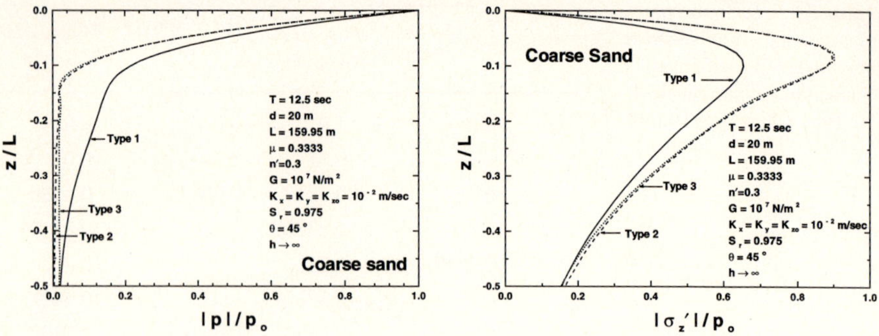

(a) $|p|/p_o$ (b) $|\sigma_z'|/p_o$

Fig. 5.3 Vertical distribution of (**a**) $|p|/p_o$ and (**b**) $|\sigma_z'|/p_o$ versus z/L in unsaturated coarse sand. ($S_r = 0.975$)

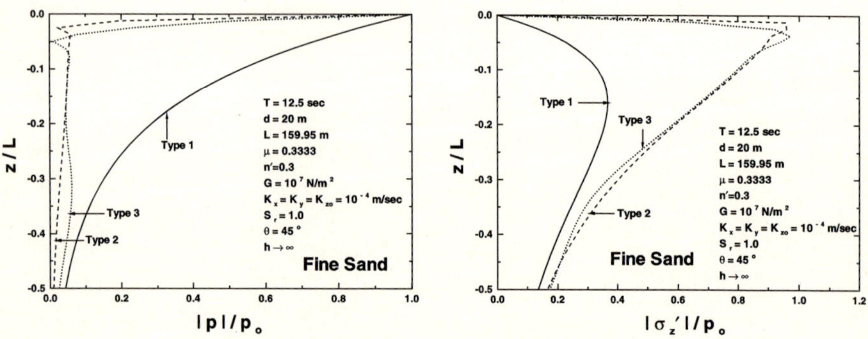

(a) $|p|/p_o$ (b) $|\sigma_z'|/p_o$

Fig. 5.4 Vertical distribution of (**a**) $|p|/p_o$ and (**b**) $|\sigma_z'|/p_o$ versus z/L in saturated fine sand. ($S_r = 1$)

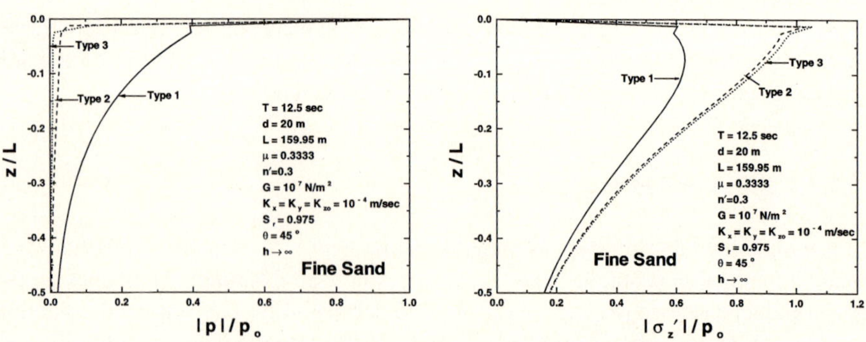

(a) $|p|/p_o$ (b) $|\sigma_z'|/p_o$

Fig. 5.5 Vertical distribution of (**a**) $|p|/p_o$ and (**b**) $|\sigma_z'|/p_o$ versus z/L in unsaturated fine sand. ($S_r = 0.975$)

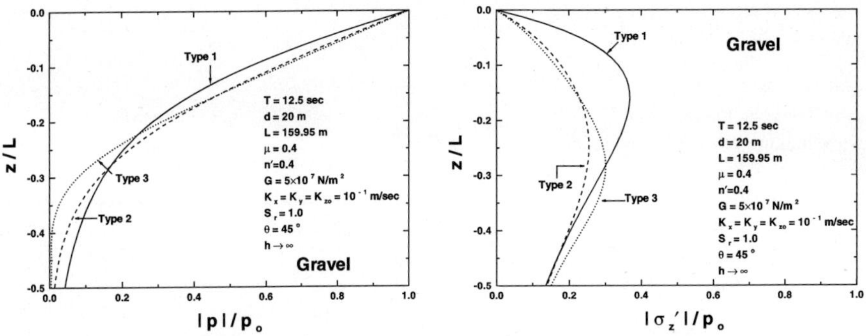

(a) $|p|/p_o$ (b) $|\sigma_z'|/p_o$

Fig. 5.6 Vertical distribution of (**a**) $|p|/p_o$ and (**b**) $|\sigma_z'|/p_o$ versus z/L in a saturated graveled seabed. ($S_r = 1$)

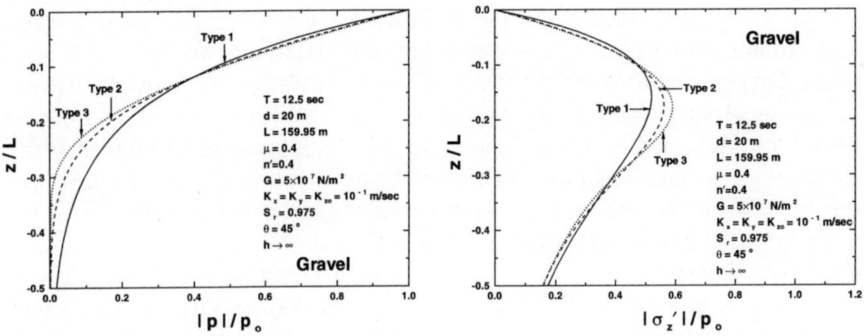

(a) $|p|/p_o$ (b) $|\sigma_z'|/p_o$

Fig. 5.7 Vertical distribution of (**a**) $|p|/p_o$ and (**b**) $|\sigma_z'|/p_o$ versus z/L in an unsaturated graveled seabed. ($S_r = 0.975$)

For coarse sand, the wave-induced pore pressure $|p|/p_o$ calculated from the solution of uniform permeability is greater than that from variable permeability, but the vertical effective normal stress $|\sigma_z'|/p_o$ is smaller in a seabed with uniform permeability (Figs. 5.2 and 5.3). The differences between these two solutions become much larger in fine sand, as shown in Figs. 5.4–5.5.

For a saturated gravel, pore pressure in the seabed with variable permeability is slightly greater than that with uniform permeability near the seabed surface (for example, $-0.22 \leq z/L \leq 0$), but becomes smaller in deeper locations (Fig. 5.6). Similar trend can be observed in unsaturated gravel, as shown in Fig. 5.7 ($S_r = 0.975$). Compared with a sandy seabed, the influences of variable permeability on pore pressure and effective stresses are insignificant in a graveled seabed.

To investigate the effect of variable permeability on the wave-induced seabed instability, Type 2 (variable permeability) and Type 1 (uniform permeability) are considered here with two different wave heights of 2.5 m and 5 m. Figure 5.8 illus-

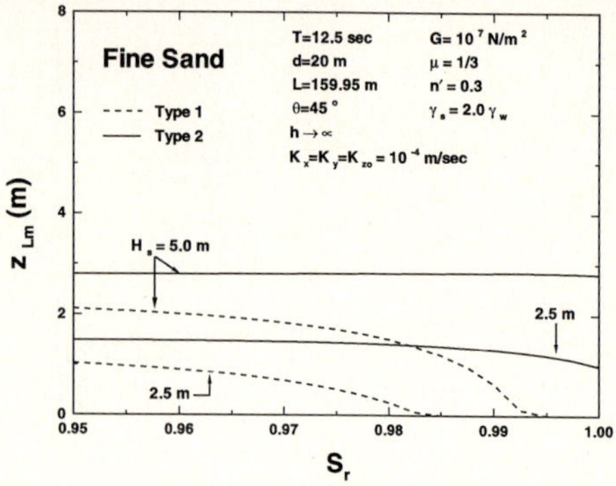

Fig. 5.8 Distribution of maximum liquefaction depth z_{Lm} versus the degree of saturation in a seabed of fine sand, showing the effect of wave height

trates the distribution of liquefaction depth z_{Lm} against the degree of saturation (S_r) in fine sand under two different wave heights. In this figure, solid lines denote results of Type 2 soil (variable permeability), and dashed lines for Type 1 (uniform permeability). This figure indicates that the maximum liquefaction depth z_{Lm} in a seabed with variable permeability is larger than that with uniform permeability. It is worth noting that liquefaction occurs in a saturated seabed with variable permeability, even under a small wave ($H_s = 2.5$ m, as shown in Fig. 5.8). This result differs from that of uniform permeability in that liquefaction does not occur in a saturated seabed under the same wave condition. As expected, no liquefaction occurs in coarse sand and gravel under the same conditions presented (figures not presented).

The distribution of the maximum shear failure depth z_{sm} versus the degree of saturation S_r are presented in Fig. 5.9 for fine sand, coarse sand and gravel, respectively. In these figures, results for the case of variable permeability are in solid lines, and those for uniform permeability in dashed lines. The maximum shear failure depth in a seabed with variable permeability is much larger than that with uniform permeability. For example, in Fig. 5.9a, the maximum shear failure depth in Type 2 soil (variable permeability) is about 7.3 m at $S_r = 0.97$ with wave height $H_s = 5$ m, but it is 4.15 m in Type 1 soil (uniform permeability) under the same conditions of soil and wave characteristics. The relative difference in the maximum shear failure depths between the cases of variable and uniform permeability becomes larger in fine sand, but is less in a graveled seabed (Figs. 5.8a and 5.9c).

5.2.3.4 Effect of the Degree of Saturation

Figures 5.2–5.7 also show the effects of the degree of saturation on the wave-induced pore pressure $|p|/p_o$ and vertical effective normal stresses $|\sigma_z'|/p_o$ for three different soil types. These figures indicate that the difference between the maximum $|p|/p_o$ and $|\sigma_z'|/p_o$ (of Type 1 and Type 2) reduces slightly as the degree of saturation S_r decreases. The differences in these two response parameters become less obvious in graveled seabeds (Figs. 5.6–5.7).

As shown in Fig. 5.8, wave-induced maximum liquefaction depth (z_{Lm}) in a seabed with uniform permeability varies with S_r, but appears rather invariant with

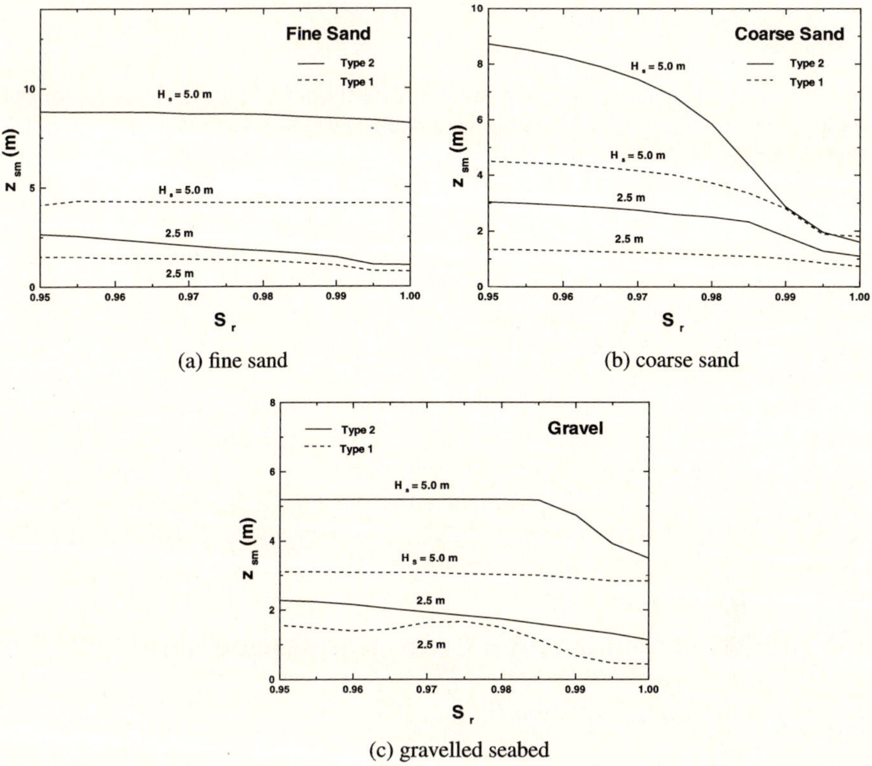

Fig. 5.9 Distribution of maximum shear failure depth z_{sm} versus the degree of saturation in (**a**) fine sand and (**b**) fine sand and (**c**) a graveled seabed, showing the effect of wave height

variable permeability for various degrees of saturation. The relative difference in z_{Lm} between variable and uniform permeabilities increases as the degree of saturation increases for a seabed of fine sand. This implies that the conventional solution with uniform permeability may under-estimate the wave-induced liquefaction potential in a saturated seabed. Thus, a saturated seabed may suffer from liquefaction, if the solution of uniform permeability is used for a seabed with variable permeability.

The wave-induced shear failure depth (z_{sm}) is also affected significantly by the degree of saturation and variable permeability in coarse sand and gravel (Fig. 5.9). For example, the maximum shear failure depth in coarse sand is about 8.66 m at $S_r = 0.95$ with $H_s = 5$ m, and it becomes 1.6 m at $S_r = 1.0$, (Fig. 5.9a). Compared with the results in coarse sand and gravel, the wave-induced shear failure depth in fine sand appears to be relatively constant over the range of S_r considered (see Fig. 5.9b). Generally speaking, the relative difference in the maximum shear failure depth z_{sm} between variable and uniform permeabilities decreases as the degree of saturation increases for three types of soil considered. This trend is different from that of liquefaction (see Fig. 5.9).

5.2.4 Summary

In this section, a series of analytical solutions for the wave-induced seabed response with variable permeability has been presented. The present solution can be simplified to yield the solution with uniform permeability, from which it is verified against the conventional solution with uniform permeability.

At least from the numerical results presented, the wave-induced seabed response is affected significantly by variable permeability. For a saturated isotropic seabed of infinite thickness, the soil response in a seabed with variable permeability is affected by soil permeability. This differs from the previous finding reported by [13, 51] for a seabed with uniform permeability.

Generally speaking, the influence of variable permeability on soil response in a sandy bed is more significant than that in gravel. The maximum depths of liquefaction and shear failure (z_{Lm} and z_{sm}) calculated from the solution of variable permeability are larger than those of uniform permeability. In addition to variable permeability, the degree of saturation affects the relative differences in the wave-induced seabed response between variable and uniform permeabilities in coarse sand.

5.3 Analytical Solution for a Cross-Anisotropic Seabed

5.3.1 Cross-Anisotropic Soil

In general, an isotropic material displays the same elastic properties irrespective of the orientation of the samples. Such a material can be described by two elastic constants, Young's modulus (E) and Poisson's ratio (μ). However, five parameters are required for describing a cross-anisotropic elastic material which possesses the same soil properties in all horizontal directions, but different in the vertical direction [32]. These are defined as

- E_v: Young's modulus in the vertical direction;
- E_h: Young's modulus in the horizontal directions;
- μ_{hh}: Poisson's ratio as the corresponding operator of lateral expansion in horizontal direction due to horizontal direct stress normal to the former;
- μ_{vh}: Poisson's ratio as the corresponding operator of lateral expansion in vertical direction due to horizontal direct stress;
- G_v: Modulus of shear deformation in a vertical plane.

The other two dependent elastic constants μ_{hv} and G_h can be inter-related by

$$\frac{\mu_{hv}}{\mu_{vh}} = \frac{E_h}{E_v} = \Omega, \qquad G_h = \frac{E_h}{2(1+\mu_{hh})}. \tag{5.71}$$

For an isotropic sand, the non-dimensional parameter Ω is equal to 1. Wiendieck [48] has carried out isotropic compressive tests for soil and proposed limits of variation of Ω as follows:

- Sand: $0.4 \le \Omega \le 1.0$,
- Normally consolidated clay: $\Omega \approx 1.0$,
- Over-consolidated clay: $1 \le \Omega \le 3$.

For the shear modulus in a vertical plane (G_v) the relations proposed by different authors are as follows:

- Wolf [49]: $\frac{E_h E_v}{E_h+(1+2\mu_{hv})E_v}$,
- Wiendieck [48]: $\frac{E_v}{\sqrt{1+\frac{E_h}{E_v}+2\mu_{hv}}}$,
- Barden [2]: $\frac{E_h}{1+\frac{E_h}{E_v}+2\mu_{hv}}$.

In summary, the shear modulus in the vertical direction (G_v) can be expressed in term of the Young's modulus (E_v) as

$$G_v = \Lambda E_z, \tag{5.72}$$

where Λ is the so-called *anisotropic constant* [10]. It is equal to $E/2(1+\mu)$ for an isotropic material.

Based on the thermodynamic considerations, Pickering [32] concluded that the strain energy of an elastic material should always be positive. This condition requires that the following inequalities should be satisfied,

$$E_h, E_v, G_v \ge 0, \tag{5.73}$$

$$\mu_{hh} \ge -1, \tag{5.74}$$

$$\frac{E_v}{E_h}(1-\mu_{hh}) - 2\mu_{vh}^2 \ge 0. \tag{5.75}$$

In (5.74), although the value of Poisson's ratio μ_{hh} in most soils is between 0 and 0.5, negative values have been reported for certain cases [10].

It is important to note that the second term of (5.75) is always negative. Therefore, the first term has to be positive. Since E_h and E_v are both positive, this requires $(1-\mu_{hh})$ to be positive, and hence $\mu_{hh} \le 1$. From (5.75), based on the same requirement for positive strain energy, Poisson's ratio μ_{vh} should satisfy the condition

$$-\sqrt{\frac{E_v(1-\mu_{hh})}{2E_h}} \le \mu_{vh} \le \sqrt{\frac{E_v(1-\mu_{hh})}{2E_h}}. \tag{5.76}$$

Similarly, the value of μ_{hv} must satisfy the condition

$$-\sqrt{\frac{E_h(1-\mu_{hh})}{2E_v}} \le \mu_{hv} \le \sqrt{\frac{E_h(1-\mu_{hh})}{2E_v}}. \tag{5.77}$$

Based on the generalized Hook's law, the relationships between elastic incremental stresses and strains in a cross-anisotropic seabed can be expressed [32] as:

$$\sigma'_x = C_{11}\frac{\partial u}{\partial x} + C_{12}\frac{\partial v}{\partial y} + C_{13}\frac{\partial w}{\partial z}, \tag{5.78}$$

$$\sigma'_y = C_{12}\frac{\partial u}{\partial x} + C_{11}\frac{\partial v}{\partial y} + C_{13}\frac{\partial w}{\partial z}, \tag{5.79}$$

$$\sigma'_z = C_{13}\frac{\partial u}{\partial x} + C_{13}\frac{\partial v}{\partial y} + C_{33}\frac{\partial w}{\partial z}, \tag{5.80}$$

and

$$\tau_{xy} = G_h\left(\frac{\partial u}{\partial y} + \frac{\partial v}{\partial x}\right) = \tau_{yx}, \tag{5.81}$$

$$\tau_{xz} = G_v\left(\frac{\partial u}{\partial z} + \frac{\partial w}{\partial x}\right) = \tau_{zx}, \tag{5.82}$$

$$\tau_{yz} = G_v\left(\frac{\partial v}{\partial z} + \frac{\partial w}{\partial y}\right) = \tau_{zy}, \tag{5.83}$$

where C_{ij} coefficients are given as

$$C_{11} = E_h(1 - \mu_{hv}\mu_{vh})/\Delta, \qquad C_{12} = E_h(\mu_{vh}\mu_{hv} + \mu_{hh})/\Delta, \tag{5.84}$$

$$C_{13} = E_h\mu_{hv}(1 + \mu_{hh})/\Delta, \qquad C_{33} = E_v\big(1 - \mu_{hh}^2\big)/\Delta, \tag{5.85}$$

$$\Delta = (1 + \mu_{hh})(1 - \mu_{hh} - 2\mu_{hv}\mu_{vh}). \tag{5.86}$$

Barden [2] and Koning [28] have also given different restrictions for elastic constants.

5.3.2 Boundary Value Problem

Considering a soil column in a porous seabed fronting a vertical reflecting wall, as depicted in Fig. 3.1, the seabed is now taken as cross-anisotropic and with thickness ranging from infinite to finite and layered. Substituting (5.78)–(5.83) into (3.10)–(3.12), the equations for force equilibrium of the pore-elastic medium can be written as

$$C_{11}\frac{\partial^2 u}{\partial x^2} + G_h\frac{\partial^2 u}{\partial y^2} + G_v\frac{\partial^2 u}{\partial z^2} + (G_h + C_{12})\frac{\partial^2 v}{\partial x\partial y} + (G_v + C_{13})\frac{\partial^2 w}{\partial x\partial z} = \frac{\partial p}{\partial x}, \tag{5.87}$$

$$C_{11}\frac{\partial^2 v}{\partial x^2} + G_h\frac{\partial^2 v}{\partial y^2} + G_v\frac{\partial^2 v}{\partial z^2} + (G_h + C_{12})\frac{\partial^2 u}{\partial x\partial y} + (G_v + C_{13})\frac{\partial^2 w}{\partial y\partial z} = \frac{\partial p}{\partial y}, \tag{5.88}$$

$$G_v\left(\frac{\partial^2 w}{\partial x^2} + \frac{\partial^2 w}{\partial y^2}\right) + C_{33}\frac{\partial^2 w}{\partial z^2} + (G_v + C_{13})\frac{\partial^2 u}{\partial x\partial z} + (G_v + C_{13})\frac{\partial^2 v}{\partial y\partial z} = \frac{\partial p}{\partial z}, \tag{5.89}$$

in the *x*-, *y*- and *z*-directions, respectively.

For a three-dimensional problem and treating a hydraulically anisotropic porous bed with permeabilities K_x, K_y and K_z in the x-, y- and z-directions, respectively, the consolidation equation is given by (3.7).

The governing equations (5.87)–(5.89) and (3.7), describing the wave-soil interaction in a cross-anisotropic seabed, can be solved with the appropriate boundary conditions at the seabed surface and rigid bottom, (5.6)–(5.10).

5.3.3 General Solutions

The boundary value problem, including governing equations (5.87)–(5.89), (3.7) and boundary conditions (5.6)–(5.10), can be used to derive wave-induced soil response in a cross-anisotropic seabed. The analytical solutions for the wave-induced pore pressure and soil displacements are obtained first, from which the effective stresses can then be determined from (5.78)–(5.83). Analytical solutions for soil response in a porous seabed of infinite and finite thickness and a layered seabed are developed systematically in this section.

5.3.3.1 Basic Framework

Since all equations are linear, it is expedient to employ complex variables in the analysis. The wave-induced pore pressure and soil displacements can be expressed as (3.25), i.e.,

$$\begin{Bmatrix} p \\ u \\ v \\ w \end{Bmatrix} = \begin{Bmatrix} P_1 \\ U_1 \\ V_1 \\ W_1 \end{Bmatrix} e^{i(mkx+nky-\omega t)} + \begin{Bmatrix} P_2 \\ U_2 \\ V_2 \\ W_2 \end{Bmatrix} e^{i(mkx-nky-\omega t)}. \tag{5.90}$$

Firstly, the reflected wave component is considered (with subscript "1"). Introducing (5.90) into governing equations (5.87)–(5.89) and (3.7), after some algebraic manipulation, renders

$$nU_1 - mV_1 = a_1 e^{\zeta_o kz} + a_2 e^{-\zeta_o kz}, \tag{5.91}$$

$$\begin{aligned} &\{G_v D^3 + (G_v + C_{13} - C_{11})k^2 D\}U_1 + \{(G_v + C_{13} - C_{33})D^2 + G_v k^2\}(imkW_1) \\ &\quad = nk^3\zeta_o\{G_v + C_{13} - G_h - C_{12}\}\big(a_1 e^{\zeta_o kz} - a_2 e^{-\zeta_o kz}\big), \end{aligned} \tag{5.92}$$

$$\begin{aligned} &\left\{G_v D^4 - (\alpha_0 G_v + C_{11}k^2)D^2 + k^2\left(\alpha_0 C_{11} - \frac{i\omega\gamma_w}{K_z}\right)\right\}U_1 \\ &\quad + \left\{(G_v + C_{13})D^3 + \left[-\alpha_0(G_v + C_{13}) + \frac{i\omega\gamma_w}{K_z}\right]D\right\}(imkW_1) \end{aligned}$$

$$= nk^2\left\{(\alpha_0 - \zeta_o^2 k^2)(G_h + C_{12}) - \frac{i\omega\gamma_w}{K_z}\right\}(a_1\, e^{\zeta_o kz} + a_2\, e^{-\zeta_o kz}), \tag{5.93}$$

where

$$\zeta_o = \sqrt{\frac{G_h}{G_v}}\alpha_0 = k^2\left[\frac{K_x}{K_z}m^2 + \frac{K_y}{K_z}n^2\right] - \frac{i\omega\gamma_w}{K_z}n'\beta. \tag{5.94}$$

Combining (5.91)–(5.93) yields

$$(\alpha_1 D^6 - \alpha_2 D^4 + \alpha_3 D^2 + \alpha_4)\left\{\begin{matrix} U_1 \\ imkW_1 \end{matrix}\right\} = n\left\{\begin{matrix} R_1(a_1 e^{\zeta_o kz} + a_2\, e^{-\zeta_o kz}) \\ R_2(a_1 e^{\zeta_o kz} - a_2\, e^{-\zeta_o kz}) \end{matrix}\right\}, \tag{5.95}$$

in which "D" denotes differential operator ($= d/dz$), and α coefficients, R_1 and R_2 are given as

$$\alpha_1 = C_{33}G_v, \tag{5.96}$$

$$\alpha_2 = k^2(C_{13} + G_v)^2 - k^2(G_v^2 + C_{33}C_{11}) - G_v C_{33}\xi_1, \tag{5.97}$$

$$\alpha_3 = k^2\{-(C_{13} + G_v)^2\xi_2 + C_{33}C_{11}\xi_3 + G_v^2\xi_4\}, \tag{5.98}$$

$$\alpha_4 = -k^4 G_v C_{11}\xi_3, \tag{5.99}$$

and

$$\begin{aligned} R_1 = k^4\{&\zeta_o^2(G_v + C_{13})^2(\zeta_o^2 k^2 - \xi_5) + \zeta_o^2 C_{33}(G_h + C_{12}) \\ &+ (\xi_1 - \xi_6) - G_v(G_h + C12)(\alpha_0 - \xi_6)\}, \end{aligned} \tag{5.100}$$

$$R_2 = k^5\zeta_o(\zeta_o^2 G_v - G_h)(\xi_5 - \zeta_o^2 k^2)(G_v + C_{13}), \tag{5.101}$$

where coefficients ξ_i $(i = 1\text{–}6)$ are given by

$$\xi_1 = k^2\left[\frac{K_x}{K_z}m^2 + \frac{K_y}{K_z}n^2\right] - \frac{i\omega\gamma_w}{K_z}\left[n'\beta + \frac{1}{C_{33}}\right], \tag{5.102}$$

$$\xi_2 = k^2\left[\frac{K_x}{K_z}m^2 + \frac{K_y}{K_z}n^2\right] - \frac{i\omega\gamma_w}{K_z}\left[n'\beta + \frac{2}{G_v + C_{13}}\right], \tag{5.103}$$

$$\xi_3 = k^2\left[\frac{K_x}{K_z}m^2 + \frac{K_y}{K_z}n^2\right] - \frac{i\omega\gamma_w}{K_z}\left[n'\beta + \frac{1}{C_{11}}\right], \tag{5.104}$$

$$\xi_4 = k^2\left[\frac{K_x}{K_z}m^2 + \frac{K_y}{K_z}n^2 + \frac{C_{11}}{G_v}\right] - \frac{i\omega\gamma_w}{K_z}\left[n'\beta + \frac{C_{11}}{C_v^2}\right], \tag{5.105}$$

$$\xi_5 = k^2\left[\frac{K_x}{K_z}m^2 + \frac{K_y}{K_z}n^2\right] - \frac{i\omega\gamma_w}{K_z}\left[n'\beta + \frac{1}{G_v + C_{13}}\right], \tag{5.106}$$

$$\xi_6 = k^2\zeta_o^2 + \frac{i\omega\gamma_w}{K_z(G_h + C_{12})}. \tag{5.107}$$

The general solution of the sixth-order ordinary differential equation (5.95) can be expressed as:

$$\begin{aligned} U_1 &= n\xi_7\left(a_1 e^{\zeta_o kz} + a_2 e^{-\zeta_o kz}\right) + \left(a_3 e^{\zeta_1 z} + a_4 e^{-\zeta_1 z}\right) \\ &\quad + \left(a_5 e^{\zeta_2 z} + a_6 e^{-\zeta_2 z}\right) + \left(a_7 e^{\zeta_3 z} + a_8 e^{-\zeta_3 z}\right), \end{aligned} \tag{5.108}$$

$$\begin{aligned} mV_1 &= \left(n^2\xi_7 - 1\right)\left(a_1 e^{\zeta_o kz} + a_2 e^{-\zeta_o kz}\right) + n\left(a_3 e^{\zeta_1 z} + a_4 e^{-\zeta_1 z}\right) \\ &\quad + n\left(a_5 e^{\zeta_2 z} + a_6 e^{-\zeta_2 z}\right) + n\left(a_7 e^{\zeta_3 z} + a_8 e^{-\zeta_3 z}\right), \end{aligned} \tag{5.109}$$

$$\begin{aligned} imkW_1 &= n\xi_8\left(a_1 e^{\zeta_o kz} - a_2 e^{-\zeta_o kz}\right) - \xi_9\left(a_3 e^{\zeta_1 z} - a_4 e^{-\zeta_1 z}\right) \\ &\quad - \xi_{10}\left(a_5 e^{\zeta_2 z} - a_6 e^{-\zeta_2 z}\right) - \xi_{11}\left(a_7 e^{\zeta_3 z} - a_8 e^{-\zeta_3 z}\right), \end{aligned} \tag{5.110}$$

$$\begin{aligned} P_1 &= \frac{1}{m}\left\{\xi_{12}\left(a_1 e^{\zeta_o kz} + a_2 e^{-\zeta_o kz}\right) + \xi_{13}\left(a_3 e^{\zeta_1 z} + a_4 e^{-\zeta_1 z}\right) \right. \\ &\quad \left. + \xi_{14}\left(a_5 e^{\zeta_2 z} + a_6 e^{-\zeta_2 z}\right) + \xi_{15}\left(a_7 e^{\zeta_3 z} + a_8 e^{-\zeta_3 z}\right)\right\}, \end{aligned} \tag{5.111}$$

where the coefficients ζ_1, ζ_2 and ζ_3 are given by

$$\zeta_1 = \frac{1}{\sqrt{3}}\sqrt{-\frac{\alpha_2}{\alpha_1} + \frac{\sqrt[3]{2}(\alpha_2^2 - 3\alpha_1\alpha_3)}{\alpha_1\sqrt[3]{\alpha_5}} + \frac{\sqrt[3]{\alpha_5}}{\sqrt[3]{2}\alpha_1}}, \tag{5.112}$$

$$\zeta_2 = \frac{1}{\sqrt{6}}\left\{-\frac{2\alpha_2}{\alpha_1} - (1 - i\sqrt{3})\left[\frac{\sqrt[3]{2}\alpha_2^2}{\alpha_1\sqrt[3]{\alpha_5}} - \frac{3\sqrt[3]{2}\alpha_3}{\sqrt[3]{\alpha_5}}\right] - (1 + i\sqrt{3})\frac{\sqrt[3]{\alpha_5}}{\sqrt[3]{2}\alpha_1}\right\}^{1/2}, \tag{5.113}$$

$$\zeta_3 = \frac{1}{\sqrt{6}}\left\{-\frac{2\alpha_2}{\alpha_1} - (1 + i\sqrt{3})\left[\frac{\sqrt[3]{2}\alpha_2^2}{\alpha_1\sqrt[3]{\alpha_5}} - \frac{3\sqrt[3]{2}\alpha_3}{\sqrt[3]{\alpha_5}}\right] - (1 - i\sqrt{3})\frac{\sqrt[3]{\alpha_5}}{\sqrt[3]{2}\alpha_1}\right\}^{1/2}, \tag{5.114}$$

and the adding coefficients ξ_j ($j = 7$–15) and α_5 are given by

$$\xi_7 = \frac{R_1}{\alpha_1\zeta_o^6 k^6 + \alpha_2\zeta_o^4 k^4 + \alpha_3\zeta_o^2 k^2 + \alpha_4}, \tag{5.115}$$

$$\xi_8 = \frac{R_2}{\alpha_1\zeta_o^6 k^6 + \alpha_2\zeta_o^4 k^4 + \alpha_3\zeta_o^2 k^2 + \alpha_4}, \tag{5.116}$$

$$\xi_9 = \frac{G_v\zeta_1^3 + k^2\zeta_1(G_v + C_{13} - C_{11})}{G_v k^2 + \zeta_1^2(G_v + C_{13} - C_{33})}, \tag{5.117}$$

$$\xi_{10} = \frac{G_v\zeta_2^3 + k^2\zeta_2(G_v + C_{13} - C_{11})}{G_v k^2 + \zeta_2^2(G_v + C_{13} - C_{33})}, \tag{5.118}$$

$$\xi_{11} = \frac{G_v\zeta_3^3 + k^2\zeta_3(G_v + C_{13} - C_{11})}{G_v k^2 + \zeta_3^2(G_v + C_{13} - C_{33})}, \tag{5.119}$$

$$\xi_{12} = \frac{nk\omega\gamma_w(\xi_7 + 1 - \zeta_o\xi_8)}{K_z(k^2\zeta_o^2 - \alpha_o)}, \tag{5.120}$$

$$\xi_{13} = \frac{\omega\gamma_w(k^2 + \zeta_1\xi_9)}{kK_z(\zeta_1^2 - \alpha_0)}, \tag{5.121}$$

$$\xi_{14} = \frac{\omega\gamma_w(k^2 + \zeta_2\xi_{10})}{kK_z(\zeta_2^2 - \alpha_0)}, \tag{5.122}$$

$$\xi_{15} = \frac{\omega\gamma_w(k^2 + \zeta_3\xi_{11})}{kK_z(\zeta_3^2 - \alpha_0)}, \tag{5.123}$$

$$\begin{aligned} \alpha_5 &= -2\alpha_2^3 + 9\alpha_1\alpha_2\alpha_3 - 27\alpha_1^2\alpha_4 \\ &\quad + 3\sqrt{3}\sqrt{-\alpha_1^2\alpha_2^2\alpha_3^2 + 4\alpha_1^3\alpha_3^3 + 4\alpha_1^2\alpha_2^3\alpha_4 - 18\alpha_1^3\alpha_2\alpha_3\alpha_4 + 27\alpha_1^4\alpha_4^2}. \end{aligned} \tag{5.124}$$

Substituting (5.108)–(5.111) into the boundary conditions, (5.6)–(5.10), analytical solutions of the wave-induced soil response in a cross-anisotropic seabed of infinite, finite thickness and a layered seabed can be further derived.

5.3.3.2 Analytical Solutions

For a porous seabed of infinite thickness, the bottom boundary condition specified by (5.8) requires coefficients, a_2, a_4, a_6 and a_8 to be zero. Thus, the wave-induced soil displacements in the x-direction can be expressed as

$$U_1 = n\xi_7 a_1\, e^{\zeta_o kz} + a_3\, e^{\zeta_1 z} + a_5\, e^{\zeta_2 z} + a_7\, e^{\zeta_3 z}. \tag{5.125}$$

Similarly, the soil displacements V_1 and W_1 in the y- and z-directions and pore pressure P_1 can also be expressed in terms of a_1, a_3, a_5 and a_7. Substituting U_1, V_1, W_1 and P_1 into (5.6) and (5.7), the solution for the reflected wave component can be obtained. Following the same procedure, the solution for the incident wave component can also be derived.

Combining the solutions of the incident and reflected wave components, a complete solution for the wave-induced soil displacements can be expressed as

$$u = mp_o\left\{A_1^\infty\, e^{\zeta_1 z} + A_2^\infty\, e^{\zeta_2 z} + A_3^\infty\, e^{\zeta_3 z}\right\}\cos nky\; e^{i(mkx-\omega t)}, \tag{5.126}$$

$$v = inp_o\left\{A_1^\infty\, e^{\zeta_1 z} + A_2^\infty\, e^{\zeta_2 z} + A_3^\infty\, e^{\zeta_3 z}\right\}\sin nky\; e^{i(mkx-\omega t)}, \tag{5.127}$$

$$w = \frac{i}{k}p_o\left\{\xi_9 A_1^\infty\, e^{\zeta_1 z} + \xi_{10}A_2^\infty\, e^{\zeta_2 z} + \xi_{11}A_3^\infty\, e^{\zeta_3 z}\right\}\cos nky\; e^{i(mkx-\omega t)}, \tag{5.128}$$

and pore pressure is given by

$$p = p_o\left\{\xi_{13}A_1^\infty\, e^{\zeta_1 z} + \xi_{14}A_2^\infty\, e^{\zeta_2 z} + \xi_{15}A_3^\infty\, e^{\zeta_3 z}\right\}\cos nky\; e^{i(mkx-\omega t)}, \tag{5.129}$$

where superscript "∞" denotes the solution of infinite thickness. It is noted that the unknown coefficient a_1 in (5.125) is equal to zero, after some algebraic process.

The coefficients A_i^∞ ($i = 1$–3) in (5.126)–(5.129) are given by

$$A_1^\infty = \frac{(\xi_{10} - \xi_{11})(C_{13}k^2 + C_{33}\zeta_2\zeta_3) + (\zeta_3 - \zeta_2)(C_{13}k^2 + C_{33}\xi_{10}\xi_{11})}{A_0^\infty}, \tag{5.130}$$

$$A_2^\infty = \frac{(\xi_{11} - \xi_9)(C_{13}k^2 + C_{33}\zeta_1\zeta_3) + (\zeta_1 - \zeta_3)(C_{13}k^2 + C_{33}\xi_9\xi_{11})}{A_0^\infty}, \tag{5.131}$$

$$A_3^\infty = \frac{(\xi_9 - \xi_{10})(C_{13}k^2 + C_{33}\zeta_1\zeta_2) - (\zeta_1 - \zeta_2)(C_{13}k^2 + C_{33}\xi_{10}\xi_9)}{A_0^\infty}, \tag{5.132}$$

and

$$\begin{aligned} A_0^\infty = {} & \xi_{13}(\xi_{10} - \xi_{11})\big(C_{13}k^2 + C_{33}\zeta_2\zeta_3\big) + \xi_{14}(\xi_{11} - \xi_9)\big(C_{13}k^2 + C_{33}\zeta_1\zeta_3\big) \\ & + \xi_{15}(\xi_9 - \xi_{10})\big(C_{13}k^2 + C_{33}\zeta_1\zeta_2\big) + \xi_{14}(\zeta_1 - \zeta_3)(C_{13} + C_{33}\xi_{11}\xi_9) \\ & + \xi_{15}(\zeta_2 - \zeta_1)(C_{13} + C_{33}\xi_{10}\xi_9) + \xi_{13}(\zeta_3 - \zeta_2)(C_{13} + C_{33}\xi_{11}\xi_{10}). \end{aligned} \tag{5.133}$$

Substituting (5.126)–(5.129) into the relationship of stresses and displacements, the wave-induced effective normal stresses can be obtained as

$$\begin{aligned} \sigma_x' = ip_o \Bigg\{ & \left[k\big(C_{11} - 2n^2G_h\big) + \frac{\zeta_1}{k}\xi_9 C_{13}\right] A_1^\infty e^{\zeta_1 z} \\ & + \left[k\big(C_{11} - 2n^2G_h\big) + \frac{\zeta_2}{k}\xi_{10} C_{13}\right] A_2^\infty e^{\zeta_2 z} \\ & + \left[k\big(C_{11} - 2n^2G_h\big) + \frac{\zeta_3}{k}\xi_{13} C_{13}\right] A_3^\infty e^{\zeta_1 z} \Bigg\} \cos nky\, e^{i(mkx - \omega t)}, \end{aligned} \tag{5.134}$$

$$\begin{aligned} \sigma_y' = ip_o \Bigg\{ & \left[k\big(C_{11} - 2m^2G_h\big) + \frac{\zeta_1}{k}\xi_9 C_{13}\right] A_1^\infty e^{\zeta_1 z} \\ & + \left[k\big(C_{11} - 2m^2G_h\big) + \frac{\zeta_2}{k}\xi_{10} C_{13}\right] A_2^\infty e^{\zeta_2 z} \\ & + \left[k\big(C_{11} - 2m^2G_h\big) + \frac{\zeta_3}{k}\xi_{13} C_{13}\right] A_3^\infty e^{\zeta_1 z} \Bigg\} \cos nky\, e^{i(mkx - \omega t)}, \end{aligned} \tag{5.135}$$

$$\begin{aligned} \sigma_z' = ip_o \Bigg\{ & \left[kC_{13} + \frac{1}{k}\zeta_1\xi_9 C_{33}\right] A_1^\infty e^{\zeta_1 z} + \left[kC_{13} + \frac{1}{k}\zeta_2\xi_{10} C_{33}\right] A_2^\infty e^{\zeta_2 z} \\ & + \left[kC_{13} + \frac{1}{k}\zeta_3\xi_{11} C_{33}\right] A_3^\infty e^{\zeta_1 z} \Bigg\} \cos nky\, e^{i(mkx - \omega t)}, \end{aligned} \tag{5.136}$$

and shear stresses are

$$\tau_{xy} = -2mnG_h p_o\{A_1^\infty e^{\zeta_1 z} + A_2^\infty e^{\zeta_2 z} + A_3^\infty e^{\zeta_3 z}\}\sin nky\, e^{i(mkx-\omega t)}, \quad (5.137)$$

$$\tau'_{xz} = mG_v p_o\{(\zeta_1 - \xi_9)A_1^\infty e^{\zeta_1 z} + (\zeta_2 - \xi_{10})A_2^\infty e^{\zeta_2 z} + (\zeta_3 - \xi_{11})A_3^\infty e^{\zeta_3 z}\}\cos nky\, e^{i(mkx-\omega t)}, \quad (5.138)$$

$$\tau'_{yz} = inG_v p_o\{(\zeta_1 - \xi_9)A_1^\infty e^{\zeta_1 z} + (\zeta_2 - \xi_{10})A_2^\infty e^{\zeta_2 z} + (\zeta_3 - \xi_{11})A_3^\infty e^{\zeta_3 z}\}\sin nky\, e^{i(mkx-\omega t)}. \quad (5.139)$$

For a seabed of finite thickness, substituting (5.108)–(5.111) into the boundary conditions, (5.6)–(5.7) and (5.9), renders a closed form expressions for the wave-induced soil displacements as

$$u = mp_o\{(A_1 e^{\zeta_1 z} + A_2 e^{-\zeta_1 z}) + (A_3 e^{\zeta_2 z} + A_4 e^{-\zeta_2 z}) + (A_5 e^{\zeta_3 z} + A_6 e^{-\zeta_3 z})\}\cos nky\, e^{i(mkx-\omega t)}, \quad (5.140)$$

$$v = inp_o\{(A_1 e^{\zeta_1 z} + A_2 e^{-\zeta_1 z}) + (A_3 e^{\zeta_2 z} + A_4 e^{-\zeta_2 z}) + (A_5 e^{\zeta_3 z} + A_6 e^{-\zeta_3 z})\}\sin nky\, e^{i(mkx-\omega t)}, \quad (5.141)$$

$$w = \frac{i}{k} p_o\{\xi_9(A_1 e^{\zeta_1 z} + A_2 e^{-\zeta_1 z}) + \xi_{10}(A_3 e^{\zeta_2 z} + A_4 e^{-\zeta_2 z}) + \xi_{11}(A_5 e^{\zeta_3 z} + A_6 e^{-\zeta_3 z})\}\cos nky\, e^{i(mkx-\omega t)}, \quad (5.142)$$

and

$$p = p_o\{\xi_{13}(A_1 e^{\zeta_1 z} + A_2 e^{-\zeta_1 z}) + \xi_{14}(A_3 e^{\zeta_2 z} + A_4 e^{-\zeta_2 z}) + \xi_{15}(A_5 e^{\zeta_3 z} + A_6 e^{-\zeta_3 z})\}\cos nky\, e^{i(mkx-\omega t)}, \quad (5.143)$$

where coefficients A_i ($i = 1$–6) are available in the appendix of [19]. It is worth noting that the unknown coefficients a_1 and a_2 in (5.108)–(5.111) are zero, after some algebraic process.

Again, the wave-induced effective stresses can be obtained as

$$\sigma'_x = ip_o\Big\{\Big[k(C_{11} - 2n^2G_h) + \frac{\zeta_1}{k}\xi_9 C_{13}\Big](A_1 e^{\zeta_1 z} + A_2 e^{-\zeta_1 z}) + \Big[k(C_{11} - 2n^2G_h) + \frac{\zeta_2}{k}\xi_{10} C_{13}\Big](A_3 e^{\zeta_1 z} + A_4 e^{-\zeta_1 z}) + \Big[k(C_{11} - 2n^2G_h) + \frac{\zeta_1}{k}\xi_{11} C_{13}\Big](A_5 e^{\zeta_1 z} + A_6 e^{-\zeta_1 z})\Big\}\cos nky\, e^{i(mkx-\omega t)}, \quad (5.144)$$

$$\sigma'_y = ip_o\Big\{\Big[k(C_{11} - 2m^2G_h) + \frac{\zeta_1}{k}\xi_9 C_{13}\Big](A_1 e^{\zeta_1 z} + A_2 e^{-\zeta_1 z})$$

$$+\left[k\left(C_{11}-2m^2G_h\right)+\frac{\zeta_2}{k}\xi_{10}C_{13}\right]\left(A_3\,e^{\zeta_1 z}+A_4\,e^{-\zeta_1 z}\right)$$
$$+\left[k\left(C_{11}-2m^2G_h\right)+\frac{\zeta_1}{k}\xi_{11}C_{13}\right]\left(A_5\,e^{\zeta_1 z}+A_6\,e^{-\zeta_1 z}\right)\Bigg\}\cos nky\; e^{i(mkx-\omega t)}, \tag{5.145}$$

$$\sigma_z'=ip_o\Bigg\{\left[kC_{13}+\frac{\zeta_1}{k}\xi_9C_{33}\right]\left(A_1\,e^{\zeta_1 z}+A_2\,e^{-\zeta_1 z}\right)$$
$$+\left[kC_{13}+\frac{\zeta_2}{k}\xi_{10}C_{33}\right]\left(A_3\,e^{\zeta_2 z}+A_4\,e^{-\zeta_2 z}\right)$$
$$+\left[kC_{13}+\frac{\zeta_3}{k}\xi_{11}C_{33}\right]\left(A_5\,e^{\zeta_3 z}+A_6\,e^{-\zeta_3 z}\right)\Bigg\}\cos nky\; e^{i(mkx-\omega t)}, \tag{5.146}$$

and

$$\tau_{xy}=-2mnG_hp_o\Big\{\left(A_1\,e^{\zeta_1 z}+A_2\,e^{-\zeta_1 z}\right)+\left(A_3\,e^{\zeta_2 z}+A_4\,e^{-\zeta_2 z}\right)$$
$$+\left(A_5\,e^{\zeta_3 z}+A_6\,e^{-\zeta_3 z}\right)\Big\}\sin nky\; e^{i(mkx-\omega t)}, \tag{5.147}$$

$$\tau_{xz}=mG_vp_o\Big\{(\zeta_1-\xi_9)\left(A_1\,e^{\zeta_1 z}-A_2\,e^{-\zeta_1 z}\right)$$
$$+(\zeta_2-\xi_{10})\left(A_3\,e^{\zeta_2 z}-A_4\,e^{-\zeta_2 z}\right)$$
$$+(\zeta_3-\xi_{11})\left(A_5\,e^{\zeta_3 z}+A_6\,e^{-\zeta_3 z}\right)\Big\}\cos nky\; e^{i(mkx-\omega t)}, \tag{5.148}$$

$$\tau_{yz}=inG_vp_o\Big\{(\zeta_1-\xi_9)\left(A_1\,e^{\zeta_1 z}-A_2\,e^{-\zeta_1 z}\right)$$
$$+(\zeta_2-\xi_{10})\left(A_3\,e^{\zeta_2 z}-A_4\,e^{-\zeta_2 z}\right)$$
$$+(\zeta_3-\xi_{11})\left(A_5\,e^{\zeta_3 z}+A_6\,e^{-\zeta_3 z}\right)\Big\}\sin nky\; e^{i(mkx-\omega t)}, \tag{5.149}$$

Following the procedure outlined previously, the wave-induced pore pressure and soil displacements in a layered seabed with variable permeability can be obtained, but not shown here.

5.3.3.3 Special Case I: An Isotropic Seabed

Since an isotropic seabed is a particular case of the cross-isotropic seabed discussed, it is logical to see whether the cross-anisotropic solution is reducible to such a condition. For an isotropic seabed, the corresponding elastic constants become

$$E_h=E_v=E,$$
$$\mu_{hh}=\mu_{vh}=\mu_{hv}=\mu, \tag{5.150}$$
$$G_h=G_v=G=\frac{E}{2(1+\mu)}.$$

Substituting (5.150) into (5.84)–(5.86), the coefficients C_{ij} can be rewritten as

$$C_{11} = C_{33} = \frac{2G(1-\mu)}{1-2\mu}, \qquad C_{12} = C_{13} = \frac{2G\mu}{1-2\mu}. \tag{5.151}$$

The plane stress-strain relationship for an isotropic soil can be written as

$$\sigma_x' = \frac{2G}{1-2\mu}\left[(1-\mu)\frac{\partial u}{\partial x} + \mu\frac{\partial v}{\partial y} + \mu\frac{\partial w}{\partial z}\right] = 2G\left[\frac{\partial u}{\partial x} + \frac{\mu}{1-2\mu}\varepsilon\right], \tag{5.152}$$

$$\sigma_y' = \frac{2G}{1-2\mu}\left[\mu\frac{\partial u}{\partial x} + (1-\mu)\frac{\partial v}{\partial y} + \mu\frac{\partial w}{\partial z}\right] = 2G\left[\frac{\partial v}{\partial y} + \frac{\mu}{1-2\mu}\varepsilon\right], \tag{5.153}$$

$$\sigma_z' = \frac{2G}{1-2\mu}\left[\mu\frac{\partial u}{\partial x} + \mu\frac{\partial v}{\partial y} + (1-\mu)\frac{\partial w}{\partial z}\right] = 2G\left[\frac{\partial w}{\partial z} + \frac{\mu}{1-2\mu}\varepsilon\right], \tag{5.154}$$

$$\tau_{xz} = G\left[\frac{\partial u}{\partial z} + \frac{\partial w}{\partial x}\right] = \tau_{zx}, \tag{5.155}$$

$$\tau_{yz} = G\left[\frac{\partial v}{\partial z} + \frac{\partial w}{\partial y}\right] = \tau_{zy}, \tag{5.156}$$

$$\tau_{xy} = G\left[\frac{\partial u}{\partial y} + \frac{\partial v}{\partial x}\right] = \tau_{yx}. \tag{5.157}$$

These equations are found to be identical to (3.13)–(3.15).

Introducing (5.150) and (5.151) into the final governing equation of soil displacement in the x-direction, (5.95), yields

$$\left[D^6 - (2k^2 - \xi^2)D^4 + (k^4 + 2k^2\delta^2)D^2 - k^4\delta^2\right]U_1 = (D^2 - k^2)^2(D^2 - \delta^2)U_1 = 0, \tag{5.158}$$

where coefficient δ was given by (3.27). It is obvious that (5.158) and (3.26) are identical. Thus, the isotropic solutions can be obtained from the present solution derived for cross-anisotropic condition, and the resultant expressions are expected to be the same as presented in Sect. 3.4.

5.3.3.4 Special Case II: A Fully Saturated Seabed

In a fully saturated seabed of infinite thickness, the wave-induced soil response has been shown to depend only on wave conditions, and not on soil characteristics [13, 51]. It is desirable to re-examine this statement from reducing the solution for a cross-anisotropic seabed. The compressibility of the pore fluid β approaches zero ($\beta = 5.0 \times 10^{-10}$ m^2/N, from (3.9)), since $S_r = 1$.

Substituting $\beta = 0$ into ξ_i coefficients, given by (5.102)–(5.107) and (5.115)–(5.123), it is found that the resultant ξ_i coefficients are found to contain cross-anisotropic parameters in terms of C_{ij} coefficients and G_v. Thus, the coefficients ζ_i ($i = 1, 2, 3$) and A_i^∞ ($i = 1, 2, 3$) also depend on the soil parameters. This implies that the wave-induced pore pressure (p) and soil displacements (u, v, and w) in a cross-anisotropic seabed depend on soil characteristics as well as wave conditions. This result differs from the conclusion for a saturated seabed under the conventional *isotropic* solutions of [13, 51]. Numerical examples are given to support this between these two solutions in the next section.

5.3.4 Results and Discussion

Based on the analytical solutions for the wave-induced soil response, together with the criteria of seabed instability given in Chap. 4, the influence of the cross-anisotropic parameters on the wave-induced seabed response can be examined.

Since detailed discussions for the relative differences between the solution of infinite, finite thickness and layered medium have been presented in Chaps. 3 and 4, only the results for an infinite seabed are demonstrated here. Furthermore, as discussed in previous chapters, the effect of the combined obliquity-permeability parameter (*KMN*) on the wave-induced seabed instability is insignificant, compared with other parameters. Therefore, only hydraulically isotropic conditions are discussed.

5.3.4.1 Comparison Between Cross-Anisotropic and Isotropic Solutions

For a fully saturated seabed of infinite thickness, it has been concluded that wave-induced seabed response depends only on the wave characteristics, implying independence of soil characteristics [13, 51]. However, as reported by Jeng and Hsu [21], this conclusion is invalid for an isotropic seabed of finite thickness under such a condition. Since an analytical solution for a cross-anisotropic seabed is now available, it is of interest to investigate the effect of the behavior of cross-anisotropic soils on the wave-induced soil response in a saturated seabed.

As a comparative example, the wave-induced pore pressure and vertical effective normal stress calculated for the *isotropic* and *cross-anisotropic* solutions are tabulated in Table 5.3. In the table, subscripts "*i*" and "*c*" denote the results calculated from isotropic and cross-anisotropic solutions, respectively. The relative differences in pore pressure, $(|p_c| - |p_i|)/p_o$, and vertical effective normal stress, $(|\sigma'_{zc}| - |\sigma'_{zi}|)/p_o$, are also included in the table. It is found that $|p_c|/p_o$ calculated from the cross-anisotropic solution is smaller than $|p_i|/p_o$ from the isotropic except for the very thin portion beneath the seabed surface, where $|\sigma'_{zc}|/p_o > |\sigma'_{zi}|/p_o$. This means that the conventional isotropic solution over-estimates the pore pressure, but under-estimates the vertical effective normal stress. Furthermore, it is important to note that the relative difference between these two solutions may reach 7.1 % of p_o

Table 5.3 Comparison of pore pressure and vertical effective normal stress obtained from cross-anisotropic and isotropic solutions[a]

z/L	Isotropic solution		Cross-anisotropic solution	
	$\vert p_i\vert/p_o$	$\vert\sigma'_{zi}\vert/p_o$	$\vert p_c\vert/p_o$ $\Delta p/p_o{}^+$	$\vert\sigma'_{zc}\vert/p_o$ $\Delta\sigma'_z/p_o{}^+$
0.0	1.0000	0.0000	1.0000 (0.0000)	0.0000 (0.0000)
–0.05	0.7304	0.2294	0.7584 (0.028)	0.2304 (0.0010)
–0.10	0.5333	0.3352	0.5232 (–0.0102)	0.3381 (0.0029)
–0.15	0.3896	0.3672	0.3648 (–0.0248)	0.3733 (0.0061)
–0.20	0.2846	0.3576	0.2427 (–0.0419)	0.3561 (–0.0015)
–0.25	0.2078	0.3265	0.1509 (–0.0569)	0.3351 (0.0086)
–0.30	0.1518	0.2862	0.0845 (–0.0673)	0.2920 (0.0058)
–0.35	0.1109	0.2438	0.0395 (–0.0714)	0.2445 (0.0007)
–0.40	0.0810	0.2035	0.0155 (–0.0655)	0.1978 (–0.0057)
–0.45	0.0591	0.1672	0.0183 (–0.0408)	0.1548 (–0.0124)
–0.50	0.0432	0.1357	0.0258 (–0.0174)	0.1174 (–0.0183)

[a]Input data for this comparison:
(1) wave conditions: $T = 12.5$ s, $d = 20$ m, $L = 159.95$ m
(2) soil conditions: $K_x = K_y = K_z = 10^{-2}$ m/s, $S_r = 1.0$, $n' = 0.2$, $h \to \infty$
(3) *anisotropic* parameters: $E_v = 10^7$ N/m^2, $\mu_{hh} = 0.4$, $\mu_{vh} = 0.3$, $\Omega = 0.8$, $\Lambda = 0.5$
(4) *isotropic* parameters: $E = 10^7$ N/m^2, $\mu = 0.3$
$^+\Delta p = (|p_c| - |p_i|)$, $\Delta = (|\sigma'_{zc}| - |\sigma'_{zi}|)$

for pore pressure at $z/L = -0.35$. This difference comes obviously from the cross-anisotropic parameters. This implies that soil response in a cross-anisotropic and saturated seabed depends on both wave and soil characteristics.

Based on the same input data presented in Table 5.3, the maximum amplitude of the wave-induced soil response is calculated, and its vertical distribution in terms of z/L are shown in Fig. 5.10. In these figures, solid lines represent the results of the cross-anisotropic solution, and dashed lines for the isotropic solution.

Figure 5.10a shows the vertical distribution of the maximum $|p|/p_o$ for the isotropic and cross-anisotropic solutions. The pore pressure $|p_i|/p_o$ is almost identical with $|p_c|/p_o$ near the seabed surface (for example, $-0.1 < z/L \le 0$ in Fig. 5.10a). However, the $|p_c|/p_o$ is smaller than $|p_i|/p_o$ for $z/L < -0.1$. The differences between them become larger in deeper locations until $z/L = -0.4$.

As shown in Fig. 5.10b, the vertical effective normal stress $|\sigma'_{zc}|/p_o$ is almost identical with $|\sigma'_{zi}|/p_o$. The horizontal effective normal stress, $|\sigma'_{xi}|/p_o$ also matches well with $|\sigma'_{xc}|/p_o$ near the seabed surface ($-0.11 \le z/L \le 0$), as shown in Fig. 5.10c, but the latter becomes slightly larger as $|z/L|$ increases. However, for

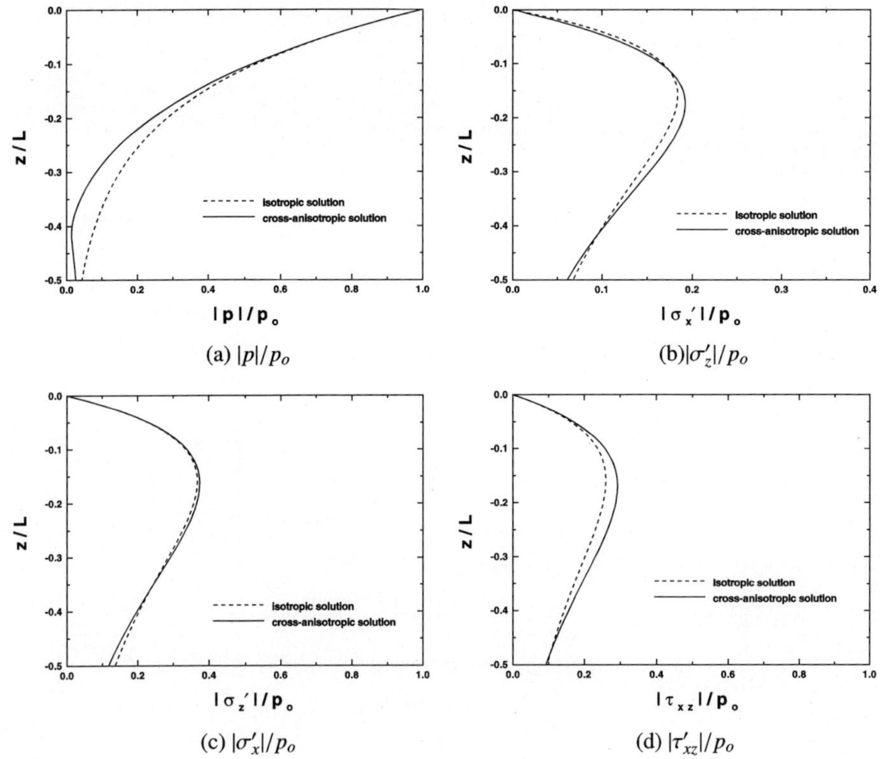

Fig. 5.10 Vertical distribution of the maximum (**a**) $|p|/p_o$, (**b**) $|\sigma_z'|/p_o$ and (**c**) $|\sigma_x'|/p_o$ and (**d**) $|\tau_{xz}'|/p_o$ for isotropic and cross-anisotropic seabed. (See footnote of Table 5.3 for input data.)

the effective shear stress $|\tau_{xz}|/p_o$, the isotropic solution is slightly greater than that of cross-anisotropic solution (Fig. 5.10d).

Comparisons on other soil response parameters (including $|\tau_{xy}|/p_o$, $|2G_hku|/p_o$ and $|2G_hkw|/p_o$) for these two solutions are presented in Figs. 5.11a–5.11c. The numerical values of the wave-induced soil displacements are also tabulated in Table 5.4. For these three physical parameters, the results from the isotropic solution are much larger than those from the cross-anisotropic solution. For example, the relative difference in the maximum soil displacement in the x-direction, $2G_hk(|u_c| - |u_i|)/p_o$, occurring at $z/L = -0.15$ is 9.15 % of p_o (Table 5.4). Similarly, the relative difference in the maximum soil displacement in the z-direction, $2G_hk(|w_c| - |w_i|)/p_o$, reaches 42 % of p_o at $z/L = 0$, and 33.2 % of p_o at $z/L = -0.15$. This implies that the effect of cross-anisotropic parameters on the wave-induced seabed response should be examined properly, as it may lead to substantial errors in some response parameters.

The other soil response parameters in the y-direction can be obtained directly from those in the x-direction through the wave obliquity θ. For example, $|2G_hkv|/p_o = |2G_hku|/p_o$, $|\tau_{yz}|/p_o = |\tau_{xz}|/p_o$ and $|\sigma_y'|/p_o = |\sigma_x'|/p_o$, since $\theta = 45°$ in this example. Thus, the results of these parameters are not presented here.

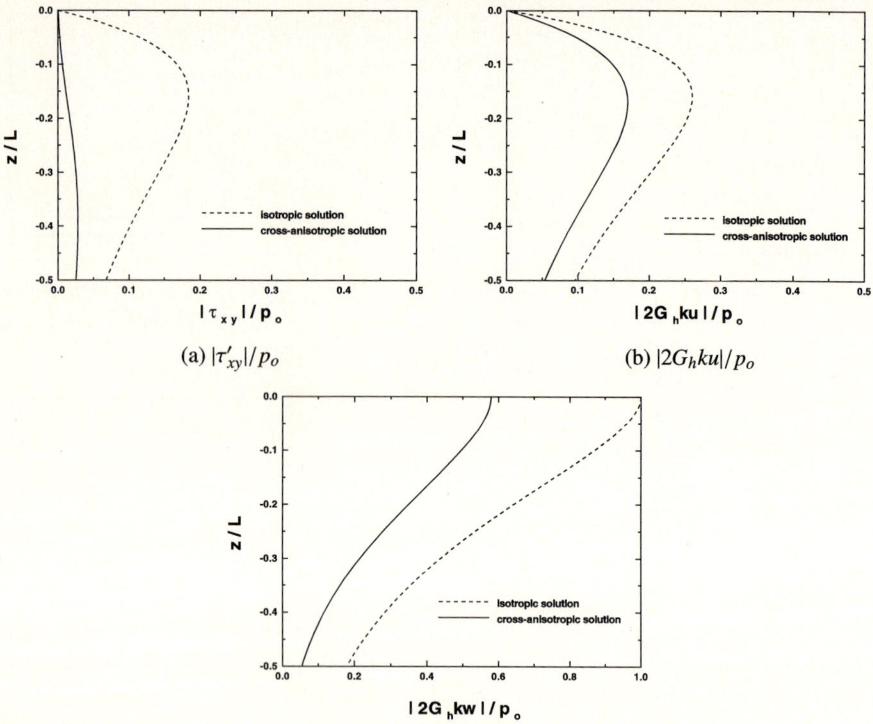

(c) $|2G_h kw|/p_o$

Fig. 5.11 Vertical distribution of the maximum (**a**) $|\tau'_{xy}|/p_o$, (**b**) $|2G_h ku|/p_o$ and (**c**) $|2G_h kw|/p_o$ versus z/L for isotropic and cross-anisotropic seabed

At least from the examples demonstrated in Figs. 5.10 and 5.11 plus Tables 5.3 and 5.4, it may be concluded that seabed response depends on soil characteristics as well as wave characteristics for a fully saturated seabed of infinite thickness.

5.3.4.2 Effect of Poisson's Ratio μ_{hh}

As discussed previously, the wave-induced soil response in a cross-anisotropic seabed depends on both wave and soil characteristics even for a saturated seabed of infinite thickness. Among the cross-anisotropic parameters, Poisson's ratio (μ_{hh}) describes the strain in one horizontal direction caused by the strain on another horizontal direction normal to the former. Its effects on the wave-induced pore pressure and effective stresses are discussed in this section.

The vertical distribution of the maximum pore pressure $|p|/p_o$ and vertical effective normal stress $|\sigma'_z|/p_o$ in a coarse sand are presented in Fig. 5.12. Generally speaking, $|p|/p_o$ increases, but $|\sigma'_z|/p_o$ decreases as μ_{hh} increases. However, the differences in magnitude are small for various values of μ_{hh}. In the region near the seabed surface (for example, $-0.1 \leq z/L \leq 0$ in Fig. 5.12), Poisson's ratio μ_{hh} affects the soil response slightly (see curves for $\mu_{hh} = 0$ and 0.45). Similar trends can

Table 5.4 Comparison of the wave-induced soil displacements from cross-anisotropic and isotropic solutions[#]

z/L	Isotropic solution		Cross-anisotropic solution	
	$\|2G_hku_i\|/p_o$	$\|2G_hkw_i\|/p_o$	$\|2G_hku_c\|/p_o$ $\Delta u/p_o{}^*$	$\|2G_hkw_c\|/p_o$ $\Delta w/p_o{}^*$
0.0	0.0000	1.0000	0.0015 (0.0015)	0.5791 (−0.4209)
−0.05	0.1623	0.9599	0.0972 (−0.0651)	0.5542 (−0.4057)
−0.10	0.2370	0.8687	0.1488 (−0.0882)	0.4964 (−0.3723)
−0.15	0.2597	0.7569	0.1672 (−0.0915)	0.4244 (−0.3325)
−0.20	0.2529	0.6423	0.1672 (−0.0857)	0.3496 (−0.2927)
−0.25	0.2309	0.5344	0.1538 (−0.0771)	0.2786 (−0.2558)
−0.30	0.2024	0.4380	0.1341 (−0.0683)	0.2152 (−0.2228)
−0.35	0.1725	0.3548	0.1122 (−0.0603)	0.1611 (−0.1937)
−0.40	0.1440	0.2846	0.0906 (−0.0534)	0.2167 (−0.1679)
−0.45	0.1183	0.2265	0.0708 (−0.0475)	0.0812 (−0.1453)
−0.50	0.0959	0.1789	0.0536 (−0.0423)	0.0539 (−0.1250)

[#]See footnote of Table 5.3 for input data

$^*\Delta u = 2G_hk(|u_c| - |u_i|)$, $\Delta w = 2G_hk(|w_c| - |w_i|)$

be observed in fine sand (Fig. 5.13). From Figs. 5.12–5.13, the influence of μ_{hh} on the wave-induced soil response in a seabed of fine sand is marginally different than in coarse sand.

Based on the criterion (d) for liquefaction presented in Chap. 4, the calculated maximum liquefaction depths (z_{Lm}) versus Poisson's ratio μ_{hh} for various degrees of saturation in fine sand are plotted in Fig. 5.14a. It is found that z_{Lm} increases gradually as the Poisson's ratio μ_{hh} increases, and as the degree of saturation decreases. This implies that a seabed fronting a breakwater may have a greater liquefaction depth z_{Lm} with a large μ_{hh} under unsaturated conditions. It is expected that no liquefaction occurs in a coarse sand in this example, because liquefaction only occurs in coarse sand when the degree of saturation S_r approaches 0.9, as seen in Fig. 4.17.

Figures 5.14b and 5.14c show the maximum shear failure depth z_{sm} versus Poisson's ratio μ_{hh} for various values of S_r in a seabed of coarse and fine sand, respectively. The maximum shear failure depth z_{sm} in coarse sand increases rapidly as μ_{hh} increases at $S_r = 0.95$, as seen in Fig. 5.14b. However, the influence of μ_{hh} on z_{sm} become less as S_r increases. On the other hand, the maximum shear failure depth z_{sm} in fine sand is affected slightly by μ_{hh} for all values of S_r (Fig. 5.14c).

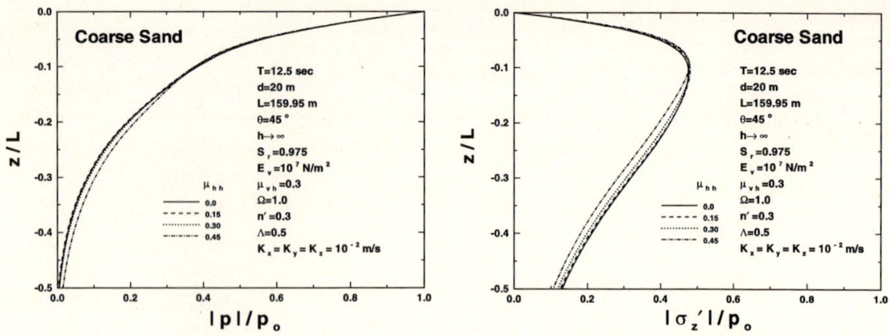

(a) $|p|/p_o$ (b) $|\sigma_z'|/p_o$

Fig. 5.12 Vertical distribution of the maximum (**a**) $|p|/p_o$ and (**b**) $|\sigma_z'|/p_o$ versus z/L for various Poisson's ratio μ_{hh} in coarse sand

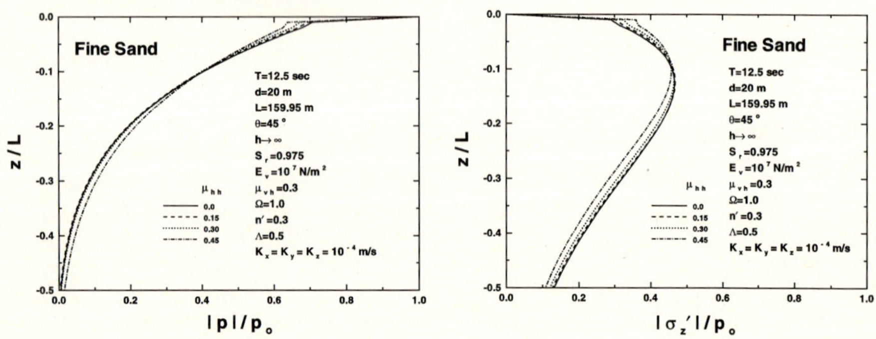

(a) $|p|/p_o$ (b) $|\sigma_z'|/p_o$

Fig. 5.13 Vertical distribution of the maximum (**a**) $|p|/p_o$ and (**b**) (**b**) $|\sigma_z'|/p_o$ versus z/L for various Poisson's ratio μ_{hh} in fine sand

5.3.4.3 Effect of Poisson's Ratio μ_{vh}

Another Poisson's ratio (μ_{vh}) describes the strain in the horizontal direction caused by the strain in the vertical direction. The vertical distributions of the maximum pore pressure $|p|/p_o$ and vertical effective normal stress $|\sigma_z'|/p_o$ in coarse sand are presented in Fig. 5.15. They indicate Poisson's ratio μ_{vh} has significant effect on the wave-induced pore pressure and effective stresses. Generally speaking, $|p|/p_o$ decreases, but $|\sigma_z'|/p_o$ increases as μ_{vh} increases. Similar trends can be observed in fine sand (Fig. 5.16).

The distribution of the maximum liquefaction depth (z_{Lm}) against Poisson's ratio μ_{vh} for various degrees of saturation in fine sand are illustrated in Fig. 5.17a. It is found that z_{Lm} increases as Poisson's ratio μ_{vh} increases. Again, no liquefaction occurs in coarse sand in this example for various values of μ_{vh}.

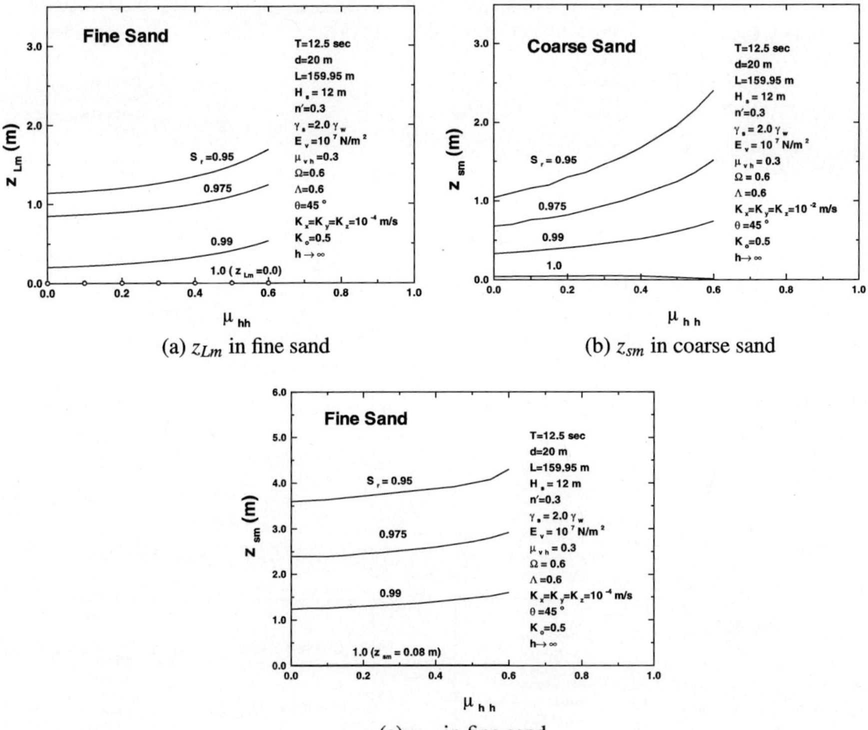

(a) z_{Lm} in fine sand

(b) z_{sm} in coarse sand

(c) z_{sm} in fine sand

Fig. 5.14 Relationship between the maximum liquefaction depth z_{Lm} and shear failure depth z_{sm} and μ_{hh} for various degrees of saturation S_r in seabeds

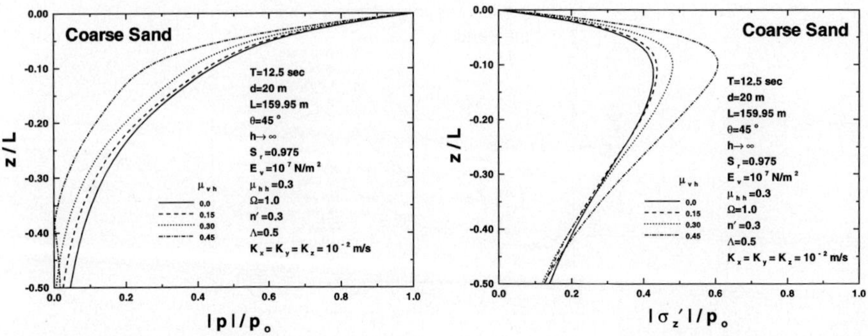

(a) $|p|/p_o$

(b) $|\sigma_z'|/p_o$

Fig. 5.15 Vertical distribution of the maximum (**a**) $|p|/p_o$ and (**b**) $|\sigma_z'|/p_o$ versus z/L for various Poisson's ratio μ_{vh} in coarse sand

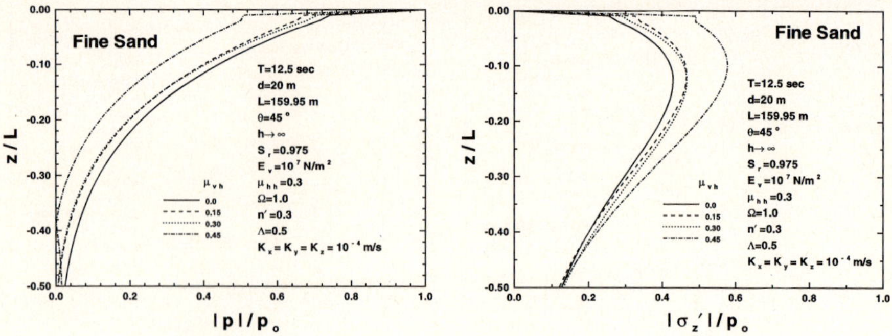

(a) $|p|/p_o$ (b) $|\sigma_z'|/p_o$

Fig. 5.16 Vertical distribution of the maximum (**a**) $|p|/p_o$ and (**b**) $|\sigma_z'|/p_o$ versus z/L for various Poisson's ratio μ_{vh} in fine sand

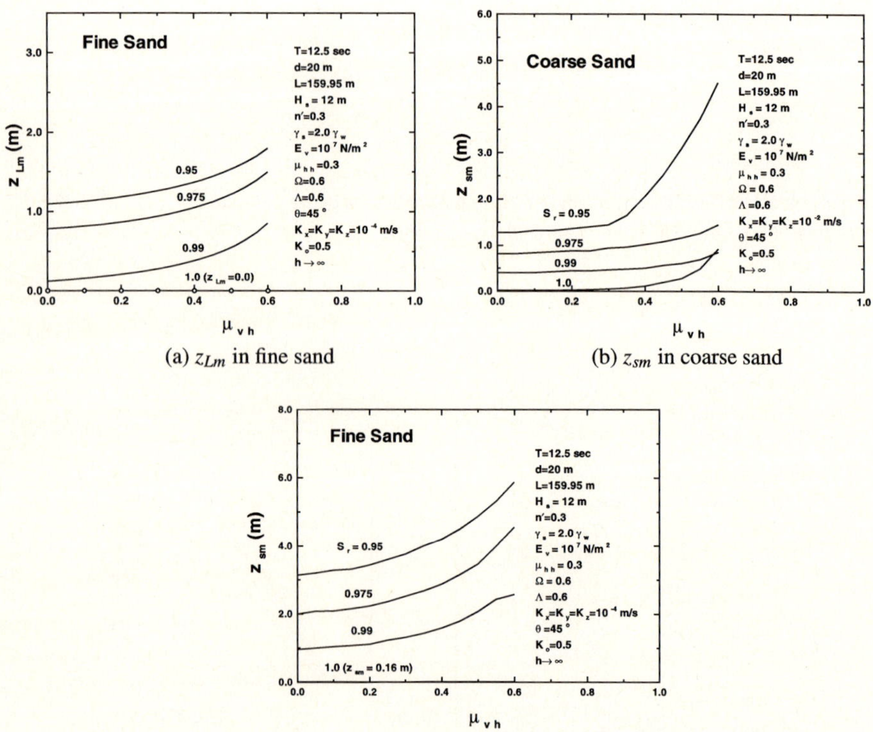

(a) z_{Lm} in fine sand (b) z_{sm} in coarse sand

(c) z_{sm} in fine sand

Fig. 5.17 Relationship between the maximum shear failure depth z_{sm} and μ_{vh} for various degrees of saturation S_r in seabeds

Figures 5.17b and 5.17c show the maximum shear failure depth z_{sm} versus Poisson's ratio μ_{vh} for various values of S_r in coarse and fine sand, respectively. For μ_{vh} between 0.0 and 0.6 and the four S_r values considered, the resultant z_{sm} values in a seabed of fine sand is generally much greater than that in a coarse sandy seabed. The gradients of z_{sm} to μ_{vh} are very orderly, except for the curve with $S_r = 0.95$ (in Fig. 5.17b) in which a steep increase is found for $\mu_{vh} > 0.4$. Unfortunately, supportive reasons for this have not been found. The maximum shear failure depth z_{sm} is constant ($z_{sm} = 0.16$ m) in fine sand under saturated condition, as shown in Fig. 5.17c.

Compared with Poisson's ratio μ_{hh}, the influence of Poisson's ratio μ_{vh} on the wave-induced seabed response is more significant, especially for the wave-induced shear failure in coarse sand.

5.3.4.4 Effect of Anisotropic Constant Ω

Besides the two Poisson's ratios mentioned (μ_{hh} and μ_{vh}) there are three other cross-anisotropic constants (E_h, E_v and G_v) to be discussed. As defined by (5.71) and (5.72), these three anisotropic parameters are related to two non-dimensional parameters, Ω and Λ. The influence of $\Omega = (E_h/E_v)$ on the wave-induced seabed response is first discussed.

The vertical distribution of the maximum pore pressure $|p|/p_o$ and vertical effective normal stress $|\sigma_z'|/p_o$ in coarse sand for four values of Ω is plotted in Fig. 5.18. These show that pore pressure $|p|/p_o$ increases and the vertical effective normal stress $|\sigma_z'|/p_o$ decreases as the anisotropic constant Ω increases after $|z/L| \geq 0.05$. It is unexpected that a reversal point appears in the vertical distribution of the pore pressure at $z/L = -0.32$ when $\Omega = 0.4$ (Fig. 5.18a).

A similar trend is found in a seabed of fine sand as in Fig. 5.19a. However, pore pressure $|p|/p_o$ increases as Ω decreases in the region near the seabed surface (for instance, $-0.075 \leq z/L \leq 0$ in Fig. 5.19a). Below this depth, it increases as Ω increases. An opposite trend can be observed in Fig. 5.19b for the vertical effective normal stress $|\sigma_z'|/p_o$.

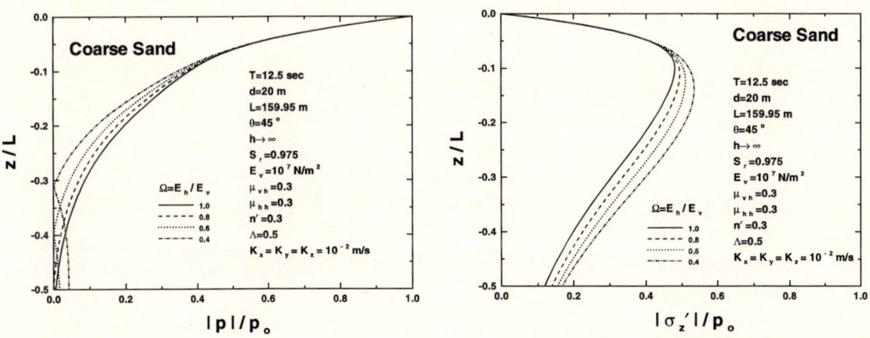

(a) $|p|/p_o$ (b) $|\sigma_z'|/p_o$

Fig. 5.18 Vertical distribution of the maximum (**a**) $|p|/p_o$ and (**b**) $|\sigma_z'|/p_o$ versus z/L for various anisotropic constant Ω in coarse sand

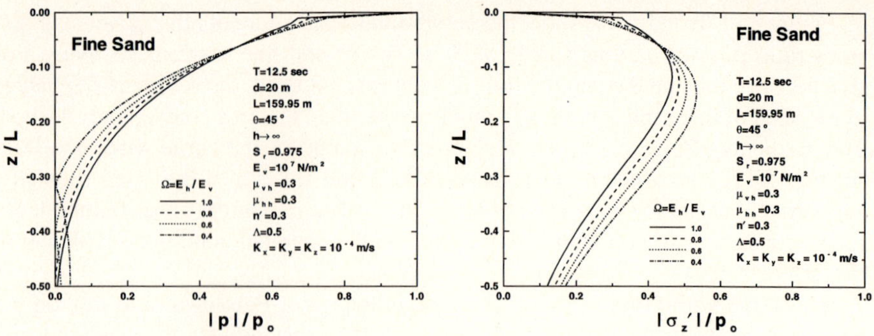

(a) $|p|/p_o$ (b) $|\sigma_z'|/p_o$

Fig. 5.19 Vertical distribution of the maximum (**a**) $|p|/p_o$ and (**b**) $|\sigma_z'|/p_o$ versus z/L for various anisotropic constant Ω in fine sand

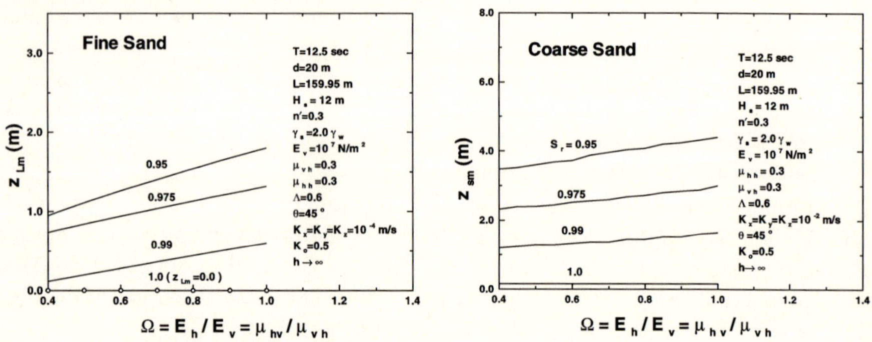

(a) z_{Lm} in fine sand (b) z_{sm} in coarse sand

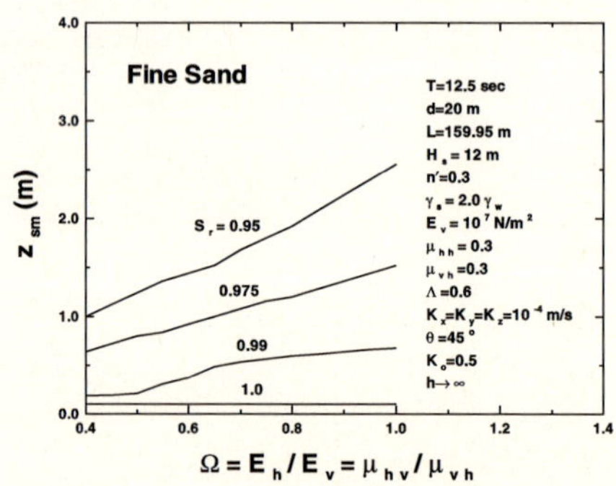

(c) z_{sm} in fine sand

Fig. 5.20 Relationship between the maximum shear failure depth z_{sm} and Ω for various degrees of saturation S_r seabeds

The distribution of the maximum liquefaction depth (z_{Lm}) versus non-dimensional parameter Ω for various degrees of saturation is depicted in Fig. 5.20a. It is found that z_{Lm} increases linearly as anisotropic parameter Ω increases.

As shown in Figs. 5.20b and 5.20c, the maximum shear failure depth z_{sm} increases with anisotropic parameter Ω in both coarse and fine sand. The rate of increase in the shear failure depth is faster in fine sand under an unsaturated condition. This implies that the influence of the anisotropic parameter Ω on the wave-induced shear failure in fine sand is more sensitive than in coarse sand.

5.3.5 *Effect of Anisotropic Constant* Λ

Figure 5.21 demonstrate the effect of the anisotropic parameter $\Lambda(= G_v/E_v)$ on the pore pressure and vertical effective normal stress in coarse sand. Generally speaking, pore pressure $|p|/p_o$ increases slightly as the anisotropic parameter Λ increases near the seabed surface ($-0.175 \leq z/L \leq 0$ in Fig. 5.21). However, the vertical effective normal stress $|\sigma_z'|/p_o$ decreases as Λ increases throughout the entire soil column (Fig. 5.21a). Similar trends can be observed in Fig. 5.22 for fine sand.

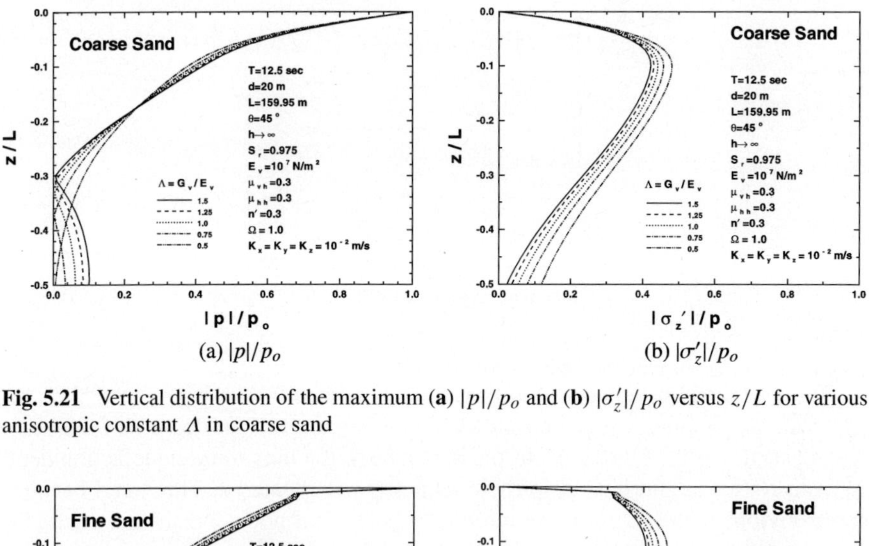

(a) $|p|/p_o$ (b) $|\sigma_z'|/p_o$

Fig. 5.21 Vertical distribution of the maximum (**a**) $|p|/p_o$ and (**b**) $|\sigma_z'|/p_o$ versus z/L for various anisotropic constant Λ in coarse sand

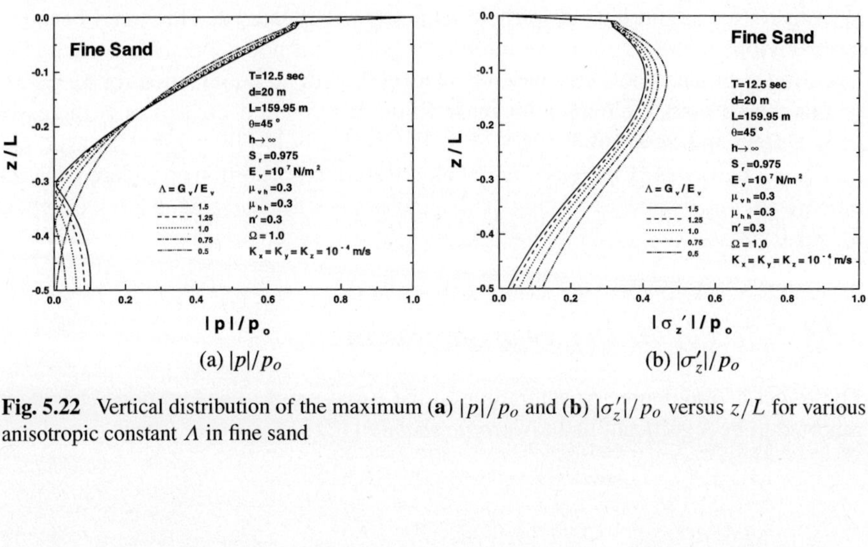

(a) $|p|/p_o$ (b) $|\sigma_z'|/p_o$

Fig. 5.22 Vertical distribution of the maximum (**a**) $|p|/p_o$ and (**b**) $|\sigma_z'|/p_o$ versus z/L for various anisotropic constant Λ in fine sand

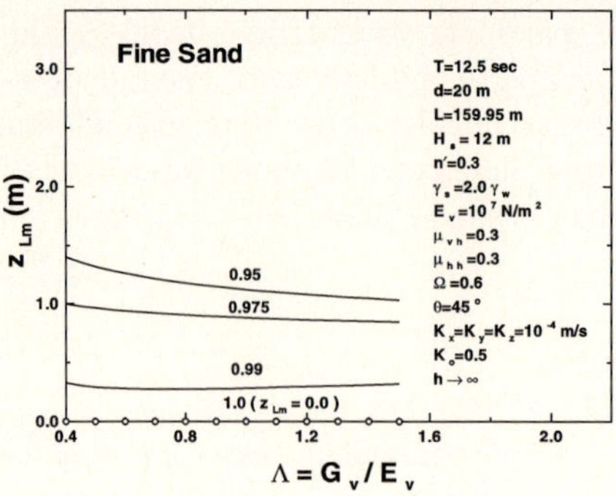

(a) z_{Lm} in fine sand

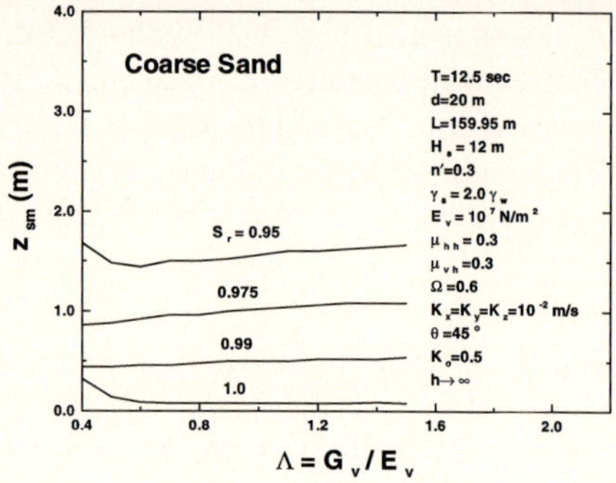

(b) z_{sm} in coarse sand

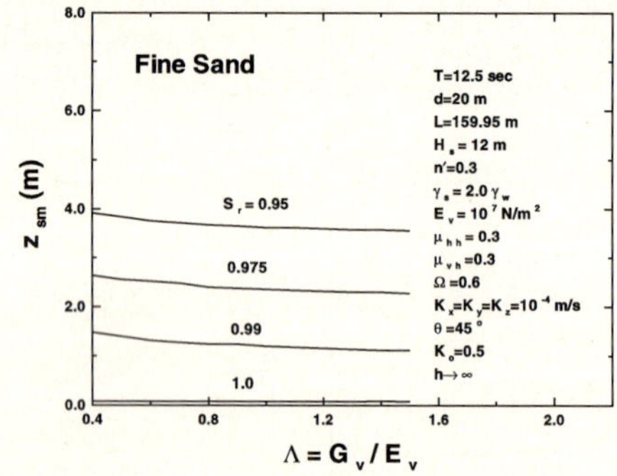

(c) z_{sm} in fine sand

Fig. 5.23 Relationship between the maximum shear failure depth z_{sm} and Λ for various degrees of saturation S_r in seabeds

The relationship between the maximum liquefaction depth z_{Lm} and the non-dimensional parameter Λ is presented in Fig. 5.23a, for various degrees of saturation in fine sand. Unlike other anisotropic parameters, the maximum liquefaction depth z_{Lm} decreases slightly as anisotropic parameter Λ increases. This trend becomes more obvious as the degree of saturation decreases. It is noted that the maximum liquefaction depth is almost constant for $S_r = 0.99$, with no liquefaction for $S_r = 1.0$.

For coarse sand, the maximum shear failure depth z_{sm} decreases as Λ increases at $S_r = 0.95$ and $S_r = 1.0$ as $0.4 \leq \Lambda \leq 0.6$, as shown in Fig. 5.23b. However, z_{sm} increases slightly as Λ increases under all unsaturated condition considered. On the other hand, z_{sm} decreases slightly as Λ increases in fine sand (Fig. 5.23c), except at $S_r = 1.0$.

5.3.6 *Effect of the Degree of Saturation*

The effect of the degree of saturation on the wave-induced soil response has been reported to be significant in an *isotropic seabed* [12, 31], also see Chaps. 3 and 4. It is desirable to examine the effects of the degree of saturation (S_r), taken from 0.95 to 1.0 (fully saturated), on the wave-induced seabed instability in a cross-anisotropic seabed.

As shown in Figs. 5.14a, 5.17a, 5.20a and 5.23a, the degree of saturation affects significantly the magnitude of the maximum liquefaction depth z_{Lm} in a cross-anisotropic seabed. The depth z_{Lm} increases as S_r decreases in a seabed of fine sand. However, the effect of the degree of saturation becomes less sensitive as S_r increases.

Shear failure depth z_{sm} also increases as the degree of saturation decreases in a cross-anisotropic seabed, as shown in Figs. 5.14b and 5.14c, 5.17b and 5.17c, 5.20b and 5.20c and 5.23b and 5.23c. It is important to point out that *the degree of saturation plays an important role on the wave-induced seabed instability in a cross-anisotropic seabed as well as in an isotropic seabed.*

5.3.7 Summary

In this section, analytical solutions of the wave-induced soil response in a cross-anisotropic seabed have been presented. Based on the general solutions derived, the effects of anisotropic parameters and the degree of saturation on the wave-induced seabed response have been discussed.

From the numerical results obtained for a fully saturated seabed of infinite thickness, it may be concluded that the influence of cross-anisotropic parameters (including μ_{hh}, μ_{vh}, Ω and Λ) on the wave-induced seabed response cannot be ignored in engineering applications. The relative difference of soil response between cross-anisotropic and isotropic solutions may be remarkable in some cases.

For an unsaturated seabed, the wave-induced pore pressure and vertical effective normal stress are almost unaffected by μ_{hh}. Furthermore, anisotropic constants, Ω and Λ, only affect the soil response slightly. However, μ_{vh} affects the pore pressure and vertical effective normal stress significantly. Generally speaking, the pore pressure ($|p|/p_o$) increases as μ_{vh} decreases, but the vertical normal stress ($|\sigma_z'|/p_o$) increases as μ_{vh} increases.

The maximum depths of liquefaction and shear failure (z_{sm} and z_{Lm}) increase as Poisson's ratios (μ_{hh} and μ_{vh}) and anisotropic constant Ω increase, while both increase as anisotropic constant Λ decreases. The degree of saturation also dominates the wave-induced soil response in cross-anisotropic seabeds.

5.4 Numerical Model for Seabed Response in Anisotropic Seabed with Variable Soil Characteristics

Besides the development of analytical solutions presented in previous section, numerical simulations have been widely applied to investigate such problems in recent years, such as the finite difference methods [52], the finite element methods

[8, 9, 22, 43, 44] and the boundary element method [35]. However, it is noted that all these investigations have only examined the soil response in a porous seabed under the action of two-dimensional progressive waves , without the presence of a marine structure.

The shortcomings of the previous models are summarized as below:

- Gatmiri's model [8, 9]: Only a fully saturated seabed with uniform soil characteristics subject to 2-D progressive waves was considered in the model. As seen in Fig. 6 in his paper [8], the $|p|/p_o$ value for $h/L = 0.2$ is greater than that for $h/L = 0.1$, and that of $h/L = 0.5$ is greater than 0.4. These are inconsistent in the trend for pore pressure distribution versus h/L, compared with the close-form analytical solution [21]. It seems difficult to provide a reason for the jump order in pore pressure for these h/L values. A possible error in the results of [8, 9] may have stemmed from the lateral boundary conditions used, which was mentioned in [9]. The lateral boundaries at $x/L = 1/4$, $5/4$ (i.e., $x/L = 0, 1$ in his paper), "$v = 0$, $p = 0$, u free", were used in his model. This assumption is only valid for a fully saturated isotropic seabed of infinite thickness, not for unsaturated seabed [31]. However, it has been proved that there is a phase lag in soil response in either a fully saturated isotropic seabed of finite thickness [21] or a fully saturated cross-anisotropic seabed of infinite thickness [19]. This implies that the lateral boundary conditions, "$v = 0$, and $p = 0$", are invalid in a saturated either isotropic seabed of finite thickness or cross-anisotropic seabed of finite thickness. Thus, the numerical results of Gatmiri's model [8, 9] seem doubtful.
- Thomas' model [43, 44]: Using two-nodal elements, only an isotropic seabed with uniform permeability and Young's modulus subject two-dimensional waves was considered in the mathematical derivation of his first paper [43]. In his second papers [44], the model verification includes the comparison between his model and the previous analytical solutions for a seabed of infinite thickness [30, 51]. Both single and two-layered seabed are considered. It is noted that the soil response in Gibson soil was mentioned in his paper [44]. However, Thomas [44] only directly modified the analytical solution for a saturated semi-infinite seabed for a semi-infinite Gibson soil by $G' = dG/dz$, without modifying the original governing equations. As presented in later section, variable modulus will result in few extra items in the governing equations. Thus, treating the soil response in Gibson soil by simply modifying G to $G' = dG/dz$ may not be a rigorous way. Furthermore, Thomas' model didn't include the influences of either variable soil characteristics (such as permeability and Young's modulus) or cross-anisotropic soil behavior, which is one of the main concerns in this study.
- Jeng's model [15, 21, 24, 26]: A series of closed form analytical solutions have been developed by the first author for the wave-induced seabed response in the vicinity of a vertical wall. In his models, both isotropic and cross-anisotropic seabeds have been considered with infinite and finite thickness, as well as a layered medium. The influence of variable permeability in an isotropic seabed has also been discussed [24, 26]. However, the influence of variable shear modulus on the soil response was not included in the previous models due to the mathe-

matical difficulties. Furthermore, the closed form analytical solution consists of variable soil characteristics and anisotropic soil behavior has not been available.

Recently, Jeng and his co-workers [23, 29] further extended their previous finite element model [22] to investigate the combined effects of cross-anisotropic soil behavior and variable soil characteristics on the wave-induced seabed response. In this section, the finite element model proposed by Jeng and Lin [23] will be outlined first. The model will be verified with previous solutions through the reduction of the model. The effects of variable soil characteristics and cross-anisotropic soil behavior on the wave-induced soil response will be detailed through a parametric study. Then, the wave-induced liquefaction potential in a non-homogeneous cross-anisotropic seabed will be investigated.

5.4.1 Boundary Value Problem

Herein, we consider a soil column of finite thickness (h) in front of a vertical reflecting wall, as depicted in Fig. 3.1. A short-crested wave is produced by fully oblique reflection of the incident waves, with obliquity angle θ measured between a wave orthogonal and the normal direction to the wall and also between a wave orthogonal and the wall itself. The resultant wave crests propagate in the positive x-direction parallel to the wall, while the z-direction is measured as positive upwards from the *seabed surface* for the wave-seabed interaction problem.

For a three-dimensional problem and treating the porous bed as a hydraulically isotropic medium with variable permeability (K) in all directions, the consolidation equation suggested by Biot [5] can be expressed as

$$K(z)\nabla^2 p + \frac{dK(z)}{dz}\frac{\partial p}{\partial z} - \gamma_w n' \beta \frac{\partial p}{\partial t} = \gamma_w \frac{\partial \epsilon}{\partial t}, \tag{5.159}$$

where p is pore pressure, $K(z)$ is the coefficient of soil permeability, γ_w is the unit weight of the pore water, n' is the soil porosity and t is the time. It is noted that the second term on the left hand side of (5.159) comes from variable permeability, and it will vanish for a seabed with uniform permeability. In (5.159), the volume strain (ϵ) and the compressibility of the pore fluid (β) are defined in (3.8) and (3.9), respectively.

The equation for overall equilibrium in a poro-elastic medium can be expressed in terms of pore pressure and effective stresses as (3.10)–(3.12). Substituting (5.78)–(5.83) into (3.10)–(3.12), by adopting Young modulus as function of z, the equations of force equilibrium, are rendered as

$$E_v(z)\left[C_{11}\frac{\partial^2 u}{\partial x^2} + G\frac{\partial^2 u}{\partial y^2} + \Lambda\frac{\partial^2 u}{\partial z^2} + (C_{12}+G)\frac{\partial^2 v}{\partial x \partial y} + (C_{13}+\Lambda)\frac{\partial^2 w}{\partial x \partial z}\right]$$
$$+ \frac{dE_v(z)}{dz}\left(\frac{\partial u}{\partial z} + \frac{\partial w}{\partial x}\right) = -\frac{\partial p}{\partial x}, \tag{5.160}$$

$$E_v(z)\left[G\frac{\partial^2 v}{\partial x^2}+C_{11}\frac{\partial^2 v}{\partial y^2}+\Lambda\frac{\partial^2 v}{\partial z^2}+(G+C_{12})\frac{\partial^2 u}{\partial x\partial y}+(C_{13}+\Lambda)\frac{\partial^2 w}{\partial y}\partial z\right]$$

$$+\Lambda\frac{dE_v(z)}{dz}\left(\frac{\partial v}{\partial z}+\frac{\partial w}{\partial y}\right)=-\frac{\partial p}{\partial y}, \tag{5.161}$$

$$E_v(z)\left[\Lambda\left(\frac{\partial^2 w}{\partial x^2}+\frac{\partial^2 w}{\partial y^2}\right)+C_{33}\frac{\partial^2 w}{\partial z^2}+(\Lambda+C_{13})\frac{\partial^2 u}{\partial x\partial z}+(\Lambda+C_{13})\frac{\partial^2 v}{\partial y\partial z}\right]$$

$$+\frac{dE_v(z)}{dz}\left(C_{13}\frac{\partial u}{\partial x}+C_{13}\frac{\partial v}{\partial y}+C_{33}\frac{\partial w}{\partial z}\right)=-\frac{\partial p}{\partial z}, \tag{5.162}$$

in the x-, y- and z-directions, respectively.

The soil displacements and pore pressure in the governing equations (5.159) and (5.160)–(5.162) can be solved with the boundary conditions at the surface of the seabed and impermeable bottom, given in (5.6)–(5.10).

The governing equations (5.159) and (5.160)–(5.162) for the wave-induced soil response in a porous seabed can be solved by using a combination of semi-analytical technique and a finite element Galerkin method [53]. The details of finite element formulation can be found in Sect. 5.6.

5.4.2 Wave-Induced Seabed Response

Based on the finite element model presented previously, the influences of cross-anisotropic soil behavior and variable soil behavior on the short-crested wave-induced seabed response are investigated in this section. To have basic understanding in the effects of variable soil characteristics, two different categories of depth function K/K_o and E_v/E_o for variable soil permeability and Young's modulus are applied to two types of marine sediments, from which pore pressure and effective stresses are calculated.

In nature, the soil permeability of marine sediments has been reported to decrease with the buried depth (z) [6]. For variable permeability, as shown in Fig. 5.24a, type K_1 is for $K = K_s$ (= constant, K_s denotes the permeability at the surface of the seabed), which represents the conventional assumption for a seabed with uniform permeability. Type K_2, $K(z) = K_s e^{\alpha_1 z/h}$, is for an exponential reduction in permeability with depth, and type K_3, $K(z) = K_s(1+\alpha_1 z/h)$, is for a linear reduction in permeability. The arbitrary coefficient α_1 is determined by the distribution of permeability versus the normalized buried depth z/h. In this study, the value of permeability at rigid bottom ($z = -h$) is assumed to be one tenth of the permeability at the seabed surface K_s.

Conversely, the Young's modulus of soil increases with burial depth in natural sedimentary seabeds. For variable Young's modulus, as depicted in Fig. 5.24b, type E_1 is for $E_v(z) = E_s$ (= constant, E_s represents the Young's modulus at the seabed surface), which is the conventional assumption of constant Young's modulus. Type

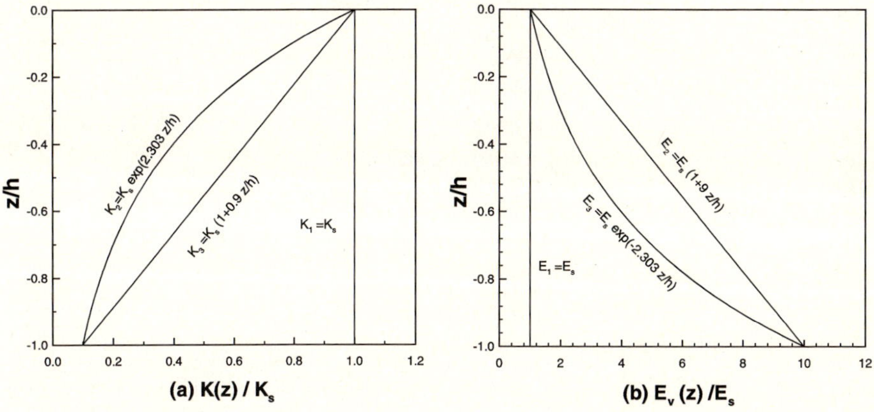

Fig. 5.24 Three different cases of depth function (**a**) $K(z)/K_s$ and (**b**) $E_v(z)/E_s$ numerically examined for soil matrix with variable permeability and Young's modulus, respectively

E_2, $E_v(z) = E_s(1 + \alpha_2 z/h)$, which is so called *Gibson Soil*, is for a linear increase in Young's modulus with burial depth, and type E_3, $E_v(z) = E_s e^{\alpha_2 z/h}$, is for an exponential increase in Young's modulus with depth. The arbitrary coefficient α_2 is determined by the distribution of Young' modulus.

To preform a parametric study, we consider a short-crested wave system with period $T = 13.5$ s, and height $H_s = 5.0$ m propagating over a uniform water depth $d = 50.0$ m on a porous seabed of thickness $h = 50.0$ m. Two marine sediments are considered here, coarse sand and fine sand, with their distinctive soil properties specified in Table 5.5.

Nine different combinations of variable soil characteristics, Type ij, which denotes the combination of type K_i and E_j, are considered here. It is worth noting that Type 11 is the conventional assumption for a *homogeneous* seabed that has been studied in the previous works.

5.4.2.1 Effects of Variable Soil Characteristics

The vertical distributions of pore pressure ($|p|/p_o$) and vertical effective normal stress ($|\sigma_z'|/p_o$) versus the relative burial soil depth (z/h) in a cross-anisotropic seabed with variable soil characteristics are depicted in Figs. 5.25–5.27 for coarse and fine sand. Both saturated ($S = 1$) and unsaturated ($S = 0.975$) conditions are considered here. As seen in Figs. 5.25–5.27, both normalized pore pressure ($|p|/p_o$) and vertical effective normal stress ($|\sigma_z'|/p_o$) are affected significantly by variable soil characteristics.

For coarse sand (Figs. 5.25 and 5.26), both variable soil permeability and Young's modulus affect the wave-induced pore pressure and vertical effective normal stress significantly. The influence of variable soil characteristics also depends on the combinations of K_i and E_j as well as the degree of saturation. For example, the effect of

Table 5.5 Input data for case study

Wave characteristics		
wave period T		13.5 s
water depth d		50 m
incident wave angle θ		45°
Soil characteristics		
soil depth h		50 m
the degree of saturation S		0.975 and 1.0
soil proposity n'		0.4
	Coarse sand (Cc)	Fine sand (Fs)
permeability K_o	5×10^{-2} m/s	5×10^{-4} m/s
Young's modulus E_o	5×10^{7} N/m^2	5×10^{7} N/m^2
Anisotropic parameters		
Poisson's ratio μ_{hh}		0.2–0.4
Poisson's ratio μ_{vh}		0.2–0.4
anisotropic constant Ω		0.8–1.0
anisotropic constant Λ		0.4–0.8

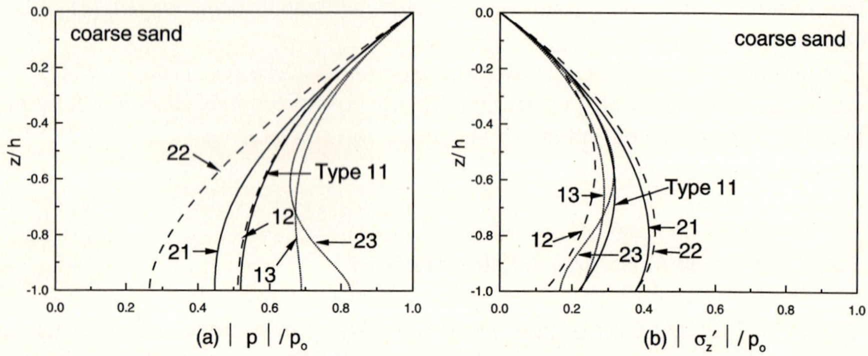

Fig. 5.25 Vertical distributions of (**a**) pore pressure ($|p|/p_o$) and (**b**) vertical effective normal stress ($|\sigma_z'|/p_o$) versus z/h in a saturated ($S_r = 1$) coarse sand with variable soil characteristics

variable Young's modulus on the wave-induced pore pressure in a saturated seabed with uniform permeability (Types 11 and 12) is less than that with variable permeability (Types 21 and 22). An opposite trend is observed in the figure for $|\sigma_z'|/p_o$. The difference between uniform and variable soil characteristics becomes less in an unsaturated seabed (Fig. 5.26).

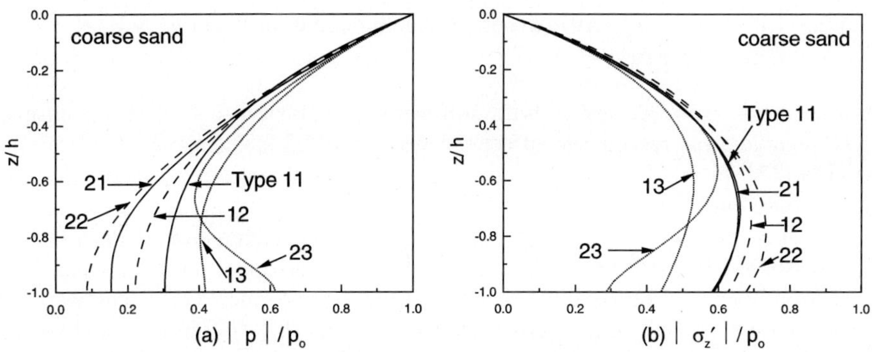

Fig. 5.26 Vertical distributions of (**a**) pore pressure ($|p|/p_o$) and (**b**) vertical effective normal stress ($|\sigma_z'|/p_o$) versus z/h in an unsaturated ($S_r = 0.975$) coarse sand with variable soil characteristics

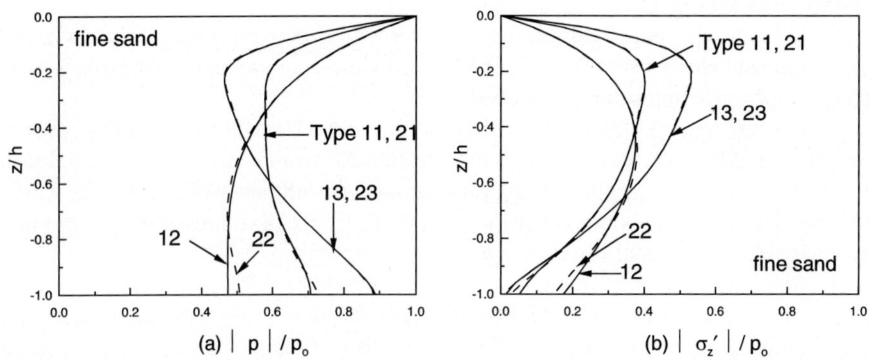

Fig. 5.27 Vertical distributions of (**a**) pore pressure ($|p|/p_o$) and (**b**) vertical effective normal stress ($|\sigma_z'|/p_o$) versus z/h in a saturated ($S_r = 1$) fine sand with variable soil characteristics

For fine sand (Fig. 5.27), variable Young's modulus plays a more important role on the evaluation of the wave-induced soil response than variable permeability. For example, variable permeability has no influence on the wave-induced soil response in a seabed with uniform Young's modulus. In the other words, the results from Types 11 and 21 are identical, as seen in Fig. 5.27. However, variable permeability still affect the soil response in a seabed with variable Young's modulus. This influence of variable characteristics will become less significant in an unsaturated seabed (graphs not shown here).

Numerical examples also conclude (graphs not shown here, but available in [23]), the wave-induced shear stresses ($|\tau_{xz}|/p_o$ and $|\tau_{xy}|/p_o$) are significantly affected by variable Young's modulus, rather than by variable permeability. It is noted that the shear stress in the vertical plane (τ_{xz}) with an increasing function of Young's modulus (E_2) is greater than that with uniform Young's modulus (E_1), while τ_{xz} with a decreasing function of Young's modulus (E_3) is less than that with E_1. However, the trend of shear stress in the horizontal plane (τ_{xy}) is opposite to that of τ_{xz}.

5.4.2.2 Effects of Cross-Anisotropic Soil Behavior with Variable Soil Characteristics

For a cross-anisotropic seabed with uniform soil characteristics, cross-anisotropic soil behavior was reported to affect the wave-induced soil response to a certain degree [17]. In this section, the influences of Poisson's ratios, μ_{hh} and μ_{vh}, and anisotropic constants, Ω and Λ, on the wave-induced pore pressure and vertical effective normal stress in a seabed with variable soil characteristics are discussed.

Poisson's ratio, μ_{hh}, describes the strain in one horizontal direction caused by the strain in another horizontal direction normal to the former. The calculated pore pressure ($|p|/p_o$) and vertical effective normal stress ($|\sigma_z'|/p_o$) in fine sand versus relative burial soil depth z/h for various values of μ_{hh} are illustrated in Figs. 5.28 and 5.29. As shown in Fig. 5.28, the maximum amplitude of pore pressure ($|p|/p_o$) decreases slightly as Poisson's ratio μ_{hh} increases in a saturated seabed. However, $|\sigma_z'|/p_o$ increases slightly as μ_{hh} increases. The influence of μ_{hh} on the soil response increases in an unsaturated seabed (Fig. 5.29).

Compared with pore pressure (p) and vertical effective normal stress (σ_z'), the wave-induced shear stresses (τ_{xz} and τ_{xy}) are almost unaffected by Poisson's ratio (μ_{hh}), especially in a saturated seabed.

A similar trend is observed in Figs. 5.30 and 5.31 for the relationship of soil response and Poisson's ratio μ_{vh}, which relates the strains in the vertical direction owing to a horizontal strain. Compared with μ_{hh}, the influence of μ_{vh} on the wave-induced soil response is more significant, especially for pore pressure ($|p|/p_o$) and vertical effective normal stress (σ_z'/p_o).

Besides the Poisson's ratios, μ_{hh} and μ_{vh}, the other three cross-anisotropic parameters, E_h, E_v and G_v, can be related to two non-dimensional parameters, Ω and Λ, as defined in (5.71) and (5.72). The influence of Ω and Λ on the wave-induced soil response are depicted in Figs. 5.32–5.34. Generally speaking, the wave-induced pore pressure ($|p|/p_o$) increases as Ω and Λ increase, but $|\sigma_z'|/p_o$ decreases. The influence of Ω and Λ on the wave-induced soil response decreases as the degree of saturation decreases (Figs. 5.33 and 5.35).

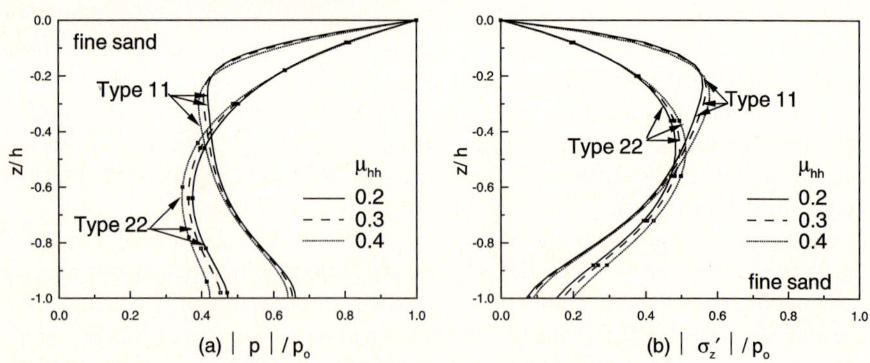

Fig. 5.28 Vertical distributions of (**a**) pore pressure ($|p|/p_o$) and (**b**) vertical effective normal stress ($|\sigma_z'|/p_o$) versus z/h in a saturated ($S_r = 1.0$) fine sand with variable soil characteristics for various μ_{hh}. ($\mu_{vh} = 0.4$, $\Omega = 0.8$ and $\Lambda = 0.4$)

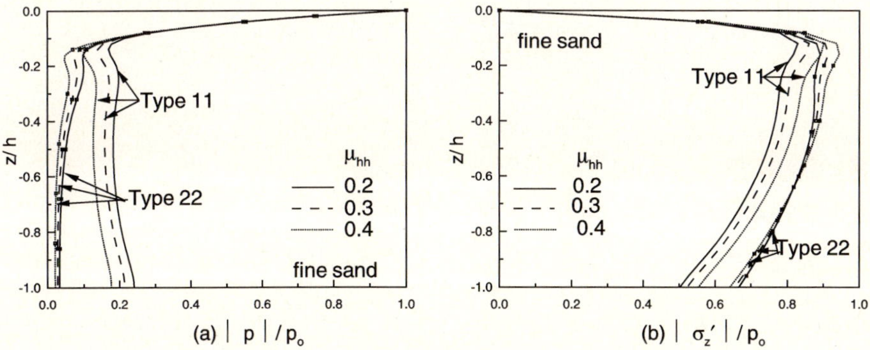

Fig. 5.29 Vertical distributions of (**a**) pore pressure ($|p|/p_o$) and (**b**) vertical effective normal stress ($|\sigma_z'|/p_o$) versus z/h in an unsaturated ($S_r = 0.975$) fine sand with variable soil characteristics for various μ_{hh}. ($\mu_{vh} = 0.4$, $\Omega = 0.8$ and $\Lambda = 0.4$)

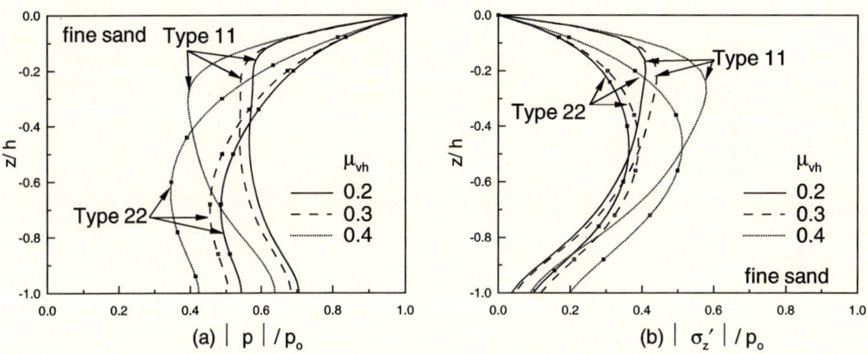

Fig. 5.30 Vertical distributions of (**a**) pore pressure ($|p|/p_o$) and (**b**) vertical effective normal stress ($|\sigma_z'|/p_o$) versus z/h in a saturated ($S_r = 1.0$) fine sand with variable soil characteristics for various μ_{vh}. ($\mu_{hh} = 0.4$, $\Omega = 0.8$ and $\Lambda = 0.4$)

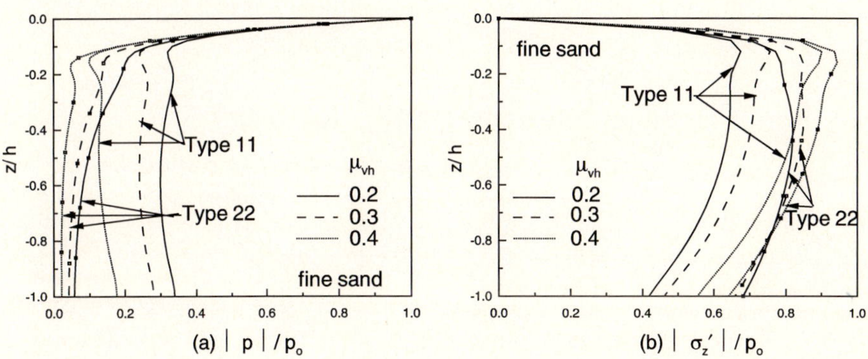

Fig. 5.31 Vertical distributions of (**a**) pore pressure ($|p|/p_o$) and (**b**) vertical effective normal stress ($|\sigma_z'|/p_o$) versus z/h in an unsaturated ($S_r = 0.975$) fine sand with variable soil characteristics for various μ_{vh}. ($\mu_{hh} = 0.4$, $\Omega = 0.8$ and $\Lambda = 0.4$)

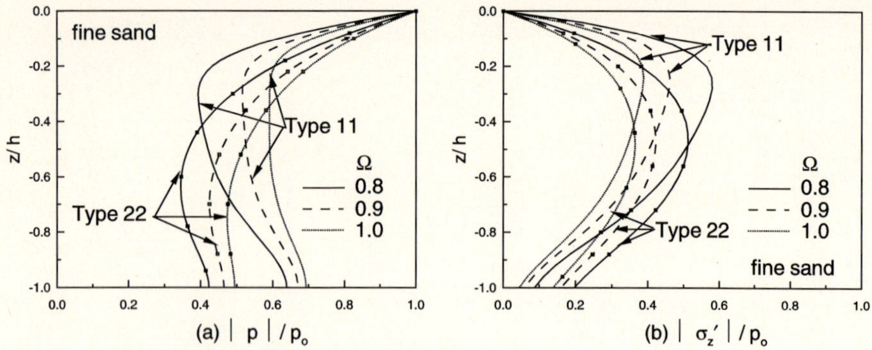

Fig. 5.32 Vertical distributions of (**a**) pore pressure ($|p|/p_o$) and (**b**) vertical effective normal stress ($|\sigma_z'|/p_o$) versus z/h in a saturated ($S_r = 1.0$) fine sand with variable soil characteristics for various Ω. ($\mu_{hh} = \mu_{vh} = 0.4$ and $\Lambda = 0.4$)

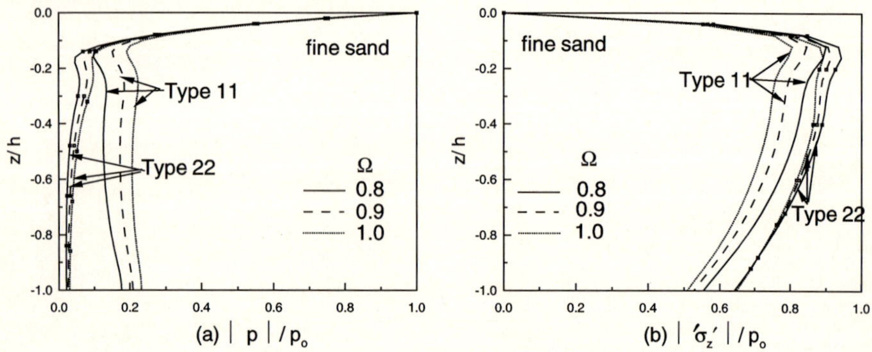

Fig. 5.33 Vertical distributions of (**a**) pore pressure ($|p|/p_o$) and (**b**) vertical effective normal stress ($|\sigma_z'|/p_o$) versus z/h in an unsaturated ($S_r = 0.975$) fine sand with variable soil characteristics for various Ω. ($\mu_{hh} = \mu_{vh} = 0.4$ and $\Lambda = 0.4$)

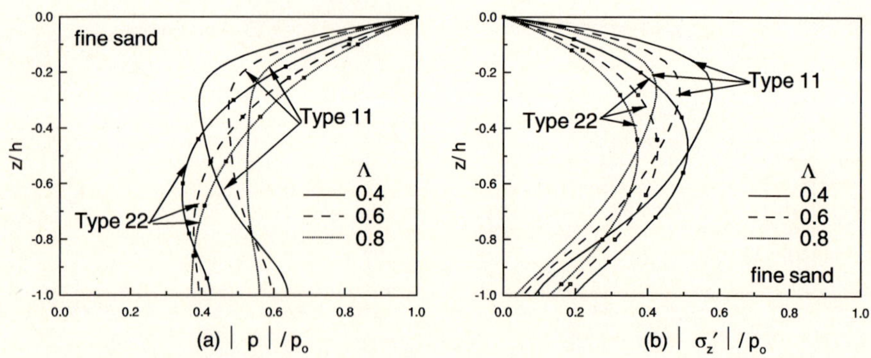

Fig. 5.34 Vertical distributions of (**a**) pore pressure ($|p|/p_o$) and (**b**) vertical effective normal stress ($|\sigma_z'|/p_o$) versus z/h in an saturated ($S = 1.0$) fine sand with variable soil characteristics for various Λ. ($\mu_{hh} = \mu_{vh} = 0.4$ and $\Omega = 0.8$)

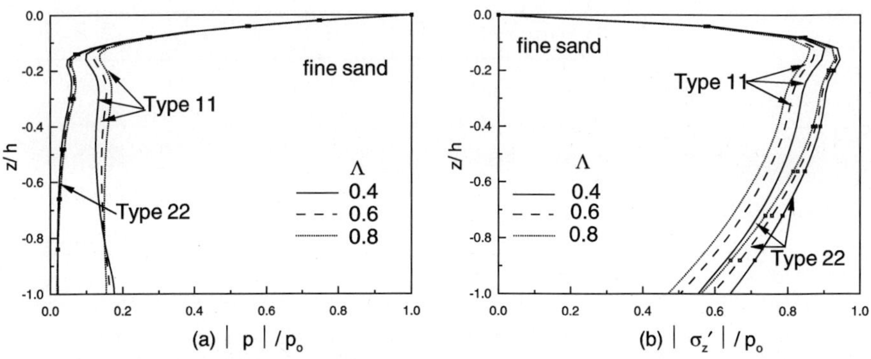

Fig. 5.35 Vertical distributions of (**a**) pore pressure ($|p|/p_o$) and (**b**) vertical effective normal stress ($|\sigma_z'|/p_o$) versus z/h in an unsaturated ($S = 0.975$) fine sand with variable soil characteristics for various Λ. ($\mu_{hh} = \mu_{vh} = 0.4$ and $\Omega = 0.8$)

Figures 5.28–5.35 also clearly show the combined effects of cross-anisotropic soil behavior and variable soil characteristics on the wave-induced soil response. Two different combinations of variable soil characteristics, Types 11 and 22, are considered in these examples. These figures indicate that cross-anisotropic soil behavior significantly affects the wave-induced soil response in a saturated seabed with either uniform or variable soil characteristics. However, the influences of cross-anisotropic soil behavior are more important in an unsaturated seabed with uniform soil characteristics than that with variable soil characteristics.

5.4.3 Wave-Induced Liquefaction

In this section, we further investigate the wave-induced liquefaction potential in a non-homogeneous anisotropic seabed. As an example, two marine sediments, coarse sand and fine sand, are considered here. The wave and soil characteristics are tabulated in Table 5.6. Nine different combinations of variable permeability and Young's modulus, Type ij, which represents the combination of type K_i and E_j, are considered here. It is noted that Type 11 is the conventional assumption for a *homogeneous* seabed that has been studied in the past.

Based on the criterion of the wave-induced liquefaction proposed by Jeng [16], the wave-induced liquefaction potential can be calculated, as depicted in Fig. 5.36. In the figure, Types 11 and 33 in both isotropic and cross-anisotropic seabeds are used as an example. The region between the curve and the surface of the seabed has potential to be liquefied. This implies that any structure embedded on this region will be damaged. It is noted that the liquefied region occurs near wave troughs, i.e., $(mkx - \omega t)/2\pi = 0.5$. This is because the wave-induced pore pressure is pushing the soil particle upward near wave troughs [52]. As seen in the Fig. 5.36, the regions for liquefaction from Type 33 for both isotropic and cross-anisotropic seabeds

Table 5.6 Input data for case study

Wave characteristics		
wave period T		12.5 s
water depth d		40 m
wavelength L		205.5 m
wave height H_s		5.0 m
incident wave angle θ		45°
Soil characteristics		
soil depth h		40 m
the degree of saturation S		0.985
soil porosity n'		0.4
	Coarse sand (Cs)	Fine sand (Fs)
permeability K_s	10^{-2} m/s	10^{-5} m/s
Young's modulus E_s	5×10^7 N/m^2	5×10^7 N/m^2
Anisotropic parameters		
Poisson's ratio μ_{hh}		0.4
Poisson's ratio μ_{vh}		0.4
anisotropic constant Ω		0.8
anisotropic constant A		0.8

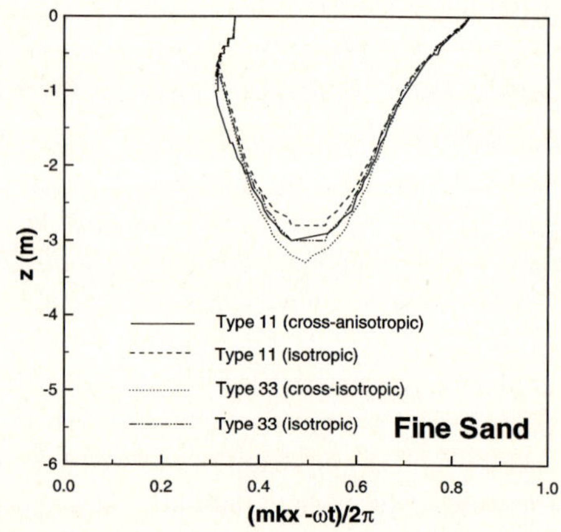

Fig. 5.36 Calculated liquefaction depth in fine sand of finite thickness ($S = 0.985$)

are deeper than that from Type 11. This implies that the conventional solution with uniform soil characteristics may under-estimate the wave-induced liquefaction potential.

Besides the variable soil characteristics, cross-anisotropic soil behavior has been reported to also affect the wave-induced soil response [15]. However, the influence

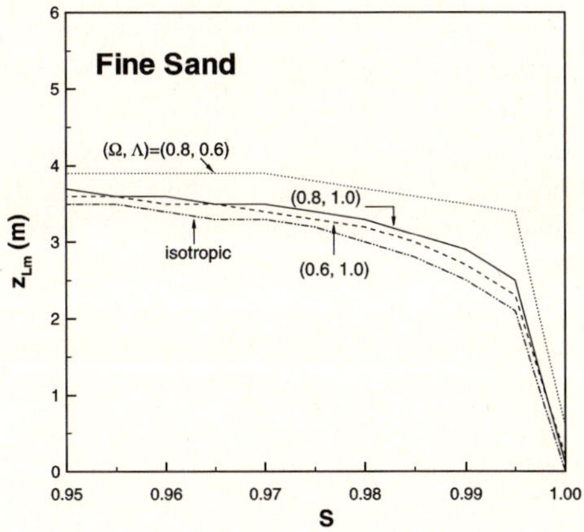

Fig. 5.37 Distribution of the maximum liquefaction depth z_{Lm} versus the degree of saturation S in fine sand

of cross-anisotropic behavior in a non-homogeneous seabed of finite thickness has not be examined.

In general, the wave-induced liquefaction potential in a cross-anisotropic seabed is larger than that in an isotropic seabed for both Types 11 and 33, as seen in Fig. 5.36. The influence of two cross-anisotropic parameters (Ω and Λ) on the maximum liquefaction depth z_{Lm} in an in-homogeneous seabed (Type 33) is presented in Fig. 5.37. The figure clearly shows that the maximum liquefaction depth z_{Lm} increases as Ω increases, but as Λ decreases. It is also noted that the result from isotropic solution always under-estimates the wave-induced liquefaction potential.

It has been reported that the wave-induced transient soil response did not cause liquefaction in a saturated seabed, because the residual soil response (i.e., the build-up pore pressure) was not taken into account [20, 34]. However, this conclusion was based on the assumption of homogeneous soil characteristics. As seen in Fig. 5.37, the soil liquefaction still occurs at $S = 1$ (i.e., fully saturation) in a cross-anisotropic seabed [for example, $(\Omega, \Lambda) = (0.8, 0.6)$] with variable characteristics (Type 33). This results is different from the conventional solution with uniform characteristics, in which no liquefaction occurs in a saturated seabed with either isotropic or cross-anisotropic soil behavior [20, 34].

Since marine sediments generally consist of three phases in the soil matrix, it is common to find gas within submarine sediments. The structure of an unsaturated marine soils (i.e. contains gas in the soils) can vary significantly depending on the relative size of the gas bubbles and the soil grains. The in-site degree of saturation of an unsaturated marine sediment normally lies on the range of 90 %–100 % [7, 33]. It is believed that most marine sediments have a degree of saturation close to unity, implying near full saturation. However, it is rare that full saturation can be attained in the field.

The degree of saturation S has been recognized as an important factor in analyzing the problem of wave-seabed interaction [12, 31]. Fig. 5.37 illustrates the vertical distribution of maximum liquefaction depth z_{Lm} versus various values of S. The value of the degree of saturation S is considered to vary from 0.95 to 1.0. The figure clearly shows that the wave-induced liquefaction potential is susceptible to the degree of saturation S, especially a nearly saturated seabed.

In order to apply the proposed model in a practical problem, the wave and soil conditions around Sines Breakwater at Portugal are used as input data here. it has been reported that the breakwater was built on a sandy bed [42]. A short-crested wave was formed at the angle of 45° in front of the breakwater. The wave conditions are: wave period—13.5 s, water depth—25 m, wave height—11 m. The soil conditions are: permeability—$K_s = 10^{-5}$ m/s, Young's modulus—$E_s = 3 \times 10^7$ N/m^2, porosity—$n' = 0.3$ and degree of saturation $S = 0.98$. Herein, we assumed the seabed to be a anisotropic medium. The anisotropic soil parameters are given as: $\mu_{vh} = \mu_{hh} = 1/3$, $(\Lambda, \Omega) = (0.8, 0.6)$. Based on the above information, the maximum wave-induced liquefaction depth z_{Lm} is 3.87 m. As seen in this example, we can further apply the proposed model to other practical cases.

5.4.4 Summary

In this section, a finite element model is proposed to investigate the wave-induced seabed response in a cross-anisotropic seabed of finite thickness with variable soil characteristics in front of a breakwater. Based on the numerical results presented previously, the following conclusions can be made:

- The wave-induced pore pressure and vertical effective normal stress in coarse sand with cross-anisotropic soil behavior are affected significantly by variable permeability and Young's modulus. However, only variable Young's modulus affects the soil response in fine sand.
- At least from the numerical results presented in this paper, the conventional solution with assumption of uniform soil characteristics under-estimates the maximum liquefaction depth (z_{Lm}). Similarly, z_{Lm} in a cross-anisotropic seabed is greater than that in isotropic seabed. It is important to note that liquefaction still occurs in a cross-anisotropic seabed with variable soil characteristics under fully saturated condition. This result differs from the conclusion based on the solutions of isotropic soil behavior and uniform soil characteristics [16].

5.5 Appendix: Exact Solutions of Linear Variable Coefficient Equations

In this appendix, the general solution of

$$\left(a^2(z)\mathcal{P}_1'\right)' - \delta^2 \mathcal{P}_1 = Q(z), \tag{5.163}$$

arising from (5.14) for a particular form of the depth function $a(z)$ is derived. The analysis is based on the result from Varley and Seymour [46] where it was shown that many linear, variable coefficient equations have exact solutions if their coefficients satisfy a particular system of nonlinear ordinary differential equations. They considered a general, linear, second order partial differential equation of the form:

$$\frac{\partial}{\partial z}\left(a^2(z)\frac{\partial f}{\partial z}\right) = \alpha\frac{\partial^2 f}{\partial t^2} + \beta\frac{\partial f}{\partial t}, \tag{5.164}$$

where α and β are arbitrary constants. Defining a new independent variable

$$R = \int^z \frac{dv}{a(v)}, \tag{5.165}$$

and regarding a and f as functions of R and t, (5.164) can be rewritten in the canonical form

$$\frac{1}{a}\frac{\partial}{\partial R}\left(a\frac{\partial f}{\partial R}\right) = \alpha\frac{\partial^2 f}{\partial t^2} + \beta\frac{\partial f}{\partial t}. \tag{5.166}$$

It was shown by Varley and Seymour [46] that if the coefficient functions $a(R)$, $e_1(R), f_1(R), \ldots, e_{N-1}(R), f_{N-1}(R)$ satisfy the system of $2N-1$ independent nonlinear ordinary differential equations

$$(\sqrt{a})' = ae_1 - f_1, \tag{5.167}$$

$$e_n + e'_{n-1} = a^{-1}f_n, \quad 1 \le n \le N, \tag{5.168}$$

$$f_n + f'_{n-1} = ae_n, \quad 1 \le n \le N, \tag{5.169}$$

where

$$f_0 = e_0^{-1} = \sqrt{a(R)}, \quad \text{and} \quad \begin{array}{l} e_N = constant = k_n, \\ f_N = constant = l_N, \end{array} \tag{5.170}$$

then any solution to (5.166) can be written in terms of the corresponding solutions F to the constant coefficient equation,

$$\alpha\frac{\partial^2 F}{\partial t^2} + \beta\frac{\partial F}{\partial t} = \frac{\partial^2 F}{\partial R^2}, \tag{5.171}$$

as

$$f = \sum_{n=0}^{N} f_n(R)\frac{\partial^{N-n}F}{\partial R^{N-n}}. \tag{5.172}$$

In summary, if the mathematical function $a(R)$ is chosen to be *any* solution to the system (5.167)–(5.170) [defined shortly as a *VS function*], then the solution of the variable coefficient problem can be written in terms of the solution to the corresponding constant coefficient problem.

Using (5.165), an ordinary differential equation of the form (5.163) transforms into

$$\frac{1}{a}(a\mathcal{P}_1')' - \delta^2 \mathcal{P}_1 = Q(z), \tag{5.173}$$

and the solution to the corresponding homogeneous equation ($Q = 0$) now has the simple form

$$\mathcal{P} = c_1 E_1(R) + c_2 E_2(R), \tag{5.174}$$

where c_1 and c_2 are arbitrary constants, and

$$E_1(R) = \sum_{n=1}^{N} \left[\delta^{N-n} f_n(R)\right] e^{\delta R}, \tag{5.175}$$

$$E_2(R) = \sum_{n=1}^{N} \left[(-\delta)^{N-n} f_n(R)\right] e^{-\delta R}. \tag{5.176}$$

For $N = 1$, the above expression can be simplified to

$$E_1(R) = \left[\lambda_1 + \ell_1 \delta / \sqrt{a(R)}\right] e^{\delta R}, \tag{5.177}$$

$$E_2(R) = \left[\lambda_1 - \ell_1 \delta / \sqrt{a(R)}\right] e^{-\delta R}. \tag{5.178}$$

For convenience, E_i is normalized so that $E_i(R_o) = 1$ for some suitable R_o.

The general solution to the non-homogeneous equation (5.163) then takes the form

$$\mathcal{P}_1\big(R(z)\big) = c_1 E_1(R) + c_2 E_2(R) + \int^{R} K(R, S) Q\big(s(S)\big)\, dS, \tag{5.179}$$

where

$$K(R, S) = \frac{E_1(R)E_2(S) - E_1(S)E_2(R)}{E_1'(S)E_2(S) - E_1(S)E_2'(S)}, \tag{5.180}$$

and $S = \int^{s} dv/a(v)$. It should be noted that for $N = 1$ the denominator in (5.180) is proportional to $a(S)$.

It was shown that many linear, variable coefficient equations will have exact solutions [46], if their coefficients (as functions of the independent variable R) satisfy the system of ordinary differential equations (5.167)–(5.170). These coefficients are defined in terms of an Nth order *VS* function, $VS_N(R; \mathbf{J}, \mathbf{P})$. Here $\mathbf{J}$ is a vector containing N integer entries, and $\mathbf{P}$ is a vector containing $2N + 1$ arbitrary constant entries. The main advantage of *VS* functions is their ability to approximate a wide variety of functions that can vary rapidly. Here only VS_1 is used; the reader is referred to [46] for the details for general N. The application of the *VS* function can also be found in [40, 41, 45, 47].

VS functions arise in the general solution to the system of (5.167)–(5.170) relating the functions $a(R)$, $e_1(R)$, $f_1(R)$, $\ldots$, $e_{N-1}(R)$, $f_{N-1}(R)$. We define the general solution for $f_0 = \sqrt{a(R)}$ as the Nth order VS function

$$f_0 = VS_N(R; \mathbf{J}, \mathbf{P}), \tag{5.181}$$

where $\mathbf{J} = (j_1, j_2, \ldots, j_N)$, $j_k = 1, 2, 3$, or 4, and $\mathbf{P} = (l_1, l_2, \ldots, l_N, R_1, R_2, \ldots, R_N)$. VS_N is defined in terms of the four base functions, $j = j_k$, where, if $\lambda_n \geq 0$,

$$z_n^{(1)} = \sqrt{\lambda_n} \tanh\bigl(\sqrt{\lambda_n}[R + R_n]\bigr), \tag{5.182}$$

or,

$$z_n^{(12} = \sqrt{\lambda_n} \coth\bigl(\sqrt{\lambda_n}[R + R_n]\bigr), \tag{5.183}$$

and if $\lambda_n \leq 0$,

$$z_n^{(3)} = -\sqrt{-\lambda_n} \tan\bigl(\sqrt{\lambda_n}[R + R_n]\bigr), \tag{5.184}$$

or,

$$z_n^{(4)} = \sqrt{-\lambda_n} \cot\bigl(\sqrt{-\lambda_n}[R + R_n]\bigr), \tag{5.185}$$

where R_n is any real constant.

For $N = 1$, the system consisting of (5.167)–(5.170) reduces to a single equation

$$(\sqrt{a})' = k_1(a - \lambda_1). \tag{5.186}$$

whose general solution for $f_0 = \sqrt{a(R)}$ can be written as:

$$VS_N(R; \mathbf{J}, \mathbf{P}) = -\frac{l_1}{\lambda_1} z_1^{(j)}(R; \lambda_1, R_1), \quad j = j_k, \tag{5.187}$$

where $\mathbf{P} = (l_1, \lambda_1, R_1)$, and $\lambda_1 = k_1 l_1$; i.e. there are *four* independent functions each containing *three* arbitrary parameters. With so many functions and parameters available, VS_1 can be used to curve-fit many monotonic functions over a large range of R by a suitable choice of $\mathbf{P}$ and j_k.

In principle, the general system (5.167)–(5.170) can be reduced to a $(2N - 1)$th order nonlinear ordinary differential equation for $f_0(R)$, containing the two arbitrary parameters k_N and l_N. Quite remarkably, the general solution to this system was constructed by [46] for *any value of N* by solving linear algebraic equations. The solution, designated as $f_0(R) = VS_N(R; \mathbf{J}, \mathbf{P})$, is given as the ratio of two determinants whose elements are defined in terms of the $z_n^{(j)}(R; \lambda_n, R_n)$, $j = j_k$, and the λ_n.

5.6 Appendix: Finite Element Formulations

The governing equations (5.159) and (5.160)–(5.162) for the wave-induced soil response in a porous seabed can be solved by using a combination of semi-analytical technique and a finite element Galerkin method [53]. The three-dimensional boundary value problem can be reduced to one-dimensional form by assuming a harmonic wave loading on the surface of the seabed. Thus, the wave-induced pore pressure and soil displacements can be expressed in a general form as

$$f(x,y,z;t)=f_a e^{i(mkx+nky-\omega t)}+f_b e^{i(mkx-nky-\omega t)}, \tag{5.188}$$

$$\begin{aligned} f &= \{p,u,v,w,\sigma'_x,\sigma'_y.\sigma'_z,\tau_{xz},\tau_{yz},\tau_{xy}\}^{\mathbf{T}}, \\ f_a &= \{p_a,u_a,v_a,w_a,\sigma'_{xa},\sigma'_{ya}.\sigma'_{za},\tau_{xza},\tau_{yza},\tau_{xya}\}^{\mathbf{T}}, \\ f_b &= \{p_b,u_b,v_b,w_b,\sigma'_{xb},\sigma'_{yb}.\sigma'_{zb},\tau_{xzb},\tau_{yzb},\tau_{xyb}\}^{\mathbf{T}}. \end{aligned} \tag{5.189}$$

The first term on the right-hand side of (5.188) represents the contribution from the reflected waves, while the second term is for the incident waves. The solutions of pore pressure, soil displacements and effective stresses can be found separately and then superimposed to yield the complete solution. Thus, the three-dimensional boundary value problem can be reduced to one-dimensional form as function of z.

For convenience, the function f_a can be separated into real and imaginary parts as

$$f_a=\big(f_r(z)+f_c(z)\big), \tag{5.190}$$

$$\begin{aligned} f_r(z) &= [P_r,U_r,V_r,W_r,S_{xr},S_{yr},S_{zr},T_{xzr},T_{yzr},T_{xyr}]^{\mathbf{T}}, \\ f_c(z) &= [P_c,U_c,V_r,W_r,S_{xc},S_{yc},S_{zc},T_{xzc},T_{yzc},T_{xyc}]^{\mathbf{T}}, \end{aligned} \tag{5.191}$$

in which f_r and f_c denote the real and imaginary parts of f_a. In (5.190), P, U, V, W, S_x, S_y, S_z, T_{xz}, T_{yz} and T_{xy} are related with p, u, v, w, σ'_x, σ'_y, σ'_z, τ_{xz}, τ_{yz} and τ_{xy}, respectively.

Substituting (5.190) into governing equations (5.159) and (5.160)–(5.162), and then directly apply the Galerkin method [53] to these equations, the finite element analytical formulations can be written in a matrix form as

$$\begin{aligned} \mathbf{F} &= \mathbf{Kq} \\ &= \int_0^{L_e}\mathbf{B}_1^{\mathbf{T}}\mathbf{D}_1\mathbf{B}_1\,dz+\int_0^{L_e}\mathbf{B}_1^{\mathbf{T}}\mathbf{D}_2\mathbf{B}_2\,dz+\int_0^{L_e}\mathbf{B}_2^{\mathbf{T}}\mathbf{D}_3\mathbf{B}_1\,dz+\int_0^{L_e}\mathbf{B}_2^{\mathbf{T}}\mathbf{D}_4\mathbf{B}_2\,dz, \end{aligned} \tag{5.192}$$

where

$$\begin{aligned} \mathbf{F}^{\mathbf{T}} = \Bigg[& -\left(\frac{dP_r}{dz}\right)_1, -\left(\frac{dP_c}{dz}\right)_1, -(T_{xzr})_1, -(T_{xzc})_1, -(T_{yzr})_1, -(T_{yzc})_1, \\ & -(S_{zr})_1, -(S_{zc})_1, \underbrace{0,0,0,\ldots,0}_{8\times(N_p-2)}, \left(\frac{dP_r}{dz}\right)_{N_p}, \left(\frac{dP_c}{dz}\right)_{N_p}, \\ & (T_{xzr})_{N_p}, (T_{xzc})_{N_p}, (T_{yzr})_{N_p}, (T_{yzc})_{N_p}, (S_{zr})_{N_p}, (S_{zc})_{N_p} \Bigg], \end{aligned} \tag{5.193}$$

$$
\begin{aligned}
\mathbf{q} &= [\mathbf{N}]\{q_e\} = [P_r, P_c, U_r, U_c, V_r, V_c, W_r, W_c]^{\mathbf{T}},\\
\{q_e\}^{\mathbf{T}} &= \left[q_1^{\mathbf{T}}, \ldots, q_{n_e}^{\mathbf{T}}\right],\\
\{q_j\}^{\mathbf{T}} &= [P_{rj}, P_{cj}, U_{rj}, U_{cj}, V_{rj}, V_{cj}, W_{rj}, W_{cj}],\\
[\mathbf{N}] &= [N_1\mathbf{I}, N_2, \mathbf{I}, \ldots, N_{n_e}\mathbf{I}] \quad j = 1, 2, 3, \ldots, n_e,
\end{aligned}
\tag{5.194}
$$

$$
\begin{aligned}
\mathbf{B}_1 &= [B_{11}, B_{12}, B_{13}, \ldots, B_{1n_e}], \qquad B_{1i} = N_i\mathbf{I},\\
\mathbf{B}_2 &= [B_{21}, B_{22}, B_{23}, \ldots, B_{2n_e}], \qquad B_{2i} = N_i\mathbf{I},
\end{aligned}
\tag{5.195}
$$

where $\mathbf{I}$ is an 8×8 identify matrix, $\mathbf{D}_i$ is a 8×8 coefficient matrix, which can be obtained from the governing equations.

Similarly, the finite element formulation for the effective normal stresses and shear stresses can be expressed in matrix form as

$$
\begin{aligned}
\mathbf{T} &= \mathbf{Hq}\\
&= \int_0^{L_e} \mathbf{H}_1^{\mathbf{T}}\mathbf{D}_{s1}\mathbf{B}_1\,dz + \int_0^{L_e} \mathbf{H}_1^{\mathbf{T}}\mathbf{D}_{s2}\mathbf{B}_2\,dz + \int_0^{L_e} \mathbf{H}_2^{\mathbf{T}}\mathbf{D}_{s3}\mathbf{B}_1\,dz + \int_0^{L_e} \mathbf{H}_2^{\mathbf{T}}\mathbf{D}_{s4}\mathbf{B}_2\,dz,
\end{aligned}
\tag{5.196}
$$

where

$$
\begin{aligned}
\mathbf{T}^{\mathbf{T}} = \big[&-(S_{xr})_1, -(S_{xc})_1, -(S_{yr})_1, -(S_{yc})_1, -(S_{zr})_1, -(S_{zc})_1,\\
&-(T_{xyr})_1, -(T_{xyc})_1, -(T_{yzr})_1, -(T_{yzc})_1, \underbrace{0, 0, 0, \ldots, 0}_{12\times(N_p-2)},\\
&-(S_{xr})_{N_p}, -(S_{xc})_{N_p}, -(S_{yr})_{N_p}, -(S_{yc})_{N_p}, -(S_{zr})_{N_p}, -(S_{zc})_{N_p},\\
&-(T_{xyr})_{N_p}, -(T_{xyc})_{N_p}, -(T_{yzr})_{N_p}, -(T_{yzc})_{N_p}\big],
\end{aligned}
\tag{5.197}
$$

$$
\begin{aligned}
\mathbf{H}_1 &= [H_{11}, H_{12}, H_{13}, \ldots, H_{1n_e}], \qquad B_{1i} = N_i\mathbf{I}_2,\\
\mathbf{H}_2 &= [H_{21}, H_{22}, H_{23}, \ldots, H_{2n_e}], \qquad B_{2i} = \frac{dN_i}{dz}\mathbf{I}_2,
\end{aligned}
\tag{5.198}
$$

where $\mathbf{I}_2$ is a 12×12 identity matrix. $\mathbf{D}_{si}$ is a 12×8 coefficient matrix, which can be obtained from the governing equations.

References

1. Atkinson, J.H.: Anisotropic elastic deformation in laboratory tests on unsaturated London clay. Géotechnique **25**(2), 357–374 (1975)
2. Barden, L.: Stresses and displacements in a cross-anisotropic soil. Géotechnique **13**, 198–210 (1963)
3. Bennett, R.H., Fischer, K.M., Li, H., Lamber, D.H., Hulbert, M.H., Yamamoto, T., Badiey, M.: In-situ porosity and permeability of selected carbonate sediments: Great Bahama Bank. Part 2. Microfabric. Mar. Geotechnol. **9**(1), 29–47 (1990)

4. Bennett, R.H., Li, H., Lamber, D.H., Fischer, K.M., Walter, D.J., Hickox, C.E., Hulbert, M.H., Yamamoto, T., Badiey, M.: In-situ porosity and permeability of selected carbonate sediments: Great Bahama Bank. Part 1: Measurements. Mar. Geotechnol. **9**(1), 1–28 (1990)
5. Biot, M.A.: General theory of three-dimensional consolidation. J. Appl. Phys. **26**(2), 155–164 (1941)
6. Bryant, W.R., Hottman, W., Trabant, P.: Permeability of unconsolidated and consolidated sediments, Gulf of Mexico. Mar. Geotechnol. **1**(1), 1–14 (1975)
7. Esrig, M.I., Kirby, R.C.: Implication of gas content for predicting the stability of submarine slopes. Mar. Geotechnol. **17**, 58–67 (1977)
8. Gatmiri, B.: A simplified finite element analysis of wave-induced effective stress and pore pressures in permeable sea beds. Géotechnique **40**(1), 15–30 (1990)
9. Gatmiri, B.: Response of cross-anisotropic seabed to ocean waves. J. Geotech. Eng. **118**(9), 1295–1314 (1992)
10. Gazetas, G.: Stresses and displacements in cross-anisotropic soils. J. Geotech. Eng. Div. **108**(4), 532–553 (1982)
11. Graham, J., Houlsby, G.T.: Anisotropic elasticity of a natural clay. Géotechnique **33**(2), 165–180 (1983)
12. Hsu, J.R.C., Jeng, D.S.: Wave-induced soil response in an unsaturated anisotropic seabed of finite thickness. Int. J. Numer. Anal. Methods Geomech. **18**(11), 785–807 (1994)
13. Hsu, J.R.C., Jeng, D.S., Tsai, C.P.: Short-crested wave-induced soil response in a porous seabed of infinite thickness. Int. J. Numer. Anal. Methods Geomech. **17**(8), 553–576 (1993)
14. Jeng, D.S.: Wave-induced liquefaction potential in a cross-anisotropic seabed. J. Chin. Inst. Eng. **19**(1), 59–70 (1996)
15. Jeng, D.S.: Soil response in cross-anisotropic seabed due to standing waves. J. Geotech. Geoenviron. Eng. **123**(1), 9–19 (1997)
16. Jeng, D.S.: Wave-induced seabed instability in front of a breakwater. Ocean Eng. **24**(10), 887–917 (1997)
17. Jeng, D.S.: Wave-induced seabed response in front of a breakwater. Ph.D. thesis, University of Western Australia (1997)
18. Jeng, D.S.: Effect of cross-anisotropic soil behaviour on the wave-induced seabed response. Géotechnique **48**(4), 555–561 (1998)
19. Jeng, D.S.: Wave-induced seabed response in a cross-anisotropic seabed in front of a breakwater: An analytical solution. Ocean Eng. **25**(1), 49–67 (1998)
20. Jeng, D.S.: Mechanism of the wave-induced seabed response in the vicinity of a breakwater: A review. Ocean Eng. **28**(5), 539–572 (2001)
21. Jeng, D.S., Hsu, J.R.C.: Wave-induced soil response in a nearly saturated seabed of finite thickness. Géotechnique **46**(3), 427–440 (1996)
22. Jeng, D.S., Lin, Y.S.: Finite element modelling for water waves-soil interaction. Soil Dyn. Earthq. Eng. **15**(5), 283–300 (1996)
23. Jeng, D.S., Lin, Y.S.: Poroelastic analysis for wave-seabed interaction problem. Comput. Geotech. **26**(1), 43–64 (2000)
24. Jeng, D.S., Seymour, B.R.: Response in seabed of finite depth with variable permeability. J. Geotech. Geoenviron. Eng. **123**(10), 902–911 (1997)
25. Jeng, D.S., Seymour, B.R.: Short-crested wave-induced seabed response with variable permeability. J. Chin. Inst. Eng. **20**(4), 377–388 (1997)
26. Jeng, D.S., Seymour, B.R.: Wave-induced pore pressure and effective stresses in a porous seabed with variable permeability. J. Offshore Mech. Arct. Eng. **119**(4), 226–233 (1997)
27. Kirkgard, M.M., Lade, P.V.: Anisotropy of normally consolidated San Francisco bay mud. Geotech. Test. J. **14**(3), 231–246 (1991)
28. Koning, K.: Stress distribution in a homogeneous anisotropic elastic semi-infinite solid. In: Fourth International Conference of Soil Mechanics and Foundations Engineering, pp. 335–338 (1957)
29. Lin, Y.S., Jeng, D.S.: Short-crested wave-induced liquefaction in porous seabed. J. Geotech. Geoenviron. Eng. **126**(5), 481–494 (2000)

30. Madsen, O.S.: Wave-induced pore pressures and effective stresses in a porous bed. Géotechnique **28**(4), 377–393 (1978)
31. Okusa, S.: Wave-induced stress in unsaturated submarine sediments. Géotechnique **35**(4), 517–532 (1985)
32. Pickering, D.J.: Anisotropic elastic parameters for soil. Géotechnique **20**(3), 271–276 (1970)
33. Pietruszczak, S., Pande, G.N.: Constitutive relations for partially saturated soils containing gas inclusions. J. Geotech. Eng. **122**(1), 50–59 (1996)
34. Rahman, M.S.: Wave-induced instability of seabed: Mechanism and conditions. Mar. Geotechnol. **10**, 277–299 (1991)
35. Raman-Nair, W., Sabin, G.C.W.: Wave-induced failure of poro-plastic seabed slopes: A boundary element study. Proc. Inst. Civ. Eng. 2. Res. Theory **91**, 771–794 (1991)
36. Rowe, P.W.: Measurement of the coefficient of consolidation of lacustrine clay. Géotechnique **9**(3), 107–118 (1959)
37. Rowe, P.W.: The stress-dilatancy relation for static equilibrium of an assembly of particles in contact. Proc. R. Soc. A **269**, 500–527 (1962)
38. Samarasinghe, A.M., Huang, Y.H., Drnevich, V.P.: Permeability and consolidation of normal consolidated soils. J. Geotech. Eng. Div. **108**(6), 835–849 (1982)
39. Seymour, B.R., Jeng, D.S., Hsu, J.R.C.: Transient soil response in a porous seabed with variable permeability. Ocean Eng. **23**(1), 27–46 (1996)
40. Seymour, B.R., Varley, E.: Exact solutions describing soliton-like interactions in a nondispersive medium. SIAM J. Appl. Math. **42**(4), 804–821 (1982)
41. Seymour, B.R., Varley, E.: Exact representations for acoustical waves when the sound speed varies in space and time. Stud. Appl. Math. **76**, 1–35 (1987)
42. Silvester, R., Hsu, J.R.C.: Sines revisited. J. Waterw. Port Coast. Ocean Eng. **115**(3), 327–344 (1989)
43. Thomas, S.D.: A finite element model for the analysis of wave induced stresses, displacements and pore pressure in an unsaturated seabed. I: Theory. Comput. Geotech. **8**(1), 1–38 (1989)
44. Thomas, S.D.: A finite element model for the analysis of wave induced stresses, displacements and pore pressure in an unsaturated seabed. II: Model verification. Comput. Geotech. **17**(1), 107–132 (1995)
45. Varley, E., Seymour, B.R.: Exact solution for large amplitude waves in dispersive and dissipative systems. Stud. Appl. Math. **72**, 241–262 (1985)
46. Varley, E., Seymour, B.R.: A method for obtaining exact solutions to partial differential equations with variable coefficients. Stud. Appl. Math. **78**, 183–225 (1988)
47. Varley, E., Seymour, B.R.: Application of exact solutions to the Navier-Stokes equations: Free shear layers. J. Fluid Mech. **274**, 267–291 (1995)
48. Wiendieck, K.: L'anisotropie des milieux pulvérulents et son influence sur les tassements. Publications Scientiques du Ministere de l'Air (1968)
49. Wolf, K.: Distribution of stress in a half plane and a half space of anisotropic material. Z. Angew. Math. Mech. **15**, 249–254 (1945)
50. Yamada, Y., Ishihara, K.: Anisotropic deformation characteristics of sand under three dimensional stress conditions. Soil Found. **19**(2), 79–94 (1979)
51. Yamamoto, T., Koning, H., Sellmeijer, H., Hijum, E.V.: On the response of a poro-elastic bed to water waves. J. Fluid Mech. **87**(1), 193–206 (1978)
52. Zen, K., Yamazaki, H.: Mechanism of wave-induced liquefaction and densification in seabed. Soil Found. **30**(4), 90–104 (1990)
53. Zienkiewicz, O.C., Taylor, R.L.: The Finite Element Method. McGraw-Hill, New York (1989)

Chapter 6
Dynamic Analysis for Wave-seabed Interaction

Abstract In this chapter, a series of analytical solutions for the wave-induced full dynamic seabed response was presented. Unlike previous chapters, the accelerations of pore fluid and solid particles were considered in the new models. Based on the new solutions, a parametristic study was performed to examine the influences of wave and seabed characteristics on the wave-induced seabed response. Finally, the applicable range of full dynamic, u–p and quasi-static models was clarified with a simple formula.

Keywords Dynamic poro-elastic theory · u–p approximation · Quasi-static model · Seabed response

6.1 Introduction

Dynamic soil behavior is another important factor in the determination of wave-induced seabed response. Most previous investigations into the wave-seabed interaction problem have treated the problem as a quasi-static state [4, 13]. Only a few researchers considered the u–p approximation [8, 9] or dynamic analysis [6, 12]. In the quasi-static state, pore fluid and soil are considered as compressible, but this approach ignores the accelerations due to the displacement of fluid and soil. In the u–p approximation, only the acceleration of solid is considered, i.e., the second-order time derivatives of the relative fluid displacement were ignored. Zienkiewicz *et al.* [14] might have been the first to propose the u–p approximation through a one-dimensional analysis. However, since the wave loading is a two-dimensional dynamic loading, which includes accelerations due to pore fluid and soil displacement, the influences of dynamic soil behavior cannot always be neglected for all cases. Also, a two-dimensional analysis for such a problem will derive a far more realistic simulation of the problem.

In this chapter, we further consider dynamic soil behavior in the poro-elastic model for wave-seabed interaction problem. The full dynamic model for such a problem proposed by Jeng and Cha [5] will be outlines first. Based on it, some special cases such as the u–p approximation will further derived. Then, the applicable range of full dynamic, u–p and quasi-static models was clarified with a simple formula.

D.-S. Jeng, *Porous Models for Wave-seabed Interactions*,
DOI 10.1007/978-3-642-33593-8_6,

6.2 Boundary Value Problem

In this section, the boundary value problem of the wave-seabed interaction is outlined. Firstly, the basic framework for the wave-induced soil response with dynamic soil behavior will be established. Then, a general solution for the wave-induced soil response with dynamic soil behavior with non-liner wave case in a seabed will be derived. With the dynamic solution, the simplified approximations, which are u–p approximation and quasi-static solutions, will be obtained.

6.2.1 Basic Ocean Wave Theory

Referring to the Stokes wave theory up to the second-order, the velocity potential (ϕ) is given as

$$\phi = \frac{gH}{2\omega}\frac{\cosh kz}{\cosh kd}\sin(kx-\omega t) + \frac{3}{8}\frac{(gH)^2}{2\omega L}\frac{\cosh(2kz)}{\sinh^4(kd)}\sin(kx-\omega t), \tag{6.1}$$

and the free surface elevation η is

$$\eta = \frac{H}{2}\cos(kx-\omega t) + \frac{h}{8}\left(\frac{gH}{L}\right)\frac{\cosh(kz)}{\sinh^3(kd)}\big[2+\cosh(2kd)\big]\cos 2(kx-\omega t) \tag{6.2}$$

where H is the wave height, k is the wave number ($k = 2\pi/L$, in which L is the wavelength), t is time and d is the water depth above the seabed surface.

The wave frequency ω ($= 2\pi/T$, where T is the wave period) satisfies the dispersion equation

$$\omega^2 = gk\tanh kd. \tag{6.3}$$

Introducing the velocity potential (ϕ) and the surface elevation (η) into the Bernoulli equation, the wave pressure is given by

$$\begin{aligned} P(x,z;t) &= \frac{\gamma_w H}{2}\frac{\cosh kz}{\cosh kd}\cos(kx-\omega t) \\ &\quad + \frac{3}{4}\gamma_w H\left(\frac{gH}{L}\right)\frac{1}{\sinh(2kd)}\left[\frac{\cosh(2kz)}{\sinh^2(kd)} - \frac{1}{3}\right]\cos 2(kx-\omega t) \end{aligned} \tag{6.4}$$

where γ_w is the unit weight of water, and P is the wave pressure.

6.2.2 Governing Equations

For a two-dimensional wave-seabed interaction problem, and treating the porous seabed as hydraulically isotropic with the same permeability. Biot [1, 2] presented a general set of governing equations, which is the behavior of liner elastic porous solid under dynamic conditions. The equations are summarized in the tensor form, as given below [3, 14]

$$\sigma_{ij,j} = \rho \ddot{u}_i + \rho_f \ddot{w}_i, \tag{6.5}$$

$$-p_{,i} = \rho_f \ddot{u} + \frac{\rho_f}{n} \ddot{w}_i + \frac{\rho_f g}{k_z} \dot{w}_i, \tag{6.6}$$

$$\dot{\varepsilon}_{ii} + \dot{w}_{ii} = -\frac{n}{K'_f} \dot{p}, \tag{6.7}$$

where p is pore pressure, n is porosity, ρ is the combined density, ρ_f is the fluid density, u and w are the displacements of solid and relative displacements pore fluid, k_z is soil permeability. K'_f is compressibility of pore fluid which can defined as

$$\frac{1}{K'_f} = \frac{1}{K_{wo}} + \frac{1 - S_r}{\gamma_w d} \tag{6.8}$$

where γ_w is unit weight of water, K_{wo} is bulk modulus of pore-water and S_r is the degree of saturation. The definition of effective stresses, σ'_{ij}, which are assumed to control the deformation of the soil skeleton, are given by the total stress (σ_{ij}) and pore pressure (p) as,

$$\sigma_{ij} = \sigma'_{ij} - \delta_{ij} p \tag{6.9}$$

where, δ_{ij} is the Delta denotation. Thus, the equation of force balance, (6.5) becomes

$$\sigma'_{ij,j} = \delta_{ij} p_{,i} + \rho \ddot{u}_i + \rho_f \ddot{w}_i. \tag{6.10}$$

Under conditions of plane strain, the relationship of plane stress-strain can be expressed as Hooke's law:

$$\sigma'_{ij} = \lambda u_{i,i} \delta_{ij} + 2G u_{i,j} \tag{6.11}$$

where G is shear modulus and μ is Poisson's ratio. $\lambda = 2G\mu/(1 - 2\mu)$ is Lames's constant.

6.2.3 Boundary Conditions

For a porous seabed of infinite or finite thickness, as shown in Fig. 3.1, the mathematical expressions for the wave-induced soil response can be derived, with appropriate boundary conditions. In general, two boundary conditions are required at a rigid impermeable bottom (**BBC**: Bottom Boundary Condition); and the seabed surface (**SBC**: Surface Boundary Condition).

(a) BBC: Boundary Condition at the Bottom ($z \to -\infty$ or $z \to -h$) For the resting on an impermeable rigid bottom, zero displacements and no vertical flow occur at the horizontal bottom. For a seabed of infinite thickness

$$u = w = p = 0 \quad \text{as } z \to -\infty. \tag{6.12}$$

For a seabed of finite thickness.

$$u = w = \frac{\partial p}{\partial z} = 0 \quad \text{at } z = -h. \tag{6.13}$$

(b) SBC: Boundary Conditions at Seabed Surface ($z = 0$) At the surface of the seabed ($z = 0$), the boundary condition imposed by the wave motion is given by

$$\begin{aligned} P_b(x, z; t) = {} & \frac{\gamma_w H}{2} \frac{1}{\cosh kd} \cos(kx - \omega t) \\ & + \frac{3}{4}\gamma_w H \left(\frac{gH}{L}\right) \frac{1}{\sinh(2kd)} \left[\frac{1}{\sinh^2(kd)} - \frac{1}{3}\right] \cos 2(kx - \omega t) \quad \text{at } z = 0 \end{aligned} \tag{6.14}$$

where "$\cos(kx - \omega t)$" denotes the spatial temporal variations in wave pressure within the two-dimensional progressive wave described above. P_b is the wave pressure at the seabed surface.

In addition, the vertical effective normal stress must vanish at the seabed surface. As for the shear stress in the z-direction, it is known that the shear stress is associated with the oscillatory flow above the seabed. However, the fluid shear stresses exerted at the seabed surface is small and may be neglected [10]. Therefore, it is reasonable to impose boundary conditions

$$\sigma_z' = \tau = 0 \quad \text{at } z = 0. \tag{6.15}$$

In (6.15), since the wave-induced transient soil response fluctuates periodically in the x-direction, thus, we keep the horizontal effective normal stresses and shear stresses in the x-direction *free* at the seabed surface without making any assumption [4].

6.3 General Solutions

6.3.1 Basic Framework

In this section, we will establish the basic framework, describe the wave-induced seabed response problem, that can be solved analytically, based on the governing equations (6.5)–(6.7) and boundary conditions (6.12)–(6.15). The basic framework for the wave-induced soil displacements and relative displacement of pore fluid can obtained. From the framework the effective stresses can then be found from the stress-strain relationship (6.9).

From the governing equation, (6.7), we can have

$$-p,_i = \frac{K'_f}{n}(\varepsilon_{ii} + w_{ii})_i \tag{6.16}$$

where n is soil porosity.

Substituting (6.16) into (6.10), the governing equation can be rewritten as

$$\frac{K'_f}{n}(\varepsilon_{ii} + w_{ii})_i = \rho_f \ddot{u}_i + \frac{\rho_f}{n}\ddot{w}_i + \frac{\rho_f g}{k_z}\dot{w}_i, \tag{6.17}$$

$$\sigma'_{ij,j} = -\frac{K'_f}{n}\delta_{ij}(\varepsilon_{jj} + w_{ii})_i + \rho \ddot{u}_i + \rho_f \ddot{w}_i. \tag{6.18}$$

Herein, re-write (6.17) in a scalar form as,

$$\frac{K'_f}{n}\left(\frac{\partial \varepsilon}{\partial x} + \frac{\partial \zeta}{\partial x}\right) = \rho_f \ddot{u}_x + \frac{\rho_f}{n}\ddot{w}_x + \frac{\rho_f g}{k_z}\dot{w}_x, \tag{6.19}$$

$$\frac{K'_f}{n}\left(\frac{\partial \varepsilon}{\partial z} + \frac{\partial \zeta}{\partial z}\right) = \rho_f \ddot{u}_z + \frac{\rho_f}{n}\ddot{w}_z + \frac{\rho_f g}{k_z}\dot{w}_z \tag{6.20}$$

in the x- and z-direction, respectively.

Similarly, (6.18) can also be re-organized in a scalar form as

$$\frac{\partial \sigma'_x}{\partial x} + \frac{\partial \tau_{xz}}{\partial z} = -\frac{K'_f}{n}\left(\frac{\partial \varepsilon}{\partial x} + \frac{\partial \zeta}{\partial x}\right) + \rho \ddot{u}_x + \rho_f \ddot{w}_x, \tag{6.21}$$

$$\frac{\partial \tau_{xz}}{\partial x} + \frac{\partial \sigma'_z}{\partial z} = -\frac{K'_f}{n}\left(\frac{\partial \varepsilon}{\partial z} + \frac{\partial \zeta}{\partial z}\right) + \rho \ddot{u}_z + \rho_f \ddot{w}_z \tag{6.22}$$

in which

$$\zeta = w_{ii} = \frac{\partial w_x}{\partial x} + \frac{\partial w_z}{\partial z}. \tag{6.23}$$

Since the wave-induced oscillatory soil response fluctuates periodically, all quantities can be replaced immediately by their complex form

$$\bar{x} = kx, \qquad \bar{z} = kz, \qquad \bar{t} = \omega t, \qquad f = \sum_{m=1}^{2} \bar{F} e^{im(\bar{x}-\bar{t})}, \tag{6.24}$$

where f denotes the wave-induced soil response parameters (such as p, σ'_x, σ'_z, w_x and w_z), and $\bar{F}$ is the corresponding parameters (such as P, S_x, S_z, T_{xz}).

$$\bar{u}_x = \bar{U}_x e^{i(\bar{x}-\bar{t})}. \tag{6.25}$$

Introducing (6.24) and (6.25) into (6.19)–(6.22), we have

$$m^2\left(\rho_f\omega^2-\frac{K'_f}{n}k^2\right)\bar{U}_x+\frac{imK'_f}{n}k^2\bar{U}'_z$$
$$+\left(\frac{\rho_f m^2}{n}\omega^2-\frac{K'_f m^2}{n}k^2+im\omega\frac{\rho_f g}{k_z}\right)\bar{W}_x+\frac{imK'_f}{n}k^2\bar{W}'_z=0, \tag{6.26}$$

$$\frac{imK'_f}{n}k^2\bar{U}'_x+\left[\frac{K'_f}{n}k^2\bar{U}''_z+\rho_f m^2\omega^2\bar{U}_z\right]+\left[\frac{imK'_f}{n}k^2\bar{W}'_x\right]$$
$$+\left[\frac{K'_f}{n}k^2\bar{W}''_z+\frac{\rho_f m^2}{n}\omega^2\bar{W}_z+im\omega\frac{\rho_f g}{k_z}\bar{W}_z\right]=0, \tag{6.27}$$

and

$$Gk^2\left[\bar{U}''_x-m^2\frac{2(1-\mu)}{1-2\mu}\bar{U}_x+m^2\frac{\rho\omega^2}{Gk^2}\bar{U}_x-m^2\frac{K'_f}{Gn}\bar{U}_x\right]$$
$$+\left(\frac{G}{1-2\mu}k^2m+\frac{K'_f}{n}k^2m\right)i\bar{U}'_z$$
$$+\left(\rho_f m^2\omega^2-\frac{K'_f}{n}k^2m^2\right)\bar{W}_x+\frac{K'_f}{n}k^2im\bar{W}'_z=0, \tag{6.28}$$

$$Gk^2\left[\frac{2(1-\mu)}{1-2\mu}\bar{U}''_z+\frac{K'_f}{Gn}\bar{U}''_z-m^2\bar{U}_z+\frac{\rho\omega^2}{Gk^2}m^2\bar{U}_z\right]$$
$$+\left(\frac{G}{1-2\mu}k^2m+\frac{K'_f}{n}k^2m\right)i\bar{U}'_x+im\frac{K'_f}{n}k^2\bar{W}'_x+\rho_f m^2\omega^2\bar{W}_z$$
$$+\frac{K'_f}{n}k^2\bar{W}''_z=0. \tag{6.29}$$

These equations can be rearranged using by new parameters defined by,

$$K_1=\frac{G}{\frac{G}{1-2\mu}+\frac{K'_f}{n}},\qquad K_2=\frac{\frac{K'_f}{n}}{\frac{G}{1-2\mu}+\frac{K'_f}{n}},$$
$$\Pi_1=\frac{\rho k_z V_c^2k^2}{\rho_f g\omega},\qquad \Pi_2=\frac{\rho\omega^2}{(\frac{G}{1-2\mu}+\frac{K'_f}{n})k^2}, \tag{6.30}$$
$$V_c^2=\frac{(\frac{G}{1-2\mu}+\frac{K'_f}{n})}{\rho},\qquad \beta=\frac{\rho_f}{\rho},$$

where K_1, K_2, Π_1 and Π_2 are non-dimensional parameters, among these parameters Π_1 and Π_2 are included as most wave and soil characteristics. Later, we will discuss these two parameters further.

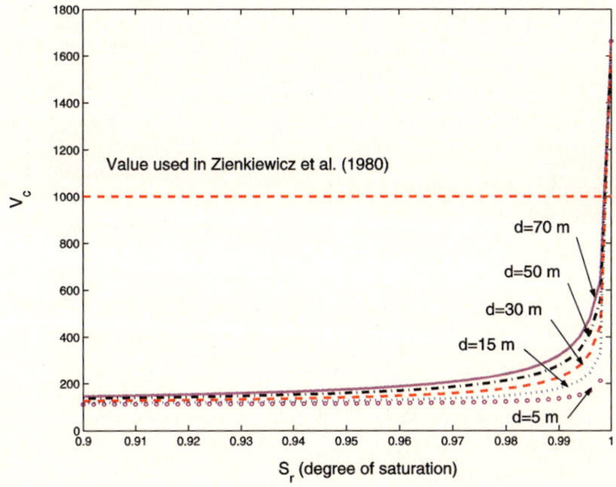

Fig. 6.1 Relationship between the speed of compressive waves V_c and the degree of saturation for various water depth

From (6.30), V_c is the speed of compressive wave, which is directly related to the degree of saturation (S_r), water depth (d), and other soil parameters (such as shear modulus (G), Poisson ratio (μ) and porosity (n)). The speed of compressive wave will depend on the degree of saturation and water depth, because of the shear modulus, Poisson ratio and porosity are almost constant for most marine sediments. Figure 6.1 illustrates the relationship between V_c and the degree of saturation for various water depths. In the example, other soil parameters are taken as $G = 5 \times 10^6\ \mathrm{N/m^2}$, $\mu = 0.35$ and $n = 0.4$. As seen in the Fig. 6.1, the speed of compressive wave (V_c) dramatically increases in near saturated seabed. The value used by Zienkiewicz *et al.* [14], i.e., $V_c = 1000\ \mathrm{m/s}$, is also indicated in the figure. From this figure, it can be concluded that the value of V_c used in [14] is invalid for unsaturated marine sediments. The reason that the value used in [14] is the case for saturated seabeds. However, the degree of saturation varies from 0.9 to 1.0 in marine sediments.

Using (6.30), (6.26)–(6.29) can be re-written as

$$(\beta\Pi_2 - K_2)m^2\bar{U}_x + imK_2D\bar{U}_z + \left(\frac{\beta\Pi_2 m^2}{n} - m^2K_2 + \frac{im}{\Pi_1}\right)\bar{W}_x + imK_2D\bar{W}_z = 0, \tag{6.31}$$

$$imK_2D\bar{U}_x + \left(K_2D^2 + \Pi_2 m^2\beta\right)\bar{U}_z + imK_2D\bar{W}_x + \left(K_2D^2 + \frac{\beta\Pi_2 m^2}{n} + \frac{im}{\Pi_1}\right)\bar{W}_z = 0, \tag{6.32}$$

$$K_1D^2\bar{U}_x - m^2\frac{2(1-\mu)}{1-2\mu}K_1\bar{U}_x + \Pi_2 m^2\bar{U}_x - K_2m^2\bar{U}_x + imD\bar{U}_z + (\Pi_2\beta - K_2)m^2\bar{W}_x + imK_2D\bar{W}_z = 0, \tag{6.33}$$

$$\left[\frac{2(1-\mu)}{1-2\mu}K_1D^2+K_2D^2-m^2K_1+m^2\Pi_2\right]\bar{U}_z$$
$$+imD\bar{U}_x+imK_2D\bar{W}_x+\left(K_2D^2+\Pi_2\beta m^2\right)\bar{W}_z=0 \tag{6.34}$$

where D denotes $\partial/\partial z$.

Equations (6.31)–(6.34) can be written in a matrix form as

$$\begin{bmatrix} A_{12} & imK_2D & K_1D^2+A_{11} & imD \\ imK_2D & K_2D^2+m^2\Pi_2\beta & imD & A_{21}D^2+A_{22} \\ A_{32} & imK_2D & A_{31} & imK_2D \\ imK_2D & K_2D^2+A_{41} & imK_2D & K_2D^2+m^2\Pi_2\beta \end{bmatrix}\begin{bmatrix}\bar{W}_x\\ \bar{W}_z\\ \bar{U}_x\\ \bar{U}_z\end{bmatrix}=\begin{bmatrix}0\\0\\0\\0\end{bmatrix}. \tag{6.35}$$

In (6.35), the A_{ij} coefficients are given by

$$A_{11}=m^2\left[\Pi_2-\frac{2(1-\mu)}{1-2\mu}K_1-K_2\right],\qquad A_{12}=m^2(\Pi_2\beta-K_2), \tag{6.36}$$

$$A_{21}=\frac{2(1-\mu)}{1-2\mu}K_1+K_2,\qquad A_{22}=m^2(\Pi_2-K_1), \tag{6.37}$$

$$A_{31}=m^2(\beta\Pi_2-K_2),\qquad A_{32}=\frac{\beta\Pi_2m^2}{n}-K_2m^2+\frac{im}{\Pi_1}, \tag{6.38}$$

$$A_{41}=\frac{\beta\Pi_2m^2}{n}+\frac{im}{\Pi_1}. \tag{6.39}$$

The characteristic equation of (6.35) can be written as

$$\left(\alpha_4D^6+\alpha_3D^4+\alpha_2D^2+\alpha_1\right)\bar{U}_x=0 \tag{6.40}$$

where α_i are derived from (6.32) and (6.34), and given as:

$$\alpha_1=-(A_{12}A_{31}-A_{11}A_{32})\left(A_{22}A_{41}-m^4\beta^2\Pi_2^2\right), \tag{6.41}$$

$$\begin{aligned}\alpha_2=\big[&K_2(-A_{31}A_{41}+A_{11}A_{22}K_2-A_{22}A_{31}K_2+A_{11}A_{41}K_2\\ &+A_{31}\Pi_2-2A_{11}K_2\Pi_2+A_{31}K_2\Pi_2)\\ &+A_{32}\big(A_{41}+A_{11}A_{21}A_{41}+A_{22}A_{41}K_1\\ &+A_{11}A_{22}K_2+A_{22}K_2^2-2K_2\Pi_2-2A_{11}K_2\Pi_2-K_1\Pi_2^2\big)\\ &-A_{12}\big(A_{21}A_{31}A_{41}+K_2(A_{22}A_{31}+A_{41}\\ &+A_{22}K_2-\Pi_2-2A_{31}\Pi_2-K_2\Pi_2)\big)\big],\end{aligned} \tag{6.42}$$

$$\begin{aligned}\alpha_3=\big[&A_{21}\big(-K_2(A_{12}A_{31}-A_{11}K_2+A_{12}K_2+A_{31}K_2)\\ &+A_{32}\big(A_{41}K_1+A_{11}K_2+K_2^2\big)\big)\end{aligned}$$

$$+ K_2\big(A_{32}(1 + A_{22}K_1 - 2K_2 - A_{11}K_2 + 2K_1\Pi_2)$$

$$+ K_2\big(1 + A_{12}A_{31} + A_{22}K_1 + A_{41}K_1 - 2K_2$$

$$- A_{11}K_2 + A_{12}K_2 + A_{31}K_2 + K_2^2 - 2K_1\Pi_2)\big)\big], \tag{6.43}$$

$$\alpha_4 = K_1(A_{21} - K_2)K_2(A_{32} + K_2). \tag{6.44}$$

The roots of the characteristic equation, (6.40), are expressed as

$$\lambda_1^2 = -\frac{\alpha_3}{3\alpha_4} + \frac{\sqrt[3]{2}(\alpha_3^2 - 3\alpha_2\alpha_4)}{3\alpha_4\alpha_5^{\frac{1}{3}}} + \frac{\alpha_5^{\frac{1}{3}}}{3\sqrt[3]{2}\alpha_4}, \tag{6.45}$$

$$\lambda_2^2 = -\frac{\alpha_3}{3\alpha_4} + \frac{(1 + i\sqrt{3})(\alpha_3^2 - 3\alpha_2\alpha_4)}{3\sqrt[3]{4}\alpha_4\alpha_5^{\frac{1}{3}}} - \frac{(1 + i\sqrt{3})\alpha_5^{\frac{1}{3}}}{6\sqrt[3]{2}\alpha_4}, \tag{6.46}$$

$$\lambda_3^2 = -\frac{\alpha_3}{3\alpha_4} + \frac{(1 - i\sqrt{3})(\alpha_3^2 - 3\alpha_2\alpha_4)}{3\sqrt[3]{4}\alpha_4\alpha_5^{\frac{1}{3}}} - \frac{(1 - i\sqrt{3})\alpha_5^{\frac{1}{3}}}{6\sqrt[3]{2}\alpha_4} \tag{6.47}$$

where

$$\alpha_5 = -2\alpha_3^3 + 9\alpha_2\alpha_3\alpha_4 - 27\alpha_1\alpha_4^2$$

$$+ \sqrt{4\left(-\alpha_3^2 + 3\alpha_2\alpha_4\right)^3 + \left(9\alpha_2\alpha_3\alpha_4 - 2\alpha_3^3 - 27\alpha_1\alpha_4^2\right)^2}. \tag{6.48}$$

Most previous studies of the wave-seabed interaction problem have been limited to the quasi-static behavior, mainly because the fully dynamic solution has complicated mathematical formulations. In this section, the analytical solution of full dynamic soil behavior will be presented.

6.3.2 A Seabed of Finite Thickness

Considering a porous seabed of finite thickness with non-linear wave loading, the soil and pore fluid displacements can be expended as;

$$\bar{U}_x = a_1 e^{\lambda_1 \bar{z}} + a_2 e^{-\lambda_1 \bar{z}} + a_3 e^{\lambda_2 \bar{z}} + a_4 e^{-\lambda_2 \bar{z}} + a_5 e^{\lambda_3 \bar{z}} + a_6 e^{-\lambda_3 \bar{z}}, \tag{6.49}$$

$$\bar{U}_z = a_1 b_1 e^{\lambda_1 \bar{z}} + a_2 b_2 e^{-\lambda_1 \bar{z}} + a_3 b_3 e^{\lambda_2 \bar{z}} + a_4 b_4 e^{-\lambda_2 \bar{z}} + a_5 b_5 e^{\lambda_3 \bar{z}} + a_6 b_6 e^{-\lambda_3 \bar{z}}, \tag{6.50}$$

$$\bar{W}_x = a_1 c_1 e^{\lambda_1 \bar{z}} + a_2 c_2 e^{-\lambda_1 \bar{z}} + a_3 c_3 e^{\lambda_2 \bar{z}} + a_4 c_4 e^{-\lambda_2 \bar{z}} + a_5 c_5 e^{\lambda_3 \bar{z}} + a_6 c_6 e^{-\lambda_3 \bar{z}}, \tag{6.51}$$

$$\bar{W}_z = a_1 d_1 e^{\lambda_1 \bar{z}} + a_2 d_2 e^{-\lambda_1 \bar{z}} + a_3 d_3 e^{\lambda_2 \bar{z}} + a_4 d_4 e^{-\lambda_2 \bar{z}} + a_5 d_5 e^{\lambda_3 \bar{z}} + a_6 d_6 e^{-\lambda_3 \bar{z}} \tag{6.52}$$

where the b_i, c_i and d_i coefficients ($i = 1, 3, 5$) can be expressed as

$$b_i = \begin{bmatrix} -im\lambda_i & im\lambda_i K_2 & \lambda_i^2 K_2 + m^2\beta\Pi_2 \\ -A_{31} & A_{32} & imK_2\lambda_i \\ -im\lambda_i K_2 & im\lambda_i K_2 & \lambda_i^2 K_2 + A_{41} \end{bmatrix} / \Delta_i, \tag{6.53}$$

$$c_i = \begin{bmatrix} A_{21}\lambda_i^2 + A_{22} & -im\lambda_i & \lambda_i^2 K_2 + m^2\beta\Pi_2 \\ im\lambda_i K_2 & -A_{31} & im\lambda_i K_2 \\ K_2\lambda_i + m^2\beta\Pi_2 & -im\lambda_i K_2 & \lambda_i^2 K_2 + A_{41} \end{bmatrix} / \Delta_i, \tag{6.54}$$

$$d_i = \begin{bmatrix} A_{21}\lambda_i^2 + A_{22} & im\lambda_i K_2 & -im\lambda_i \\ im\lambda_i K_2 & A_{32} & -A_{31} \\ K_2\lambda_i + m^2\beta\Pi_2 & im\lambda_i K_2 & -im\lambda_i K_2 \end{bmatrix} / \Delta_i \tag{6.55}$$

where

$$\Delta_i = \begin{bmatrix} A_{21}\lambda_i^2 + A_{22} & imK_2 & K_2\lambda_i^2 + m^2\beta\Pi_2 \\ im\lambda_i K_2 & A_{32} & im\lambda_i K_2 \\ K_2\lambda_i^2 + m^2\beta\Pi_2 & im\lambda_i K_2 & K_2\lambda_i^2 + A_{41} \end{bmatrix} \tag{6.56}$$

and the b_i, c_i, d_i coefficients ($i = 2, 4, 6$) can be written as

$$b_i = \begin{bmatrix} im\lambda_i & -im\lambda_i K_2 & \lambda_i^2 K_2 + m^2\beta\Pi_2 \\ -A_{31} & A_{32} & -imK_2\lambda_i \\ im\lambda_i K_2 & -im\lambda_i K_2 & \lambda_i^2 K_2 + A_{41} \end{bmatrix} / \Delta_i, \tag{6.57}$$

$$c_i = \begin{bmatrix} A_{21}\lambda_i^2 + A_{22} & im\lambda_i & \lambda_i^2 K_2 + m^2\beta\Pi_2 \\ -im\lambda_i K_2 & -A_{31} & -imK_2\lambda_i \\ K_2\lambda_i + m^2\beta\Pi_2 & im\lambda_i K_2 & \lambda_i^2 K_2 + A_{41} \end{bmatrix} / \Delta_i, \tag{6.58}$$

$$d_i = \begin{bmatrix} A_{21}\lambda_i^2 + A_{22} & -im\lambda_i K_2 & im\lambda_i \\ -im\lambda_i K_2 & A_{32} & -A_{31} \\ K_2\lambda_i + m^2\beta\Pi_2 & -im\lambda_i K_2 & im\lambda_i K_2 \end{bmatrix} / \Delta_i \tag{6.59}$$

where

$$\Delta_i = \begin{bmatrix} A_{21}\lambda_i^2 + A_{22} & -im\lambda_i K_2 & K_2\lambda_i^2 + m^2\beta\Pi_2 \\ -im\lambda_i K_2 & A_{32} & -im\lambda_i K_2 \\ K_2\lambda_i^2 + m^2\beta\Pi_2 & -im\lambda_i K_2 & K_2\lambda_i^2 + A_{41} \end{bmatrix}. \tag{6.60}$$

Then, the wave-induced pore pressure and effective normal stresses can be further expressed as,

$$\begin{aligned} p = & -\frac{K_f'}{n} k\big[(im + b_1\lambda_1 + imc_1 + d_1\lambda_1)a_1\, e^{\lambda_1 \bar{z}} \\ & + (im - b_2\lambda_1 + imc_2 - d_2\lambda_1)a_2\, e^{-\lambda_1 \bar{z}} + (im + b_3\lambda_2 + imc_3 + d_3\lambda_2)a_3\, e^{\lambda_2 \bar{z}} \\ & + (im - b_4\lambda_2 + imc_4 - d_4\lambda_2)a_4\, e^{-\lambda_2 \bar{z}} + (im + b_5\lambda_3 + imc_5 + d_5\lambda_3)a_5\, e^{\lambda_3 \bar{z}} \\ & + (im - b_6\lambda_3 + imc_6 - d_6\lambda_3)a_6\, e^{-\lambda_3 \bar{z}}\big], \end{aligned} \tag{6.61}$$

$$\sigma'_x = \frac{2Gk}{1-2\mu}\{[(1-\mu)im + \mu\lambda_1 b_1]a_1 e^{\lambda_1\bar{z}} + [(1-\mu)im - \mu\lambda_1 b_2]a_2 e^{-\lambda_1\bar{z}}$$
$$+ [(1-\mu)im + \mu\lambda_2 b_3]a_3 e^{\lambda_2\bar{z}} + [(1-\mu)im - \mu\lambda_2 b_4]a_4 e^{-\lambda_2\bar{z}}$$
$$+ [(1-\mu)im + \mu\lambda_3 b_5]a_5 e^{\lambda_3\bar{z}} + [(1-\mu)im - \mu\lambda_3 b_3]a_6 e^{-\lambda_3\bar{z}}\}, \quad (6.62)$$

$$\sigma'_z = \frac{2Gk}{1-2\mu}\{[im\mu + (1-\mu)\lambda_1 b_1]a_1 e^{\lambda_1\bar{z}} + [im\mu - (1-\mu)\lambda_1 b_2]a_2 e^{-\lambda_1\bar{z}}$$
$$+ [im\mu + (1-\mu)\lambda_2 b_3]a_3 e^{\lambda_2\bar{z}} + [im\mu - (1-\mu)\lambda_2 b_4]a_4 e^{-\lambda_2\bar{z}}$$
$$+ [im\mu + (1-\mu)\lambda_3 b_5]a_5 e^{\lambda_3\bar{z}} + [im\mu - (1-\mu)\lambda_3 b_6]a_6 e^{-\lambda_3\bar{z}}\}, \quad (6.63)$$

$$\tau_{xz} = Gk[(\lambda_1 + imb_1)a_1 e^{\lambda_1\bar{z}} - (\lambda_1 - imb_1)a_2 e^{-\lambda_1\bar{z}}$$
$$+ (\lambda_2 + imb_3)a_3 e^{\lambda_2\bar{z}} - (\lambda_2 - imb_4)a_4 e^{-\lambda_2\bar{z}}$$
$$+ (\lambda_3 + imb_5)a_5 e^{\lambda_3\bar{z}} - (\lambda_3 - imb_6)a_6 e^{-\lambda_3\bar{z}}]. \quad (6.64)$$

Substituting the above equations into the boundary conditions at the seabed surface (6.14) and (6.15), the wave-induced soil response parameters can be calculated.

6.3.3 A Seabed of Infinite Thickness

Considering the infinite thickness seabed with non-linear wave, which the bottom boundary condition ($z \to -\infty$), gives

$$a_2^\infty = a_4^\infty = a_6^\infty = 0 \quad (6.65)$$

from finite thickness solutions (6.49)–(6.52), which can be simplified as,

$$\bar{U}_x = a_1^\infty e^{\lambda_1\bar{z}} + a_3^\infty e^{\lambda_2\bar{z}} + a_5^\infty e^{\lambda_3\bar{z}}, \quad (6.66)$$

$$\bar{U}_z = a_1^\infty b_1^\infty e^{\lambda_1\bar{z}} + a_3^\infty b_3^\infty e^{\lambda_2\bar{z}} + a_5^\infty b_5^\infty e^{\lambda_3\bar{z}}, \quad (6.67)$$

$$\bar{W}_x = a_1^\infty c_1^\infty e^{\lambda_1\bar{z}} + a_3^\infty c_3^\infty e^{\lambda_2\bar{z}} + a_5^\infty c_5^\infty e^{\lambda_3\bar{z}}, \quad (6.68)$$

$$\bar{W}_z = a_1^\infty d_1^\infty e^{\lambda_1\bar{z}} + a_3^\infty d_3^\infty e^{\lambda_2\bar{z}} + a_5^\infty d_5^\infty e^{\lambda_3\bar{z}} \quad (6.69)$$

where b_i^∞, c_i^∞ and d_i^∞ coefficients can be expressed as

$$b_i^\infty = \begin{bmatrix} -im\lambda_i & im\lambda_i K_2 & \lambda_i^2 K_2 + m^2\beta\Pi_2 \\ -A_{31} & A_{32} & imK_2\lambda_i \\ -im\lambda_i K_2 & im\lambda_i K_2 & \lambda_i^2 K_2 + A_{41} \end{bmatrix} / \Delta_i^\infty, \quad (6.70)$$

$$c_i^\infty = \begin{bmatrix} A_{21}\lambda_i^2 + A_{22} & -im\lambda_i & \lambda_i^2 K_2 + m^2\beta\Pi_2 \\ im\lambda_i K_2 & -A_{31} & im\lambda_i K_2 \\ K_2\lambda_i + m^2\beta\Pi_2 & -im\lambda_i K_2 & \lambda_i^2 K_2 + A_{41} \end{bmatrix} / \Delta_i^\infty, \quad (6.71)$$

$$d_i^\infty = \begin{bmatrix} A_{21}\lambda_i^2 + A_{22} & im\lambda_i K_2 & -im\lambda_i \\ im\lambda_i K_2 & A_{32} & -A_{31} \\ K_2\lambda_i + m^2\beta\Pi_2 & im\lambda_i K_2 & -im\lambda_i K_2 \end{bmatrix} / \Delta_i^\infty \tag{6.72}$$

where,

$$\Delta_i^\infty = \begin{bmatrix} A_{21}\lambda_i^2 + A_{22} & imK_2 & K_2\lambda_i^2 + m^2\beta\Pi_2 \\ im\lambda_i K_2 & A_{32} & im\lambda_i K_2 \\ K_2\lambda_i^2 + m^2\beta\Pi_2 & im\lambda_i K_2 & K_2\lambda_i^2 + A_{41} \end{bmatrix}. \tag{6.73}$$

Then, the wave-induced pore pressure and stresses can be furthers expressed as,

$$\begin{aligned} p^\infty = -\frac{K_f'}{n}k\big[&\big(im + b_1^\infty\lambda_1 + imc_1^\infty + d_1^\infty\lambda_1\big)a_1^\infty e^{\lambda_1\bar{z}} \\ &+ \big(im + b_3^\infty\lambda_2 + imc_3^\infty + d_3^\infty\lambda_2\big)a_3^\infty e^{\lambda_2\bar{z}} \\ &+ \big(im + b_5^\infty\lambda_3 + imc_5^\infty + d_5^\infty\lambda_3\big)a_5^\infty e^{\lambda_3\bar{z}}\big], \end{aligned} \tag{6.74}$$

$$\begin{aligned} \sigma_x'^\infty = \frac{2Gk}{1-2\mu}\big\{&\big[(1-\mu)im + \mu\lambda_1 b_1^\infty\big]a_1^\infty e^{\lambda_1\bar{z}} \\ &+ \big[(1-\mu)im + \mu\lambda_2 b_3^\infty\big]a_3^\infty e^{\lambda_2\bar{z}} + \big[(1-\mu)im + \mu\lambda_5 b_3^\infty\big]a_5^\infty e^{\lambda_3\bar{z}}\big\}, \end{aligned} \tag{6.75}$$

$$\begin{aligned} \sigma_z'^\infty = \frac{2Gk}{1-2\mu}\big\{&\big[im\mu + (1-\mu)\lambda_1 b_1^\infty\big]a_1^\infty e^{\lambda_1\bar{z}} \\ &+ \big[im\mu + (1-\mu)\lambda_2 b_3^\infty\big]a_3^\infty e^{\lambda_2\bar{z}} + \big[im\mu + (1-\mu)\lambda_5 b_5^\infty\big]a_5^\infty e^{\lambda_3\bar{z}}\big\}, \end{aligned} \tag{6.76}$$

$$\tau_{xz}^\infty = Gk\big\{\big(\lambda_1 + imb_1^\infty\big)a_1^\infty e^{\lambda_1\bar{z}} + \big(\lambda_2 + imb_3^\infty\big)a_3^\infty e^{\lambda_2\bar{z}} + \big(\lambda_3 + imb_5^\infty\big)a_5 e^{\lambda_3\bar{z}}\big\}. \tag{6.77}$$

The solution given in this section was included in Jeng and Cha [5].

6.4 Simplified Solution

6.4.1 u–p Approximation

By neglecting the terms of involving the second time derivatives of the relative fluid displacement, a solution of u–p approximation can be obtained. Therefore, if $\ddot{w}_i$ is ignored, re-organizing the governing equation, will render to

$$\sigma_{ij,j}^{up\prime} = \delta_{ij} p,_i{}^{up} + \rho\ddot{u}_i^{up}, \tag{6.78}$$

$$-p,_i{}^{up} = \rho_f \ddot{u}^{up} + \frac{\rho_f g}{k_z}\dot{w}_i^{up}, \tag{6.79}$$

$$\dot{\varepsilon}_{ii}^{up} + \dot{w}_{ii}^{up} = -\frac{n}{K_f'}\dot{p}^{up} \tag{6.80}$$

where the superscript "up" denotes the u–p approximation.

The u–p approximation can be solved in a similar way to the fully dynamic solution, that is, the coupled equation will be reduced to

$$\begin{bmatrix} K_1 D^2 + A_{11}^{up} & imD & -m^2 K_2 & imK_2 D \\ imD & A_{21}^{up} D^2 + A_{22}^{up} & imK_2 D & K_2 D^2 \\ A_{31}^{up} & imK_2 D & A_{32}^{up} & imK_2 D \\ imK_2 D & K_2 D^2 + A_{41}^{up} & imK_2 D & K_2 D^2 + A_{42}^{up} \end{bmatrix} \begin{bmatrix} \bar{U}_x^{up} \\ \bar{U}_z^{up} \\ \bar{W}_x^{up} \\ \bar{W}_z^{up} \end{bmatrix} = \begin{bmatrix} 0 \\ 0 \\ 0 \\ 0 \end{bmatrix} \tag{6.81}$$

where

$$A_{11}^{up} = m^2\left[\Pi_2 - \frac{2(1-\mu)}{1-2\mu}K_1 - K_2\right], \tag{6.82}$$

$$A_{21}^{up} = \frac{2(1-\mu)}{1-2\mu}K_1 - K_2, \qquad A_{22}^{up} = \Pi_2 - K_1, \tag{6.83}$$

$$A_{31}^{up} = \beta\Pi_2 - K_2, \qquad A_{32}^{up} = \frac{im}{\Pi_1} - m^2 K_2, \tag{6.84}$$

$$A_{41}^{up} = \beta m^2 \Pi_2, \qquad A_{42}^{up} = \frac{im}{\Pi_1}. \tag{6.85}$$

Similarly, the characteristic equation derived from (6.81) is identical to (6.40). Thus, we will have the same roots for the characteristics equation as (6.45)–(6.47). The differences between the full dynamic solution and the u–p approximation are the α_i coefficients, which are given by

$$\alpha_1^{up} = \left(A_{11}^{up} A_{22}^{up} A_{32}^{up} A_{42}^{up} + A_{22}^{up} A_{31}^{up} A_{42}^{up} K_2\right), \tag{6.86}$$

$$\begin{aligned} \alpha_2^{up} = {} & \big[A_{32}^{up}\big(A_{42}^{up}\big(1 + A_{11}^{up} A_{21}^{up} + A_{22}^{up} K_1\big) + K_2\big(A_{11}^{up}\big(A_{22}^{up} - A_{41}^{up}\big) \\ & - A_{41}^{up} + A_{22}^{up} K_2\big)\big) + K_2\big(-A_{41}^{up} + A_{42}^{up} \\ & + A_{11}^{up}\big(A_{22}^{up} - A_{41}^{up} + A_{42}^{up}\big) + A_{22}^{up} K_2\big)\big], \end{aligned} \tag{6.87}$$

$$\begin{aligned} \alpha_3^{up} = {} & -\big[K_2\big(A_{32}^{up}\big(-1 - A_{22}^{up} K_1 + A_{41}^{up} K_1 + 2K_2 + A_{11}^{up} K_2 + D^2 K_1 K_2\big) \\ & + K_2\big(-1 - A_{22}^{up} K_1 + A_{41}^{up} K_1 + A_{41}^{up} K_1 \\ & - A_{42}^{up} K_1 + 2K_2 + A_{11}^{up} K_2 + D^2 K_1 K_2\big)\big) \\ & + A_{21}^{up}\big(K_2^2\big(A_{11}^{up} + D^2 K_1 + K_2\big) \\ & + A_{32}^{up}\big(A_{42}^{up} K_1 + A_{11}^{up} K_2 + D^2 K_1 K_2 + K_2^2\big)\big)\big], \end{aligned} \tag{6.88}$$

$$\alpha_4^{up} = K_1\big(A_{21}^{up} - K_2\big)K_2\big(A_{32}^{up} + K_2\big). \tag{6.89}$$

The expressions of coefficients b_i^{up}, c_i^{up} and d_i^{up} for the u–p approximation are identical to the dynamic solution with different α_i coefficients. This solution with infinite thickness under linear wave loading is identical to the solution proposed by Jeng *et al.* [8]. An alternative derivation but with the same results for a seabed of finite thickness can be found in Jeng and Rahman [7].

6.4.2 Quasi-Static Approximation

Further simplifying from u–p approximation, neglecting the terms of involving the second time derivatives of the relative solid displacement, the solution to the quasi-static model can be obtained. Hence, if $\ddot{u}_i$ and $\ddot{w}_i$ are ignored, we have a more comprehensible governing equation than the u–p approximation model.

$$\sigma_{ij,j}^{st\prime} = \delta_{ij} p,_i{}^{st}, \tag{6.90}$$

$$-p,_i{}^{up} = \frac{\rho_f g}{k_z} \dot{w}_i^{st}, \tag{6.91}$$

$$\dot{\varepsilon}_{ii}^{st} + \dot{w}_{ii}^{st} = -\frac{n}{K_f'} \dot{p}^{st} \tag{6.92}$$

where the superscript "st" denotes the quasi-static case.

Substituting (6.91) into (6.92), we have

$$\frac{k_z}{\rho_f g} p_{,ii}^{st} - \frac{n}{K_f'} \dot{p}^{st} = \dot{\varepsilon}_{ii}^{st}. \tag{6.93}$$

From the above governing equations (6.90) and (6.93), we can obtain the wave-induced pore pressure, soil and fluid displacement as,

$$\begin{bmatrix} p \\ u \\ w \end{bmatrix} = \sum_{m=1}^{2} \begin{bmatrix} P_{1(z)} \\ U_{1(z)} \\ W_{1(z)} \end{bmatrix} e^{im(kx-\omega t)}. \tag{6.94}$$

The right hand side, with subscript 1, of (6.94) represents the incident component. The solution of P_1, U_1 and W_1 will be expressed as

$$\left(D^2 - k_1^2\right)\left(U_1' - ik_1 W_1\right) = 0, \tag{6.95}$$

$$G\left(D^2 - k_1^2\right)U_1 + \frac{ik_1 G}{1-2\mu}(ik_1 U_1 + W_1) = ik_1 P_1, \tag{6.96}$$

$$\left(D^2 - k_1^2 + \frac{i\omega_1 \gamma_w n_e \beta}{K_z}\right) P_1 = -\frac{i\omega_1 \gamma_w}{K_z}\left(ik_1 U_1 + W_1'\right) \tag{6.97}$$

where $D(=d/dz)$, $k_1 = mk$, $\omega_1 = m\omega$.

From the above equations, the characteristic equation can be expressed as

$$\left(D^2 - k^2\right)^2 \left(D^2 - \delta^2\right) U_x = 0 \tag{6.98}$$

where

$$\delta^2 = k_1^2 - \frac{im\omega\gamma_w}{k_z}\left[\frac{n}{K_f'} + \frac{1-2\mu}{2G(1-\mu)}\right]. \tag{6.99}$$

If we consider the linear wave loading (i.e., $m = 1$), the above solution is identical to the previous solution presented in Chap. 3.

6.5 Numerical Results and Discussions

To have a basic understanding of the mechanism of the wave-seabed interaction, a parametric study will be performed in this chapter. As observed in the analytical solution presented in Chap. 3, two groups of parameters are involved in the whole wave-seabed interaction problem: they are:

- Soil parameters: the degree of saturation (S_r), soil permeability (k_z), Poisson' ratio (μ), soil porosity (n) and seabed thickness (h). Among these, the degree of saturation and soil permeability are the two most dominant factors in the analysis of the wave-seabed interaction, while the other soil parameters nearly remain as constant values for most marine sediments, at least from the quasi-static solution [4].
- Wave parameters: wave period (T), water depth (d) and wave height(H), which lead to the wavelength through the wave dispersion equation. Since previous studies have been limited to the linear wave condition, we will focus on the non-linear wave condition including the effects of wave height.

In this chapter, the influences of the above parameters on wave-induced soil response will be investigated through a parametric study.

6.5.1 *Effects of Dynamic Soil Behavior*

Two cases are considered to investigate the influence of dynamic soil behavior on wave-induced soil response. The wave and soil characteristics for the case study are tabulated in Table 6.1.

The vertical distributions of the wave-induced pore pressure ($|p|/p_o$), effective stresses ($|\sigma_z'|/p_o$), shear stress ($|\tau_{xz}|/p_o$) and soil displacements versus the soil depth (z/H) in a porous seabed are presented in Fig. 6.2. Here, p_o ($= \frac{\gamma_w H}{2\cosh kd}$) is the amplitude of linear wave. In the figures, the solid lines represent the present solution with dynamic soil behavior, while the dotted lines denote the previous solution with

Table 6.1 Input data for demonstration of effects of dynamic soil behavior

Parameters	Case 1	Case 2
Wave period T	4.0 s	10.0 s
Water depth d	50 m	20 m
Wave height H/L	0.125	0.125
Degree of saturation S_r	0.975	0.975
Seabed thickness h/L	0.25	0.25
Poisson's ratio μ	0.4	0.4
Soil porosity n	0.35	0.35
Soil permeability k_z	10^{-2} m/s	10^{-2} m/s
Dynamic parameters		
K_1	0.062119	0.10565
K_2	0.6894	0.47175
V_c	197.610	151.631
Π_1	3.3085	0.20644
Π_2	0.0010	0.00643

quasi-static soil behavior [4]. The results from u–p approximation are identical to the dynamic soil solution in this example. The figures clearly indicate that the influence of dynamic soil behavior is insignificant for the wave-induced pore pressure for both cases. However, the difference between the dynamic and quasi-static solutions in other soil response parameters, such as effective stresses and shear stress, are observable for Case 2, while it is insignificant for Case 1. Thus, it can be concluded that dynamic soil behavior is only important in the wave-induced seabed response for certain combinations of wave and soil characteristics. The applicable ranges of dynamic soil behavior, u–p approximation and quasi-static solutions, will be discussed in detail in Sect. 6.4.

6.5.2 *Effects of Soil Characteristics*

In this section, the effects of several important soil parameters, including the degree of saturation, soil depth and soil permeability, on the wave-induced pore pressure, effective stresses and shear stresses in a porous seabed will be investigated. Three solutions with dynamic, u–p approximation and quasi-static soil behavior will be included in the discussions performed in this section. The input data of the wave and soil characteristics for the following numerical examples are given in Table 6.2.

Generally speaking, it is common to observe gas in marine sediments [11]. It has been reported that the degree of saturation is an important factor in the evaluation of the wave induced seabed response [4, 11]. In this sub-section, the effects of degree of saturation on the pore pressure, effective stress, and shear stress are examined.

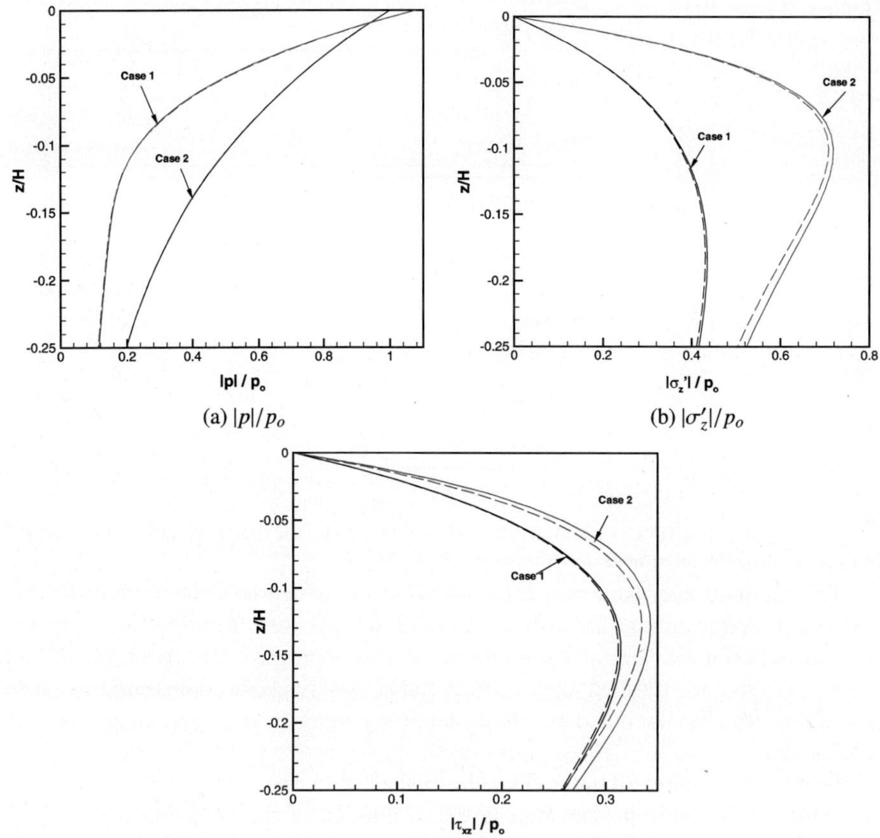

(a) $|p|/p_o$ (b) $|\sigma_z'|/p_o$

(c) $|\tau_{xz}|/p_o$

Fig. 6.2 Vertical distributions of the wave-induced (**a**) pore pressure ($|p|/p_o$), (**b**) vertical effective normal stresses ($|\sigma_z'|/p_o$) and (**c**) shear stress ($|\tau_{xz}|/p_o$) versus the soil depth (z/h) for Case 1 and Case 2. *Solid lines* are dynamic solution and *dashed lines* are quasi-static solution

As shown in Fig. 6.3, the degree of saturation significantly affects the wave-induced pore pressure. The influence of dynamic soil behavior increases as the degree of saturation increases. Additionally, the distribution of pore pressure in coarse sand ($k_z = 10^{-2}$ m/s) changes smoothly, while it changes dramatically near the seabed surface in fine sand ($k_z = 10^{-4}$ m/s). However, the influence of S_r on the relative difference of the wave-induced pore pressure among with dynamic, u–p approximation and quasi-static solutions are insignificant, except where $S_r = 1$ in coarse sand ($k_z = 10^{-2}$ m/s).

Figure 6.4 illustrates a comparison of dynamic, u–p approximation and quasi-static analytical solutions on the vertical effective normal stress for various values of degree of saturation with different soil types. As seen in the figure, the degree of saturation significantly affects the effective stress in the seabed with both soil types. Its influence increases as the degree of saturation decreases. It is observed in

Table 6.2 Input data for numerical examples

Wave characteristics	
Wave period T	10 s or various
Wave height H/L	0.125 or various
Water depth d	20 m or various
Soil characteristics	
Degree of saturation S_r	0.95 or various
Seabed thickness h/L	0.5 or various
Porosity n	0.35
Poisson ratio μ	0.4
Permeability k_z	10^{-2} m/s
	10^{-4} m/s
Shear modulus G	5.0×10^6 N/m^2

Fig. 6.4 that the effects of dynamic soil behavior on the vertical effective normal stress is more important than that on pore pressure.

The distribution of shear stress for various values of degree of saturation with different soil permeabilities are presented in Fig. 6.5. As seen from Fig. 6.5a, the difference between the dynamic (solid line), the u–p approximation (dash-dotted line) and the quasi-static (dashed line) solution can be observed. By comparing Figs. 6.5a and 6.5b, it can be concluded that the dynamic soil behavior is important in saturated coarse sand.

Besides the degree of saturation, the seabed thickness is another soil parameter which plays an important role in the evaluation of the wave-induced soil re-

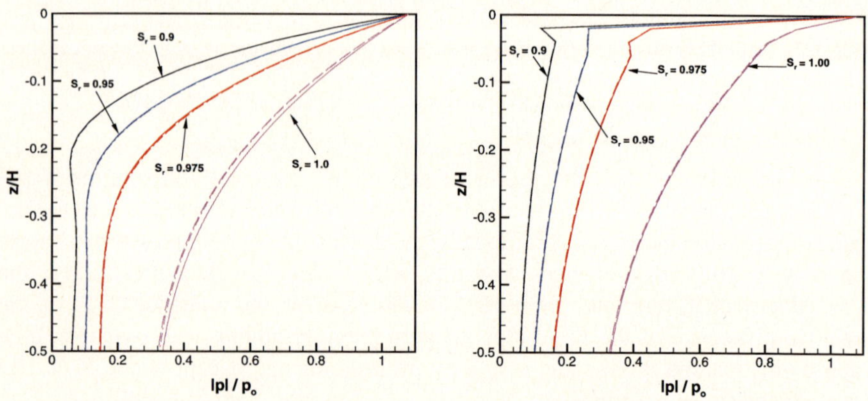

(a) Coarse sand ($k_z = 10^{-2}$ m/sec) (b) Fine sand ($k_z = 10^{-4}$ m/sec)

Fig. 6.3 Vertical distributions of the wave-induced pore pressure ($|p|/p_o$) versus the soil depth for various values of degree of saturation (S_r) in a seabed with different permeability, (**a**) coarse sand ($k_z = 10^{-2}$ m/s) and (**b**) fine sand ($k_z = 10^{-4}$ m/s). *Solid lines* are dynamic solution, *dash-dotted lines* are u–p approximation and *dashed lines* are quasi-static solution

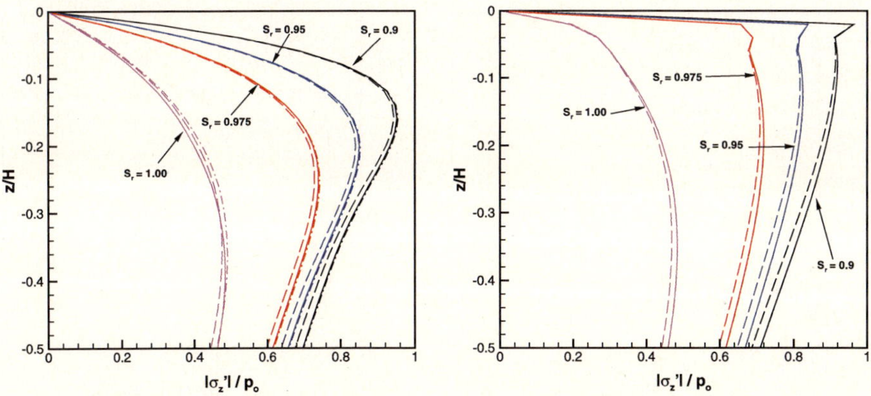

(a) Coarse sand ($k_z = 10^{-2}$ m/sec) (b) Fine sand ($k_z = 10^{-4}$ m/sec)

Fig. 6.4 Vertical distributions of the vertical effective normal stresses ($|\sigma_z'|/p_o$) versus the soil depth for various values of degree of saturation (S_r) in a seabed with different permeability (**a**) coarse sand ($k_z = 10^{-2}$ m/s) and (**b**) fine sand ($k_z = 10^{-4}$ m/s). *Solid lines* are dynamic solution, *dash-dotted lines* are u–p approximation and *dashed lines* are quasi-static solution

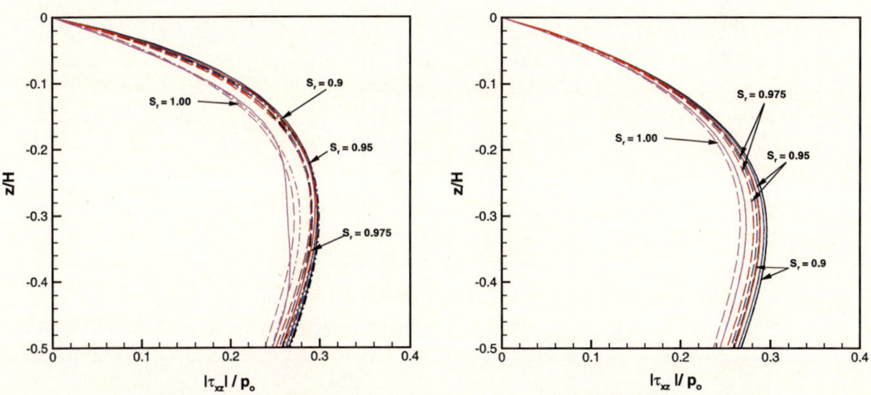

(a) Coarse sand ($k_z = 10^{-2}$ m/sec) (b) Fine sand ($k_z = 10^{-4}$ m/sec)

Fig. 6.5 Vertical distributions of the shear stresses ($|\tau_{xz}|/p_o$) versus the soil depth for various values of degree of saturation (S_r) in a seabed with different permeability, (**a**) coarse sand ($k_z = 10^{-2}$ m/s) and (**b**) fine sand ($k_z = 10^{-4}$ m/s). *Solid lines* are dynamic solution, *dash-dotted lines* are u–p approximation and *dashed lines* are quasi-static solution

sponse [4]. In this sub-section, the effects of various seabed thickness on the pore pressure, effective stress and shear stress are examined.

As shown in Fig. 6.6, the seabed thickness significantly affects the wave-induced pore pressure. Its influence increases as seabed thickness decreases. However, the influence of seabed thickness on the relative difference of the wave-induced pore pressure between dynamic, u–p approximation and quasi-static solutions are not

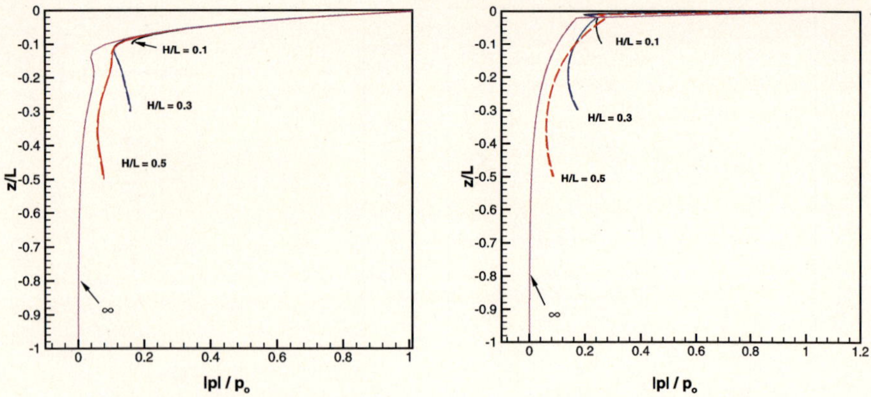

(a) Coarse sand ($k_z = 10^{-2}$ m/sec) (b) Fine sand ($k_z = 10^{-4}$ m/sec)

Fig. 6.6 Vertical distributions of the wave-induced pore pressure ($|p|/p_o$) versus the soil depth for various values of seabed thickness(h/L) in a seabed with different permeability, (**a**) coarse sand ($k_z = 10^{-2}$ m/s) and (**b**) fine sand ($k_z = 10^{-4}$ m/s). *Solid lines* are dynamic solution, *dash-doted lines* are u–p approximation and *dashed lines* are quasi-static solution

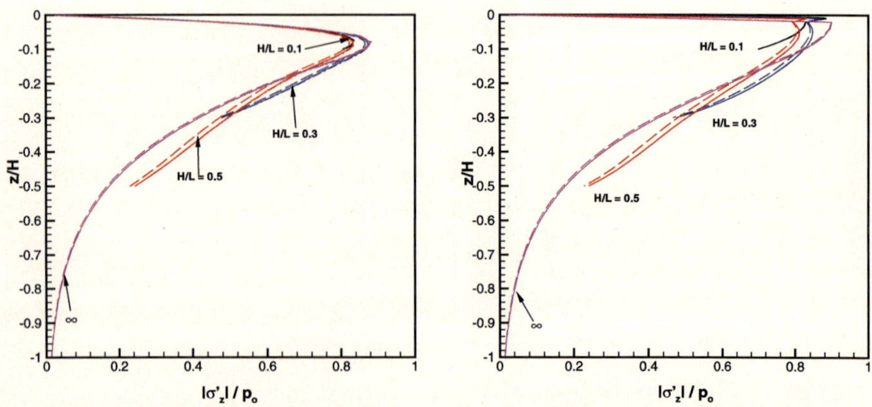

(a) Coarse sand ($k_z = 10^{-2}$ m/sec) (b) Fine sand ($k_z = 10^{-4}$ m/sec)

Fig. 6.7 Vertical distribution of vertical effective normal stresses ($|\sigma_z'|/p_o$) versus the soil depth for various values of seabed thickness (h/L) in a seabed with different permeability, (**a**) coarse sand ($k_z = 10^{-2}$ m/s) and (**b**) fine sand ($k_z = 10^{-4}$ m/s). *Solid lines* are dynamic solution, *dash-dotted lines* are u–p approximation and *dashed lines* are quasi-static solution

significant. The figure also shows that the distribution of pore pressure is different between coarse sand ($k_z = 10^{-2}$ m/s) and fine sand ($k_z = 10^{-4}$ m/s).

The distribution of vertical effective stresses for various seabed thickness with different soil types are presented in Fig. 6.7. The figure demonstrates the influences of seabed thickness on the vertical effective stress in the both soil types. The distribution of vertical effective stress in the fine sand ($k_z = 10^{-4}$ m/s) changes dramatically in near seabed surfaces, while it changes smoothly in coarse sand ($k_z = 10^{-2}$ m/s). The results of the dynamic (solid line) and the u–p approximation (dash-dotted line) are identical, but there are differences between the dynamic and quasi-static (dashed line) solutions. Similar trends have been found for the wave-induced shear stresses (graphs not shown here).

In the numerical example presented in Figs. 6.3–6.7, two different materials with different permeability, coarse sand ($k_z = 10^{-2}$ m/s) and fine sand ($k_z = 10^{-4}$ m/s), are considered. As shown in the figures, the influence of soil behavior on the wave-induced soil response in coarse sand are more significant than that in fine sand. This is because the pore fluid transfers faster and the acceleration due to solid and pore fluid is more important in coarse sand.

6.5.3 *Effects of Wave Characteristics*

Wave period, wave steepness and water depth are important wave parameters, which directly affect wavelength and other wave characteristics. Thus, it is interesting to examine the effects of wave period, wave steepness and water depth on wave-induced seabed response. Figures 6.8–6.10 present the vertical distribution of the wave-induced pore pressure, effective normal stress and shear stress for various values of the wave period (T) in coarse sand and fine sand. In general, the maximum amplitude of the wave-induced pore pressure increases as the wave period increases, as shown in Fig. 6.8. However, the wave period does not affect the relative difference of the wave induced pore pressure between dynamic and quasi-static soil behavior (at least from the examples presented in Fig. 6.8).

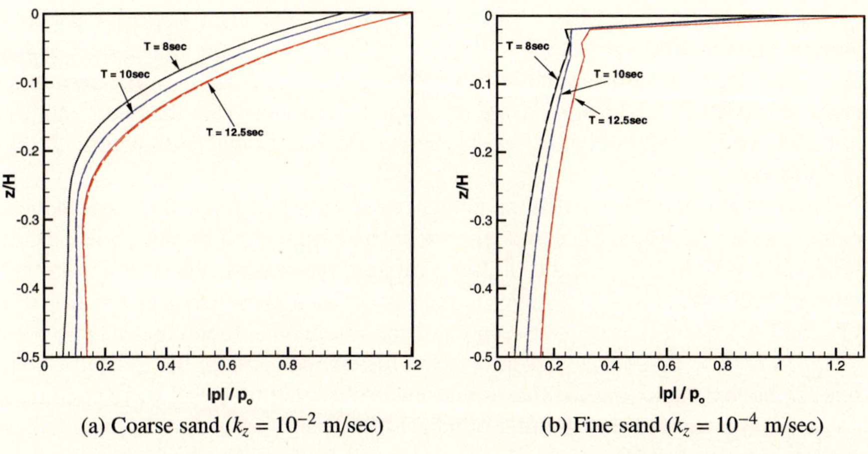

Fig. 6.8 Vertical distributions of the wave-induced pore pressure ($|p|/p_o$) versus the soil depth for various values of wave period (T) in seabeds with different permeability (**a**) coarse sand ($k_z = 10^{-2}$ m/s) and (**b**) fine sand ($k_z = 10^{-4}$ m/s). *Solid lines* are dynamic solution, *dash-dotted lines* are u–p approximation and *dashed lines* are quasi-static solution

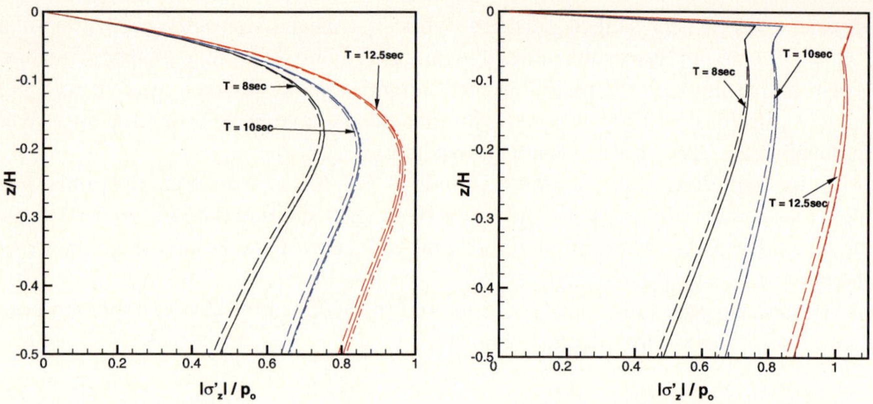

(a) Coarse sand ($k_z = 10^{-2}$ m/sec) (b) Fine sand ($k_z = 10^{-4}$ m/sec)

Fig. 6.9 Vertical distributions of the effective stresses ($|\sigma_z'|/p_o$) versus the soil depth for various values of wave period (T) in seabeds with different permeability (**a**) coarse sand ($k_z = 10^{-2}$ m/s) and (**b**) fine sand ($k_z = 10^{-4}$ m/s). *Solid lines* are dynamic solution, *dash-dotted lines* are u–p approximation and *dashed lines* are quasi-static solution

Figure 6.9 illustrates the vertical effective stresses with various wave periods in different soil types. Compared with the pore pressure, significant effects of the wave period on the effective stress are observed. For example, in the case of a wave period of 12.5 s in coarse sand ($k_z = 10^{-2}$ m/s), the differences are clearly shown with the dynamic (solid line), u–p approximation (dash-dotted line) and quasi-static (dashed line) solutions. However, in fine sand, the differences between the dynamic and quasi-static solutions are only observed. This figure concludes that the difference between full-dynamic and u–p approximation may be ignored.

Figure 6.10 illustrates the vertical distribution of the shear stress for different wave periods in different types of soil. As seen, the figures in the coarse sand seabed ($k_z = 10^{-2}$ m/s), the magnitude of the shear stress increases as wave period decreases. It is also found that there are significant differences with the three solutions at 12.5 s, while there is only a difference between dynamic and quasi-static solutions in fine sand.

Numerical studies found that the influences of wave heights on the wave-induced pore pressure between three models are insignificant (graphs not shown here). Compared with the wave-induced pore pressure, the wave height has a more significant influence on the effective normal stress in both soil types. As shown in Fig. 6.11 (a), the difference between the three solutions is found under large wave loading ($H/L = 0.125$). However, only the difference between the dynamic (solid line) and quasi-static (dashed line) solutions is observed for small wave conditions ($H/L = 0.05$ and 0.0725). Similar trends can be found in the distribution of shear stresses (see Fig. 6.12).

Similar to the cases with various wave periods, the influence of water depth on the pore pressure calculated from three models are insignificant (graphs not shown). Figure 6.13 presents the vertical distribution of effective stress for various water

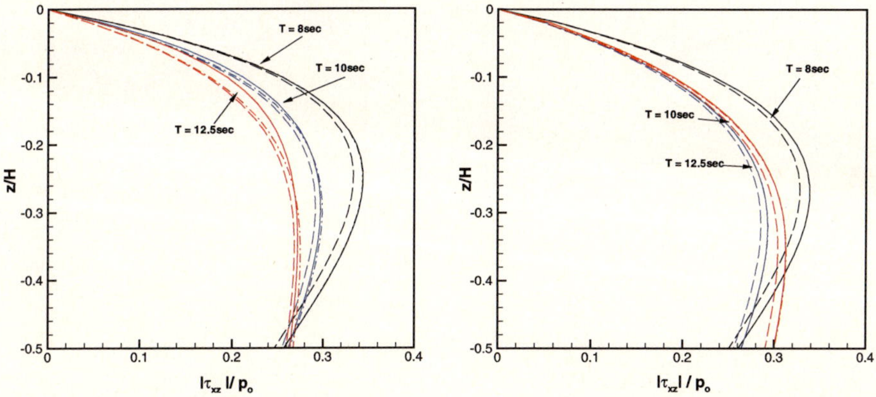

(a) Coarse sand ($k_z = 10^{-2}$ m/sec) (b) Fine sand ($k_z = 10^{-4}$ m/sec)

Fig. 6.10 Vertical distributions of the shear stresses ($|\tau_{xz}|/p_o$) versus the soil depth for various values of wave period (T) in seabeds with different permeability (**a**) coarse sand ($k_z = 10^{-2}$ m/s) and (**b**) fine sand ($k_z = 10^{-4}$ m/s). *Solid lines* are dynamic solution, *dash-dotted lines* are *u–p* approximation and *dashed lines* are quasi-static solution

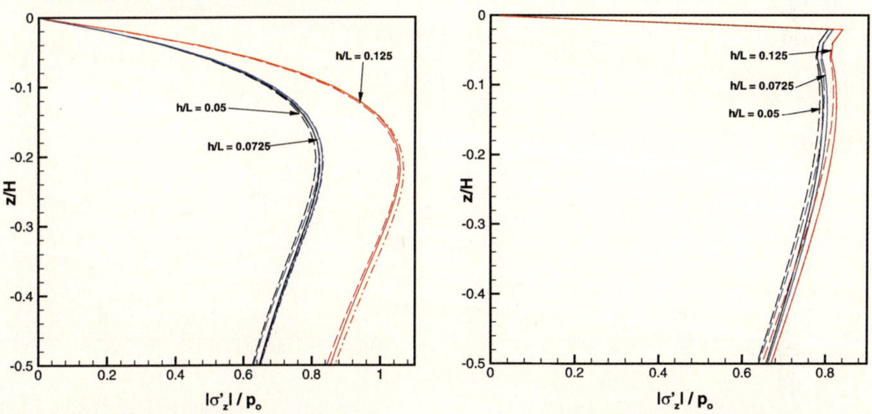

(a) Coarse sand ($k_z = 10^{-2}$ m/sec) (b) Fine sand ($k_z = 10^{-4}$ m/sec)

Fig. 6.11 Vertical distributions of the effective stresses ($|\sigma_z'|/p_o$) versus the soil depth for various values of wave height (H/L) in seabeds with different permeability (**a**) coarse sand ($k_z = 10^{-2}$ m/s) and (**b**) fine sand ($k_z = 10^{-4}$ m/s). *Solid lines* are dynamic solution, *dash-dotted lines* are *u–p* approximation and *dashed lines* are quasi-static solution

depth in seabeds with different soil types. The figure clearly indicates that water depth significantly affects the vertical effective normal stress. The effective stresses increase while water depth decreases. It is also observed that slight differences between three models can be observed in the case of 20 m water depth, while only differences between dynamic and quasi-static solution on are observed.

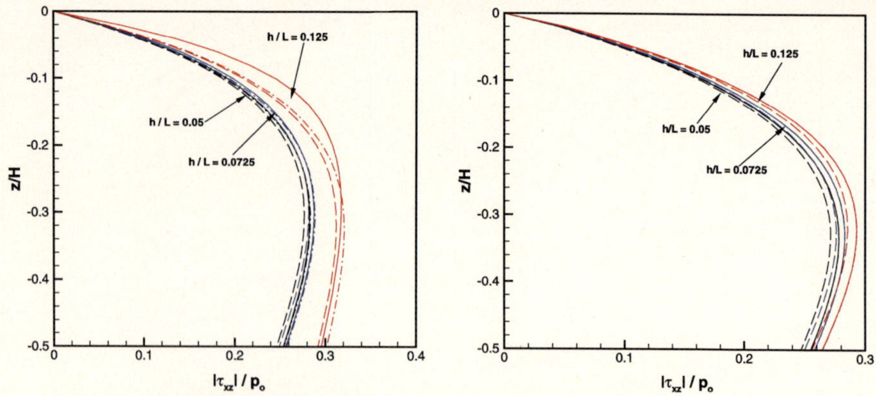

(a) Coarse sand ($k_z = 10^{-2}$ m/sec) (b) Fine sand ($k_z = 10^{-4}$ m/sec)

Fig. 6.12 Vertical distributions of the shear stresses ($|\tau_{xz}|/p_o$) versus the soil depth for various values of wave period (H/L) in seabeds with different permeability (**a**) coarse sand ($k_z = 10^{-2}$ m/s) and (**b**) fine sand ($k_z = 10^{-4}$ m/s). *Solid lines* are dynamic solution, *dash-dotted lines* are u–p approximation and *dashed lines* are quasi-static solution

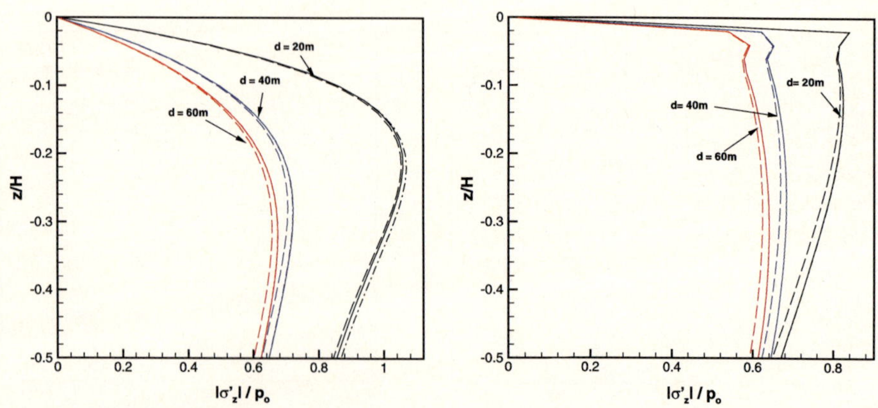

(a) Coarse sand ($k_z = 10^{-2}$ m/sec) (b) Fine sand ($k_z = 10^{-4}$ m/sec)

Fig. 6.13 Vertical distributions of the effective stresses ($|\sigma'_z|/p_o$) versus the soil depth for various values of water depth (d) in seabeds with different permeability (**a**) coarse sand ($k_z = 10^{-2}$ m/s) and (**b**) fine sand ($k_z = 10^{-4}$ m/s). *Solid lines* are dynamic solution, *dash-dotted lines* are u–p approximation and *dashed lines* are quasi-static solution

The vertical distribution of shear stress for various water depth as different soil types are presented in Fig. 6.14. Again, the differences between full dynamic and u–p approximation can only be observed in the water depth of 20 m.

Based on the above numerical examples, it can be concluded that the dynamic soil behavior needs to be considered in some combination of wave and soil parameters such as coarse sand ($k_z = 10^{-2}$ m/s) with shallow water.

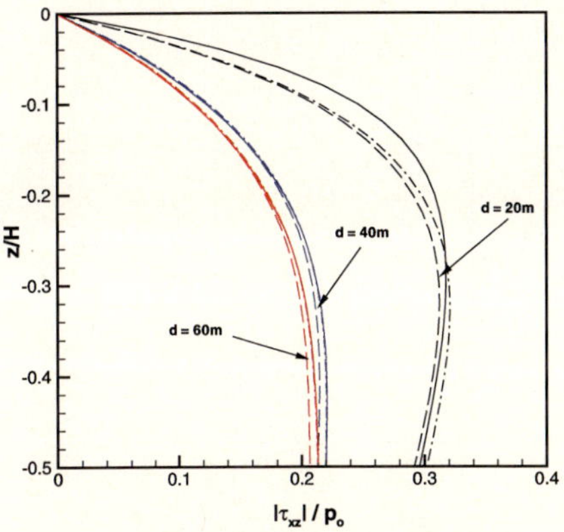

(a) Coarse sand ($k_z = 10^{-2}$ m/sec)

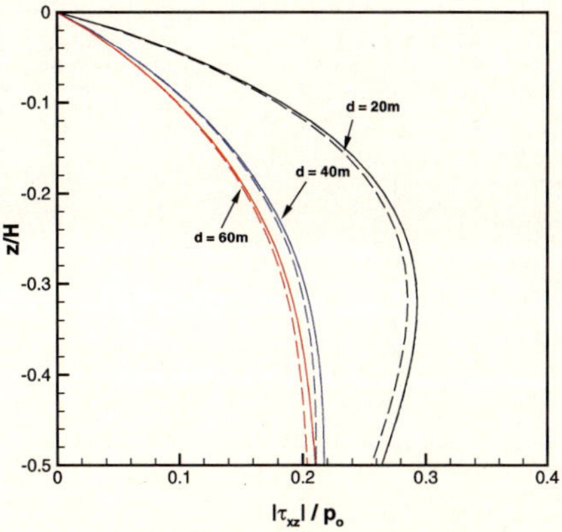

(b) Fine sand ($k_z = 10^{-4}$ m/sec)

Fig. 6.14 Vertical distributions of the shear stresses ($|\tau_{xz}|/p_o$) versus the soil depth for various values of water depth (d) in seabeds with different permeability (**a**) coarse sand ($k_z = 10^{-2}$ m/s) and (**b**) fine sand ($k_z = 10^{-4}$ m/s). *Solid lines* are dynamic solution, *dash-dotted lines* are u–p approximation and *dashed lines* are quasi-static solution

6.6 When Should Dynamic Soil Behavior Be Considered?

As demonstrated in previous numerical examples, it can be concluded that the effects of dynamic soil behavior on the wave-induced soil response needs to be considered under certain combinations of wave and soil parameters. Thus, an interesting question arises: "When should we consider dynamic soil behavior?". It is a particularly important question for practical engineers to know when the conventional quasi-static solution (u–p approximation solution is only considered in some special cases, such as in shallow break water conditions) is applicable.

As shown in the analytical solution derived in Sect. 6.3, most wave and soil characteristics can be represented in two non-dimensional parameters, Π_1 and Π_2. Thus, these two non-dimensional parameters will be a more useful way to investigate the applicable range of the quasi-static solution. Since the results of u–p approximation are almost identified to dynamic solutions for most cases, we only compared, dynamic and quasi-static soil behavior. This approach was first suggested by Zienkiewicz *et al.* [14] with a one-dimensional analysis.

Figure 6.15a illustrates the boundary between the dynamic and the quasi-static solution, with a 3 % difference in terms of non-dimensional parameters Π_1 and Π_2 for various soil permeability. To simplify the problem, linear wave loading is considered here. The input data are given in Table 6.3. Herein, we use Cases 1–4 as an example. In the figure, point "$\star$" is for Case 1, point "$\diamond$" is indicates Case 2, point "Δ" indicates Case 3, and point "$\circ$" is for Case 4. Note that the soil permeability for all cases are 10^{-2} m/s, as mentioned previously. For the point with given Π_1 and Π_2, located below the curve in Fig. 6.15a, the dynamic soil behavior must be considered, if 3 % of relative difference is acceptable (for example, Cases 2 and 4).

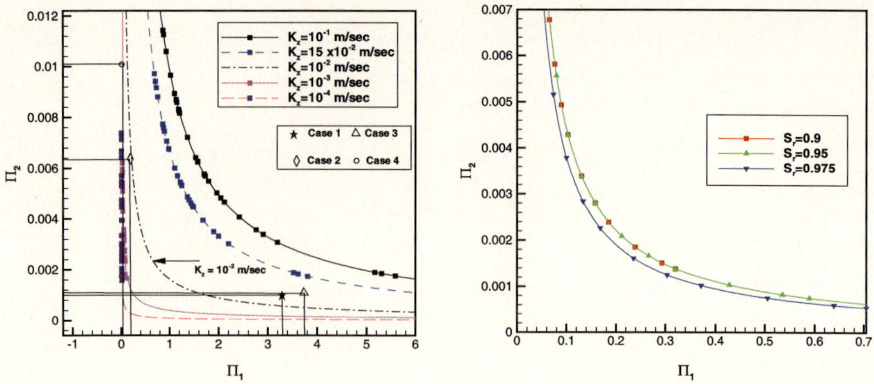

(a) permeability (b) degree of saturation

Fig. 6.15 The relationship between Π_1 and Π_2 for various (**a**) soil permeability and (**b**) degree of saturation

Table 6.3 Input data for demonstration of relationship Π_1 and Π_2 in linear wave loading

Parameters	Case 3	Case 4
Wave period T	3.0 s	10.0 s
Water depth d	50 m	30 m
Degree of saturation S_r	0.9	0.9
Poisson's ratio μ	0.4	0.4
Soil porosity n	0.35	0.35
Soil permeability k_z	10^{-2} m/s	10^{-2} m/s
Dynamic parameters		
K_1	0.12829	0.14973
K_2	0.35855	0.025134
V_c	137.133	126.9343
Π_1	3.7973	0.11366
Π_2	0.0011657	0.011684

For the point above the curve, quasi-static soil behavior is acceptable (for example, Cases 1 and 3).

Although the degree of saturation plays an important role in the analysis of the wave-induced soil response, its influences on the boundary between dynamic and quasi-static soil behavior is less significant than permeability, as shown in Fig. 6.15b. After preliminary tests for various ranges of water depths and wave periods, we find that the boundary between dynamic and quasi-static soil behavior can be considered as a constant with the same permeability.

Referring to Fig. 6.15a, it is apparent that, when the point located above a curve with particular value of soil permeability, we can use the quasi-static solution with

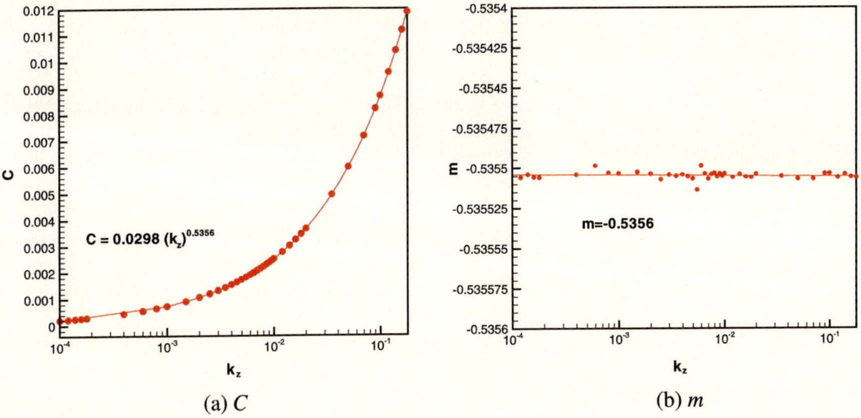

Fig. 6.16 The relationship coefficient (**a**) C and (**b**) m and permeability k_z

less than a 3 % relative difference. To further help practical engineers in determining the boundary between dynamic and quasi-static soil behavior, we further examined the relationship between boundary and soil permeability.

From the Fig. 6.15, it can be seen that the relationship between the two Π values can be represented as

$$\Pi_2 = C\Pi_1^m. \tag{6.100}$$

Thus, in the Fig. 6.16, the coefficient C can be represented in terms of soil permeability as

$$C = 0.0298(k_z)^{0.5356} \tag{6.101}$$

the coefficient m can be considered as a constant ($m \cong -0.5356$).

Now, Fig. 6.16 and (6.100) can, therefore, be applied to determine the boundary between dynamic and quasi-static soil behavior using the following procedure:

- Calculate the dynamic constants, Π_1 and Π_2, with the given wave and soil parameters.
- Based on the soil permeability given, determine C coefficient from Fig. 6.16.
- Substituting the C and m coefficients into (6.100), determine the boundary of the dynamic and quasi-static soil behavior.
- Based on the calculated Π_1 and Π_2, determine if quasi-static soil behavior is applicable or not.

References

1. Biot, M.A.: General theory of three-dimensional consolidation. J. Appl. Phys. **26**(2), 155–164 (1941)
2. Biot, M.A.: Theory of propagation of elastic waves in a fluid-saturated porous solid, Part I: Low frequency range. J. Acoust. Soc. Am. **28**, 168–177 (1956)

3. Cha, D.H., Jeng, D.S., Rahman, M.S., Sekiguchi, H., Zen, K., Yamazaki, H.: Effects of dynamic soil behaviour on the wave-induced seabed response. Int. J. Ocean Eng. Technol. **16**(21–33), 5 (2002)
4. Jeng, D.S.: Wave-induced seabed response in front of a breakwater. Ph.D. thesis, University of Western Australia (1997)
5. Jeng, D.S., Cha, D.H.: Effects of dynamic soil behavior and wave non-linearity on the wave-induced pore pressure and effective stresses in porous seabed. Ocean Eng. **30**(16), 2065–2089 (2003)
6. Jeng, D.S., Lee, T.L.: Dynamic response of porous seabed to ocean waves. Comput. Geotech. **28**(2), 99–128 (2001)
7. Jeng, D.S., Rahman, M.S.: Effective stresses in a porous seabed of finite thickness: Inertia effects. Can. Geotech. J. **37**(4), 1388–1397 (2000)
8. Jeng, D.S., Rahman, M.S., Lee, T.L.: Effects of inertia forces on wave-induced seabed response. Int. J. Offshore Polar Eng. **9**(4), 307–313 (1999)
9. Lee, T.L., Tsai, C.P., Jeng, D.S.: Analytical solution of wave-induced seabed response: Effects of inertia forces. Ocean Eng. **29**(12), 1577–1601 (2002)
10. Madsen, O.S.: Wave-induced pore pressures and effective stresses in a porous bed. Géotechnique **28**(4), 377–393 (1978)
11. Okusa, S.: Wave-induced stress in unsaturated submarine sediments. Géotechnique **35**(4), 517–532 (1985)
12. Yamamoto, T.: On the response of a Coulomb-damped poro-plastic bed to water waves. Mar. Geotechnol. **5**(2), 93–130 (1983)
13. Yamamoto, T., Koning, H., Sellmeijer, H., Hijum, E.V.: On the response of a poro-elastic bed to water waves. J. Fluid Mech. **87**(1), 193–206 (1978)
14. Zienkiewicz, O.C., Chang, C.T., Bettess, P.: Drained, undrained, consolidating and dynamic behaviour assumptions in soils. Géotechnique **30**(4), 385–395 (1980)

Chapter 7
Wave Propagation over Coulomb-Damped Seabed

Abstract In this chapter, the Coulomb-damping friction is considered in the problem of the wave-seabed interaction in a porous seabed. A closed-form analytical solution will be presented here. In the model, we couple the water waves and porous seabed with a complex wave number. Thus, the effects of soil properties on the wave characteristics can be examined. The numerical results indicate that the Coulomb-damping friction significantly affects the wave-induced pore pressure in a finer seabed (such as fine sand and clay). Furthermore, numerical examples show that, the wave characteristics have been significantly affected by the wave-induced seabed response.

Keywords Coulomb friction · Wave-seabed interaction · Pore pressure · Effective stresses · Wave loading

7.1 Introduction

A marine soil is a mixture of three phases: a solid phase, a liquid phase and a gas phase. Among these, the solid phase forms an interlocking skeletal frame, the liquid phase occupies a major portion of pore space, and the gas phase sometimes occupies a small portion of pore space. When ocean waves propagate over a seabed, they exert pore pressure gradient in the pore that induces a porous flow of pore fluid (water and gas) relative to skeletal frame. Through the fluid viscosity and pressure gradient, the pore fluid transmits the force to the skeletal frame in the form of effective stresses. Then, the wave-induced effective stresses cause deformation of the skeletal frame. Because the skeletal frame has both rigidity and compressibility, two kinds of stress waves (i.e., shear waves and compressional waves) transmit through the skeletal frame. Furthermore, since the pore fluid is compressible, compressional waves transmit through the pore fluid. Thus, ocean wave energy is dissipated by the viscous friction of fluid and the solid-to-solid friction (a Coulomb damping) at the points of contact between grains.

In marine environments, the wave characteristics, which are important in ocean engineering, are also affected by the characteristics of sedimentary beds. Traditional ocean wave theories have assumed a rigid impermeable seabed, although they are porous materials. It has been well known that the wavelength of a water wave is

D.-S. Jeng, *Porous Models for Wave-seabed Interactions*,
DOI 10.1007/978-3-642-33593-8_7,

also modified by the seabed response in a porous seabed. In such a condition, it is necessary to couple the wave field with porous flow and provide a better estimation of wave characteristics.

Since the 1980s, a few researchers attempted to consider the effects of internal friction in the model. Among these, Dalrymple and Liu [1] treated the porous seabed as incompressible viscous fluid, and assumed the soil friction to be proportional to the strain rate. Stoke *et al.* [11] determine the shear modulus and material damping of soils by directly measuring the shear stress and shear strain. Their experimental results indicate that the shear energy dissipation mechanism in soil be due to Coulomb damping (or solid-solid friction), but not due to viscous damping of fluid-solid friction. Yamamoto [14] might have been the first to consider the Coulomb damping friction in the waves/seabed interaction problem. However, he only considered a seabed of infinite thickness. Yamamoto [13] further extended his model to the seabed of finite thickness with a semi-analytical framework. Later, Yamamoto's approach was further extended to a seabed of finite thickness by the author and his co-worker [8].

In this chapter, closed-form analytical solutions for the ocean wave propagating over a Coulomb-damping seabed will be presented first. The influences of Coulomb friction and fluid acceleration on the wave-induced soil response will then be clarified; and the effects of soil properties on the wave characteristics will be examined.

7.2 Coulomb-Damping Poro-Elastic Seabed

In general, when water wave transfers into the soil matrix, two different stress waves (shear waves and compressional waves) will be generated between solid grains. This is due to the compressibility of soils. Meanwhile, another type of compressional wave is also generated when water waves transmit between pore spaces. These three stress waves will dissipate due to the internal friction of pore fluid and solids. On the other hand, the wave field is also modified by the seabed. Thus, the wave energy is affected by the viscous friction of pore fluid and Coulomb-damping friction (i.e., solid-solid friction). Therefore, the whole wave-seabed interaction problem can be considered as the energy transmission between water wave field and porous seabed. Both materials affect each other.

Although considerable amount of experiment investigations have been done on dynamic properties of marine sediments in the areas of earthquake engineering and underwater acoustics, only a few amount of experimental investigations have been done on the dynamic properties of sediments influenced by ocean waves.

Using a resonant column device dynamic shear modulus and material internal damping have been measured by many earthquake engineers [3, 11]. The shear modulus of the soil is determined from the resonance frequency at first-mode torsional frequency of cylindrical soil sample and material damping from the free-vibration decay curve. Stokoe *et al.* [11] also measured shear modulus and material damping

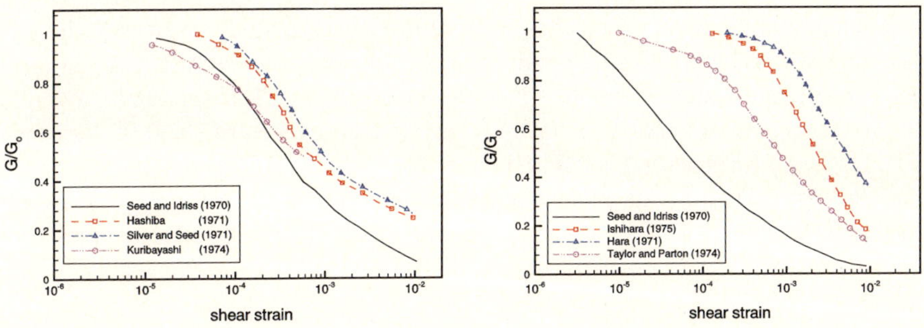

(a) G_r/G_0 (b) $\delta = G_i/G_0$

Fig. 7.1 Relationship between (**a**) shear modulus (G_r/G_0) and (**b**) Coulomb energy loss parameter ($\delta = G_i/G_0$) and shear strain in sand

of soils at a lower frequency range (1 to 15 Hz) by directly measuring the shear stress and shear strain. They concluded that the shear modulus and material damping per cycle are functions of shearing strain amplitude, magnitude of confining pressure, duration of confinement, and number of cycles of loading, but are nearly independent of loading frequency. The experimental results suggest that the shear energy dissipation mechanism in soil be due to Coulomb damping or solid-solid friction, but not due to viscous damping of fluid-solid friction. Since the friction between soil particles behaviors weakly non-linear decaying characteristics, the mathematical treatment for such a problem is very complicated. Thus, the previous studies [8, 13, 14] analyzed the phenomenon of wave propagating over a porous seabed by employing the concept of linearizion.

Shear modulus, defined as the proportional of shear stresses and shear strain, is considered as a complex constant, $G = G_r + iG_i$, due to the slightly inelastic nature of the soil. The real part, G_r, is the dynamic shear modulus of the skeletal frame of soil, while the imaginary part, G_i, is the linearized expression of the non-linear Coulomb damping due to the grain-to-grain friction. Thus, the complex shear modulus of soil may be defined in terms of the dynamic shear modulus ($G_r = G_o$) and energy loss parameter (δ) as

$$G = G_r + iG_i = G_0(1 + i\delta). \tag{7.1}$$

Note that the non-linear stress-strain relationship exists in this problem. However, such material is similar to a visco-elastic material under cyclic stresses, which can be approximately represented by introducing complex elastic module [14]. Thus, it is reasonable to use (7.2)–(7.6) to approximate the non-linear stress-strain relations.

In general, the shear modulus of clay varies between 10^5 and 10^7 N/m^2, while sand varies between 10^6 and 10^8 N/m^2. Based on previous experimental investigations, non-linear relationships between shear modulus, energy loss parameter (δ) and soil strain (Fig. 7.1). As seen in the figure, shear modulus decreases when the soil strain increases. The values of δ for clay and sand vary between 0 and 0.5, and the maximum shear strain reaches 0.01. However, in another set of experimental data conducted by [16] for clay indicates that the maximum shear strain can reach 0.1, and the maximum energy loss parameter can reach 1.27. In this study, to have a more reasonable estimation of wave-seabed interaction, we use $\delta = 0.05$ for coarse sand, $\delta = 0.4$ for fine sand and $\delta = 0.8$ for clay.

Under the condition of plane strain, the effective stresses and pore pressure can be expressed in terms of displacements of the skeletal frame of soil **u** and pore fluid **w** as

$$\sigma'_x = (H + C)\epsilon - 2G\epsilon_z - (C_M)\nabla \bullet \mathbf{w}, \tag{7.2}$$

$$\sigma'_z = (H + C)\epsilon - 2G\epsilon_x - (C - M)\nabla \bullet \mathbf{w}, \tag{7.3}$$

$$\tau_{xz} = G\left(\frac{\partial u_z}{\partial x} + \frac{\partial u_x}{\partial z}\right), \tag{7.4}$$

$$p = C\epsilon - M\nabla \bullet \mathbf{w}, \tag{7.5}$$

$$\epsilon = \epsilon_x + \epsilon_z, \quad \epsilon_x = \frac{\partial u_x}{\partial x}, \ \epsilon_z = \frac{\partial u_z}{\partial z} \tag{7.6}$$

where u_x and u_z are the components of the displacement vector (**u**). σ'_x and σ'_z are the effective normal stresses in the x- and z-directions, respectively, and τ_{xz} is the shear stress. p is pore fluid pressure. $\mathbf{w} = (w_x, w_z)$ represents the flow of the fluid relative to the solid, but measured in terms of volume per unit area of the bulk medium, which can bed defined by $\mathbf{w} = n'(\mathbf{U} - \mathbf{u})$, in which **U** represents the displacement vector of the pore fluid.

In (7.2)–(7.6), Biot's elastic modulus, H, C and M can be determined by G and other basic physical properties of the soil as

$$H = \frac{[K_r - \frac{2}{3}\frac{1+\mu}{1-2\mu}G]^2}{K_r[1 + n'(K_r\beta - 1)] - \frac{2}{3}\frac{1+\mu}{1-2\mu}G} + \frac{2(1-\mu)}{1-2\mu}, \tag{7.7}$$

$$C = \frac{[K_r - \frac{2}{3}\frac{1+\mu}{1-2\mu}G]}{K_r[1 + n'(K_r\beta - 1)] - \frac{2}{3}\frac{1+\mu}{1-2\mu}G}, \tag{7.8}$$

$$M = \frac{K_r^2}{K_r[1 + n'(K_r\beta - 1)] - \frac{2}{3}\frac{1+\mu}{1-2\mu}G} \tag{7.9}$$

where K_r is the bulk modulus of the grain, and β is the compressibility of pore fluid, which can be defined by

$$\beta = \frac{1}{K_w} + \frac{1 - S_r}{P_{wo}} \tag{7.10}$$

in which K_w is true bulk modulus of pore water, S_r is the degree of saturation and P_{wo} is the absolute pore pressure.

7.3 Boundary Value Problem

7.3.1 Governing Equations

In this study, periodic water waves are considered to propagate over a poro-elastic bed of finite thickness h in water depth d, as seen in Fig. 3.1. The wave crests are assumed to propagate in the positive x-direction, while the vertical z-axis is upward from the seabed surface.

Considering an incompressible, irrotational flow for the wave filed, the velocity potential satisfies Laplace equation as

$$\nabla^2 \phi = 0. \tag{7.11}$$

The water wave is assumed to propagate in the positive x-direction, and the surface fluctuation, η is give as

$$\eta = a\mathrm{Re}\left\{e^{i(\lambda x - \omega t)}\right\} \tag{7.12}$$

where a is the amplitude of wave, and ω is the wave frequency ($= 2\pi/T$, T is wave period). λ ($= \lambda_r + i\lambda_i$) represents the complex wave number (please note that the notation of the wave number in this chapter is different from other chapters). The real part, λ_r is related to the wavelength L ($= 2\pi/\lambda_r$), while the imaginary part λ_i represents the damping of the water waves. Unlike conventional investigations [5, 15], the wave number is an unknown parameter to be determined in this study.

The displacement components of water particles and wave pressures are given as

$$\frac{\partial \mathbf{U}}{\partial t} = \nabla \phi, \quad \text{and} \quad p_w = -\rho_f \frac{\partial \phi}{\partial t} \tag{7.13}$$

where ρ_f is the density of pore fluid.

In this study, skeletal frame and pore fluid of seabed are assumed to be compressible. The motion of pore fluid relative the skeletal is assumed to obey Darcy's law, together with inertia forces as

$$\nabla^2 p + \frac{\gamma_w}{K_s}\frac{\partial \mathbf{w}}{\partial t} = \frac{\partial^2(\rho_f \mathbf{u} + m\mathbf{w})}{\partial t^2}, \tag{7.14}$$

where K_s is hydraulic coefficient of the permeability of soil [m/s], γ_w is the unit weight of water. It is noted that the right-hand-side of (7.14) comes from the inertia forces.

In (7.14), the virtual mass of the skeletal frame (m) in the accelerated flow field is given as

$$m = (1+\alpha)\rho_f/n', \tag{7.15}$$

in which α is the added mass coefficient of the skeletal frame, and n' is porosity of soil.

Including inertia forces, equations of motion of the skeletal frame can be expressed as

$$\frac{\partial \sigma_x'}{\partial x} + \frac{\partial \tau_{xz}}{\partial z} = \frac{\partial p}{\partial x} + \rho \frac{\partial^2 u_x}{\partial t^2} + \rho_f \frac{\partial^2 w_x}{\partial t^2}, \tag{7.16}$$

$$\frac{\partial \tau_{xz}}{\partial x} + \frac{\partial \sigma_Z'}{\partial z} = \frac{\partial p}{\partial z} + \rho \frac{\partial^2 u_z}{\partial t^2} + \rho_f \frac{\partial^2 w_z}{\partial t^2}, \tag{7.17}$$

where $\rho = (1 - n')\rho_r + n'\rho_f$, in which ρ is the bulk density of soil and ρ_r is the density of grain material.

Substituting (7.2)–(7.6) into (7.14), (7.16) and (7.17), we have

$$G\nabla^2 \mathbf{u} + (H - G)\nabla(\nabla \bullet \mathbf{u}) + C\nabla(\nabla \bullet \mathbf{w}) = \rho \frac{\partial^2 \mathbf{u}}{\partial t^2} + \rho_f \frac{\partial^2 \mathbf{w}}{\partial t^2}, \tag{7.18}$$

$$\nabla(C\nabla \bullet \mathbf{u} - M\nabla \bullet \mathbf{w}) = \left(\rho_f \frac{\partial^2 \mathbf{u}}{\partial t^2} + m \frac{\partial^2 \mathbf{w}}{\partial t^2}\right) + \frac{1}{k} \frac{\partial^2 \mathbf{w}}{\partial t^2}. \tag{7.19}$$

Since (7.18)–(7.19) are linear, the principle of superposition can be applied. That is, the compressional waves and shear waves can be considered separately and then added later on, i.e.,

$$\mathbf{u} = \mathbf{u}_c + \mathbf{u}_s \quad \text{and} \quad \mathbf{w} = \mathbf{w}_c + \mathbf{w}_s, \tag{7.20}$$

where the subscript "c" denotes the components from compressional waves, and "s" is for shear waves.

For compressional waves, herein, we define a displacement potential ψ_1 and ψ_2, which satisfy

$$\mathbf{u}_c = \nabla \psi_1 \quad \text{and} \quad \mathbf{w}_c = \nabla \psi_2. \tag{7.21}$$

Substituting (7.21) into (7.18) and (7.19), we have

$$\nabla^2 (H\psi_1 + C\psi_2) = \frac{\partial^2}{\partial t^2}(\rho \psi_1 + \rho_f \psi_2). \tag{7.22}$$

Similarly, for shear waves, the displacements potential satisfies

$$\mathbf{u}_s = \nabla \times \psi_3 \quad \text{and} \quad \mathbf{w}_s = \nabla \times \psi_4. \tag{7.23}$$

Then, the governing equations (7.18) and 7.19) can be re-organized as

$$G\nabla^2 \psi_3 = \frac{\partial^2}{\partial t^2}(\rho \psi_3 + \rho_f \psi_4), \tag{7.24}$$

$$-\frac{\gamma}{K_s \eta_f} \frac{\partial \psi_4}{\partial t} = \frac{\partial^2}{\partial t^2}(\rho_f \psi_3 + m\psi_4). \tag{7.25}$$

7.3.2 Boundary Condition

The governing equations (7.22)–(7.25), can be solved with some appropriate boundary conditions at the rigid impermeable bottom, seabed surface and water free surface.

Firstly, at the free surface, the velocity potential must satisfy the linearized free surface boundary conditions for a small amplitude wave,

$$\eta = -\frac{1}{g}\frac{\partial \phi}{\partial t} \quad \text{and} \quad \frac{\partial \phi}{\partial z} = \frac{\partial \eta}{\partial t} \quad \text{at } z = d, \tag{7.26}$$

where g is the acceleration of gravity.

Secondly, at the seabed surface, the mass of the fluid must be conserved. Thus,

$$\frac{\partial \phi}{\partial z} = \frac{\partial u_z}{\partial t} + \frac{\partial w_z}{\partial t} \quad \text{at } z = 0. \tag{7.27}$$

Note that the above boundary condition is based on the continuity of velocity, which will be used to determine the complex wave number latter one. This was not used in the previous sections for un-coupled models, as the wave number was given in the previous models.

Thirdly, the vertical effective normal stress and shear stress vanish at the seabed surface, while the fluid pressure is transmitted continuously from the seabed to the pores in the seabed, Thus,

$$p_w(z=0) = p(z=0) = -\rho_f \frac{\partial}{\partial t}, \qquad \sigma_z' = \tau_{xz} = 0 \quad \text{at } z = 0. \tag{7.28}$$

Finally, zero displacements are assumed for the soil resting on an impermeable rigid bottom and no vertical flow occurs across the horizontal boundary, i.e.,

$$\mathbf{u} = \frac{\partial p}{\partial z} = 0, \quad \text{at } z = -h. \tag{7.29}$$

7.4 General Solutions

7.4.1 Analytical Solution for a Seabed of Finite Thickness

The governing equations (7.22)–(7.25) can be solved with boundary conditions (7.26)–(7.29). In this section, the closed-form solution for the wave-seabed interaction problem in a porous seabed of finite thickness will be derived in this section.

Since the free surface fluctuation given by (7.11) is harmonic in both x-direction and time, and the boundary conditions (7.26) and (7.27) are all homogeneous, the velocity potential in the area of $(0 \le z \le d)$ can be expressed as

$$\phi = -\frac{iag}{\omega}\left\{\cosh\lambda(z-d) + \frac{\omega^2}{g\lambda}\sinh\lambda(z-d)\right\} e^{i(\lambda x - \omega t)}. \tag{7.30}$$

It is noted that the wave number in the above equation is an unknown, which will be determined from the boundary condition at the surface of seabed after obtaining the general solution for the seabed response.

Similarly, the displacement potential functions ψ_i $(i=1\text{–}4)$ can be expressed as

$$\psi_i = \Psi_i(z)\, e^{i(\lambda x - \omega t)}. \tag{7.31}$$

Substituting (7.31) into (7.22)–(7.25), after some algebraic manipulation, the final form for the governing equations can be derived as

$$\begin{aligned}&\left\{(HM - C^2)\left(\lambda^2 - \frac{\partial^2}{\partial z^2}\right)^2 - \omega^2(\rho M + m_1 H - 2\rho_f C)\left(\lambda^2 - \frac{\partial^2}{\partial z^2}\right)\right.\\ &\quad \left. + (\rho m_1 - \rho_f^2)\omega^4\right\}\begin{Bmatrix}\Psi_1(z)\\ \Psi_2(z)\end{Bmatrix} = 0,\end{aligned} \tag{7.32}$$

$$\left\{G m_1 \omega^2\left(\lambda^2 - \frac{\partial^2}{\partial z^2}\right)^2 - (\rho m_1 - \rho_f^2)\omega^4\right\}\begin{Bmatrix}\Psi_3(z)\\ \Psi_4(z)\end{Bmatrix} = 0 \tag{7.33}$$

where $m_1 = m + i/\lambda\omega$.

The general solutions of (7.32)–(7.33) are

$$\begin{Bmatrix}\Psi_1(z)\\ \Psi_2(z)\end{Bmatrix} = \begin{Bmatrix}1\\ \delta_1\end{Bmatrix}\left(a_1 e^{\lambda_1 z} + a_2 e^{-\lambda_1 z}\right) + \begin{Bmatrix}1\\ \delta_2\end{Bmatrix}\left(a_3 e^{\lambda_1 z} + a_4 e^{-\lambda_1 z}\right), \tag{7.34}$$

$$\begin{Bmatrix}\Psi_3(z)\\ \Psi_4(z)\end{Bmatrix} = \begin{Bmatrix}1\\ \delta_3\end{Bmatrix}\left(a_5 e^{\lambda_3 z} + a_6 e^{-\lambda_3 z}\right) \tag{7.35}$$

where

$$\delta_i = -\frac{V_i^2 - H/\rho}{V_i^2 - C/\rho_f} \quad (i = 1, 2) \quad \text{and} \quad \delta_3 = -\frac{\rho_f}{m_1}, \tag{7.36}$$

$$\begin{aligned}\lambda_1 &= \lambda\sqrt{1 - V_w^2/V_1^2}, \qquad \lambda_2 = \lambda\sqrt{1 - V_w^2/V_2^2},\\ \lambda_3 &= \lambda\sqrt{1 - V_w^2/V_3^2}, \quad V_w = \frac{\omega}{\lambda},\end{aligned} \tag{7.37}$$

$$V_1 = \sqrt{\frac{(\rho M + m_1 H - 2\rho_f C) + \sqrt{\delta_m}}{2(HM - C^2)}}, \tag{7.38}$$

$$V_2 = \sqrt{\frac{(\rho M + m_1 H - 2\rho_f C) - \sqrt{\delta_m}}{2(HM - C^2)}}, \tag{7.39}$$

$$V_3 = \sqrt{\frac{G/\rho}{1 - \rho_f^2/\rho m_1)}}, \tag{7.40}$$

where $\delta_m = (\rho M + m_1 H - 2\rho_f C)^2 - 4(HM - C^2)(\rho m_1 - \rho_f^2)$.

Introducing (7.34)–(7.35) into (7.20), we have the final solution of displacements as

$$u_x = \gamma_w a\{i\lambda(a_1 e^{\lambda_1 z} + a_2 e^{-\lambda_1 z}) + i\lambda(a_3 e^{\lambda_2 z} + a_4 e^{-\lambda_2 z}) \\ - \lambda_3(a_5 e^{\lambda_3 z} - a_6 e^{-\lambda_3 z})\} \times \text{chd}, \tag{7.41}$$

$$u_z = \gamma_w a\{\lambda_1(a_1 e^{\lambda_1 z} - a_2 e^{-\lambda_1 z}) + \lambda_2(a_3 e^{\lambda_2 z} - a_4 e^{-\lambda_2 z}) \\ + i\lambda(a_5 e^{\lambda_3 z} + a_6 e^{-\lambda_3 z})\} \times \text{chd}, \tag{7.42}$$

$$w_x = \gamma_w a\{i\lambda\delta_1(a_1 e^{\lambda_1 z} + a_2 e^{-\lambda_1 z}) + i\lambda\delta_2(a_3 e^{\lambda_2 z} + a_4 e^{-\lambda_2 z}) \\ - \lambda_3\delta_3(a_5 e^{\lambda_3 z} - a_6 e^{-\lambda_3 z})\} \times \text{chd}, \tag{7.43}$$

$$w_z = \gamma_w a\{\lambda_1\delta_1(a_1 e^{\lambda_1 z} - a_2 e^{-\lambda_1 z}) + \lambda_2\delta_2(a_3 e^{\lambda_2 z} - a_4 e^{-\lambda_2 z}) \\ + i\lambda\delta_3(a_5 e^{\lambda_3 z} + a_6 e^{-\lambda_3 z})\} \times \text{chd}, \tag{7.44}$$

$$\text{chd} = \left[-\cosh\lambda d + \frac{\omega^2}{g\lambda}\sinh\lambda d\right] e^{i(\lambda x - \omega t)}. \tag{7.45}$$

Then, the wave-induced effective stresses and pore pressure can be expressed as

$$\sigma'_x = \gamma_w a\{[(H + \delta_1 C)s_1^2 + 2G\lambda_1^2](a_1 e^{\lambda_1 z} + a_2 e^{-\lambda_1 z}) \\ - [(H + \delta_2 C)s_2^2 + 2G\lambda_2^2](a_3 e^{\lambda_2 z} + a_4 e^{-\lambda_2 z}) \\ - 2iG\lambda\lambda_3(a_5 e^{\lambda_3 z} - a_6 e^{-\lambda_3 z})\} \times \text{chd}, \tag{7.46}$$

$$\sigma'_z = \gamma_w a\{-[(H + \delta_1 C)s_1^2 + 2G\lambda_1^2](a_1 e^{\lambda_1 z} + a_2 e^{-\lambda_1 z}) \\ - [(H + \delta_2 C)s_2^2 + 2G\lambda_2^2](a_3 e^{\lambda_2 z} + a_4 e^{-\lambda_2 z}) \\ - 2iG\lambda\lambda_3(a_5 e^{\lambda_3 z} - a_6 e^{-\lambda_3 z})\} \times \text{chd}, \tag{7.47}$$

$$\tau_{xz} = \gamma_w a G\{2i\lambda\lambda_1(a_1 e^{\lambda_1 z} - a_2 e^{-\lambda_1 z}) - 2i\lambda\lambda_2(a_3 e^{\lambda_2 z} - a_4 e^{-\lambda_2 z}) \\ - (2\lambda^2 - s_3^2)(a_5 e^{\lambda_3 z} - a_6 e^{-\lambda_3 z})\} \times \text{chd}, \tag{7.48}$$

$$p = \gamma_w a\{s_1^2(C + \delta_1 M)(a_1 e^{\lambda_1 z} - a_2 e^{-\lambda_1 z}) \\ - s_2^2(C + \delta_2 M)(a_3 e^{\lambda_2 z} - a_4 e^{-\lambda_2 z})\} \times \text{chd} \tag{7.49}$$

where $s_i^2 = \lambda_i^2 - \lambda^2$ $(i = 1, 2, 3)$. In (7.41)–(7.49), coefficients a_i $(i = 1–6)$ can be expressed in explicit form as seen in Sect. 7.7. It is noted that the wave number λ in the solution is another unknown.

Now, we still have one unknown k and the boundary condition at the seabed surface (7.27), which leads to a new wave dispersion equation as

$$\tanh\lambda d = \frac{\omega^2 - g\lambda\mathcal{F}}{g\lambda - \omega^2\mathcal{F}}, \tag{7.50}$$

where

$$\mathcal{F} = -\frac{\rho_f \omega^2}{\lambda}\left\{\lambda_1(1+\delta_1)(a_1-a_2)+\lambda_2(1+\delta_2)(a_3-a_4)+ik\lambda_3(1+\delta_3)(a_5+a_6)\right\}. \tag{7.51}$$

The complex wave number λ can be obtained from the above dispersion (7.50) by Newton's iteration method. Once the complex wave number λ is obtained, the coefficient a_1–a_6 can also be calculated through the coefficients listed in Sect. 7.7.

7.4.2 Analytical Solution for an Infinite Seabed

With the similar framework outlined in Sect. 7.4.1, we can derive the solution of infinite seabed by setting $a_2 = a_4 = a_6 = 0$, which leads to a simplified form for seabed response in a seabed of infinite thickness as

$$u_x^{\infty} = \gamma_w a\left\{i\lambda a_1^{\infty} e^{\lambda_1 z} + i\lambda a_3^{\infty} e^{\lambda_2 z} - \lambda_3 a_5^{\infty} e^{\lambda_3 z}\right\} \times \text{chd}, \tag{7.52}$$

$$u_z^{\infty} = \gamma_w a\left\{\lambda_1 a_1^{\infty} e^{\lambda_1 z} + \lambda_2 a_3^{\infty} e^{\lambda_2 z} + i\lambda a_5^{\infty} e^{\lambda_3 z}\right\} \times \text{chd}, \tag{7.53}$$

$$w_x^{\infty} = \gamma_w a\left\{i\lambda\delta_1 a_1^{\infty} e^{\lambda_1 z} + i\lambda\delta_2 a_3^{\infty} e^{\lambda_2 z} - \lambda_3\delta_3 a_5^{\infty} e^{\lambda_3 z}\right\} \times \text{chd}, \tag{7.54}$$

$$w_z^{\infty} = \gamma_w a\left\{\lambda_1\delta_1 a_1^{\infty} e^{\lambda_1 z} + \lambda_2\delta_2 a_3^{\infty} e^{\lambda_2 z} + i\lambda\delta_3 a_5^{\infty} e^{\lambda_3 z}\right\} \times \text{chd}. \tag{7.55}$$

Then, the wave-induced effective stresses and pore pressure can be expressed as

$$\begin{aligned}\sigma_x'^{\infty} = \gamma_w a\big\{&\left[(H+\delta_1 C)s_1^2 + 2G\lambda_1^2\right]a_1^{\infty} e^{\lambda_1 z} - \left[(H+\delta_2 C)s_2^2 + 2G\lambda_2^2\right]a_3^{\infty} e^{\lambda_2 z}\\ &- 2iG\lambda\lambda_3 a_5^{\infty} e^{\lambda_3 z}\big\} \times \text{chd},\end{aligned} \tag{7.56}$$

$$\begin{aligned}\sigma_z'^{\infty} = \gamma_w a\big\{&-\left[(H+\delta_1 C)s_1^2 + 2G\lambda_1^2\right]a_1^{\infty} e^{\lambda_1 z} - \left[(H+\delta_2 C)s_2^2 + 2G\lambda_2^2\right]a_3^{\infty} e^{\lambda_2 z}\\ &- 2iG\lambda\lambda_3 a_5^{\infty} e^{\lambda_3 z}\big\} \times \text{chd},\end{aligned} \tag{7.57}$$

$$\tau_{xz}^{\infty} = \gamma_w a G\left\{2i\lambda\lambda_1 a_1^{\infty} e^{\lambda_1 z} - 2i\lambda\lambda_2 a_3^{\infty} e^{\lambda_2 z} - \left(2\lambda^2 - s_3^2\right)a_5^{\infty} e^{\lambda_3 z}\right\} \times \text{chd}, \tag{7.58}$$

$$p = \gamma_w a\left\{s_1^2(C+\delta_1 M)a_1^{\infty} e^{\lambda_1 z} - s_2^2(C+\delta_2 M)a_3^{\infty} e^{\lambda_2 z}\right\} \times \text{chd}. \tag{7.59}$$

The coefficients a_i^{∞} $(i = 1, 3, 5)$ are given by

$$a_1^{\infty} = -\left(B\left(2\lambda^2 - s_3^2\right) + 4\lambda^2\lambda_2\lambda_3 G\right)/\Delta, \tag{7.60}$$

$$a_3^{\infty} = \left(A\left(2\lambda^2 - s_3^2\right) - 4\lambda^2\lambda_1\lambda_3 G\right)/\Delta, \tag{7.61}$$

$$a_5^{\infty} = -i\lambda(\lambda_2 A + \lambda_1 B)/\Delta, \tag{7.62}$$

$$\begin{aligned}\Delta = &-s_1^2(C+\delta_1 M)\left[4\lambda^2\lambda_2\lambda_3 G + B\left(2\lambda^2 - s_3^2\right)\right]\\ &+ s_2^2(C+\delta_2 M)\left[4\lambda^2\lambda_1\lambda_3 G - A\left(2\lambda^2 - s_3^2\right)\right]\end{aligned} \tag{7.63}$$

in which $A = (H+\delta_1 C)s_1^2 + 2G\lambda_1^2$ and $B = (H+\delta_2 C)s_2^2 + 2G\lambda_2^2$.

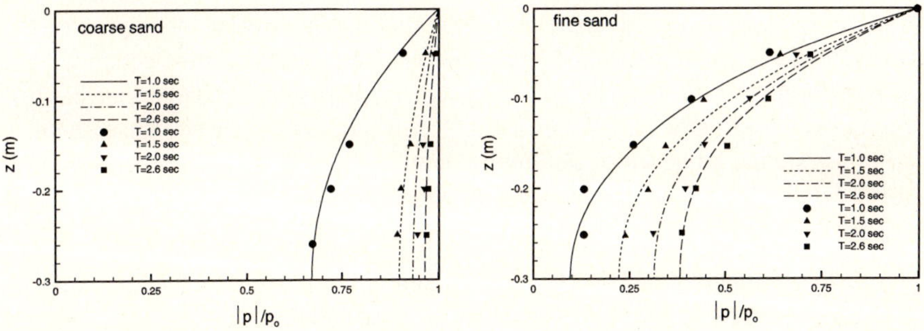

Fig. 7.2 Comparison of the wave-induced pore pressure ($|p|/p_o$) in sandy beds. *Lines* denote the present solution, and symbols are experimental data [14]

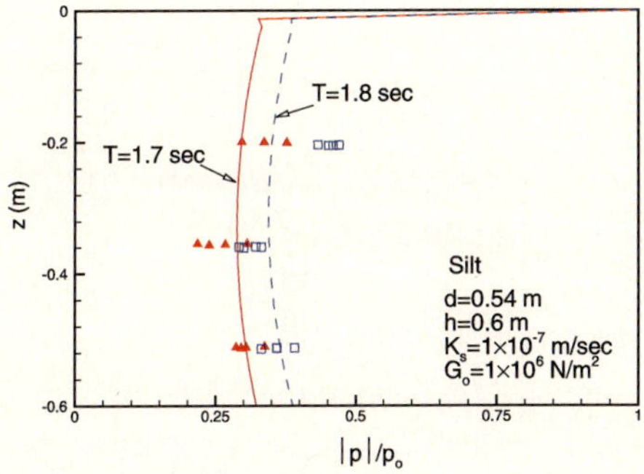

Fig. 7.3 Comparison of the wave-induced pore pressure ($|p|/p_o$) in sandy beds. *Lines* denote the present solution, and symbols are experimental data [12]

7.4.3 Verification

Since the present solution for the wave-seabed interaction is new, it is necessary to be verified with previous experimental data. In this section, we compare the present solution with the previous experimental data [12, 15]. All these two sets of experiments were conducted in the seabed of finite thickness.

The comparison between the experimental data in sandy beds [15] and the present solution is presented in Fig. 7.2 for various values of wave periods. Two different sandy seabeds (coarse sand and fine sand) are used in their experiments. As seen in the figure, the present solution overall agrees with the previous experiments [14] for sandy beds.

Figure 7.3 presents a set of comparison between the present analytical solution and the previous experimental data in silt [12]. It is noted that a silt bed was used in the wave tank experiments. As shown in the figure, the present solution overall agrees with experimental data.

7.5 Results and Discussions

This chapter is aimed at two goals: (1) the inclusion of Coulomb-damping friction and fluid acceleration and (2) the response of seabed to ocean waves. Although the semi-analytical analysis for the first component has been proposed by Yamamoto [14], a more detailed discussion will be provided in this section with the closed-form solution. The second component, response of seabed to ocean, has been rarely discussed in the past. Thus, a detailed discussion will be conducted here. To demonstrate the influence of Coulomb-damping friction and fluid acceleration, the input data in Table 1 is used for the following numerical examples. In this chapter, we consider three different solutions:

(1) Jeng's model [5]: Neither Coulomb-damping friction nor fluid acceleration was considered in the model. This is a quasi-static solution.
(2) JRL model [7]: Only fluid acceleration was considered. The Coulomb-damping friction was excluded.
(3) The present model: Both Coulomb-damping friction and fluid acceleration are considered.

7.5.1 *Effects of Coulomb-Damping Friction and Fluid Acceleration*

Figures 7.4–7.5 illustrate the vertical distribution of the wave-induced pore pressure ($|p|/p_o$) and vertical effective stress ($|\sigma_z'|/p_o$) against the relative soil depth (z/L_o, where $L_o = 1.56T^2$ represents the wavelength in deep water) in both coarse sand and fine sand. In the example, the wave period, water depth, the wavelength and seabed thickness are taken as 10 s, 25 m, 125 m and 25 m, respectively. In the figure, solid lines represent the present results with Coulomb-damping friction and fluid acceleration (the present solution), the dashed lines are the results without Coulomb-damping friction but including fluid acceleration (JRL model), and the dashed-dotted lines are the results of quasi-static solution (Jeng's model). Two different degrees of saturation are considered in this example. The figure clearly

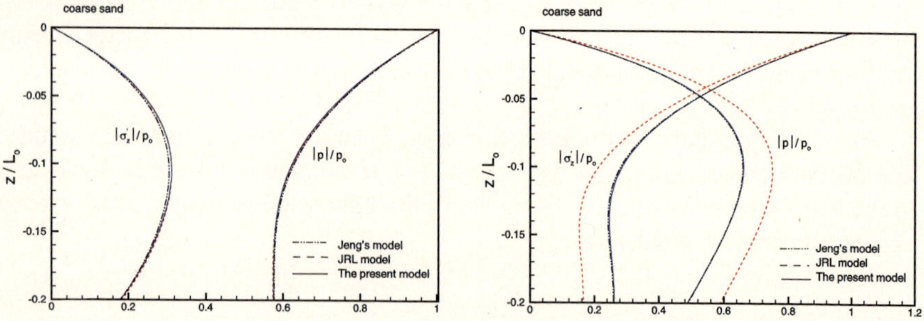

Fig. 7.4 Vertical distribution of the wave-induced pore pressure ($|p|/p_o$) and vertical effective stresses (σ_z') versus soil depth (z/L_o)in a saturated seabed ($S_r = 1$). *Solid lines* denote the present solution with Coulomb-damping friction, and *dashed lines* are the results without Coulomb-damping friction

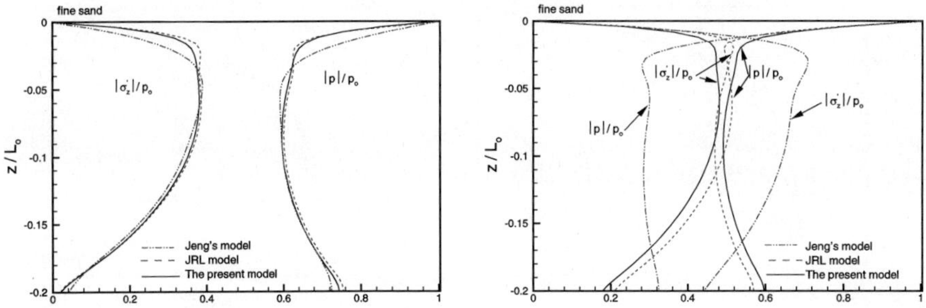

Fig. 7.5 Vertical distribution of the wave-induced pore pressure ($|p|/p_o$) and vertical effective stresses (σ_z') versus soil depth (z/L_o)in a unsaturated seabed ($S_r = 0.98$). *Solid lines* denote the present solution with Coulomb-damping friction, and *dashed lines* are the results without Coulomb-damping friction

demonstrates the significant effects of Coulomb-damping friction on the soil response in an unsaturated seabed. For example, the maximum differences of pore pressure between three solutions reach 10 % of p_o in coarse sand (Figs. 7.4 and 7.5). It is also observed that the Coulomb-damping friction is more important in a softer seabed (such as fine sand in Figs. 7.4–7.5).

The degree of saturation is involved in the determination of compressibility of pore fluid. This also included in the Biot's elastic constants, H, C, and M, which directly affects the coefficients given in the general solution. The significant influence of the degree of saturation on the wave-induced soil response is also observed in Fig. 7.5. In general, a lower degree of saturation will enhance the effects of Coulomb damping friction on the wave-induced soil response (Fig. 7.5). A possible explanation for this result is that unsaturated seabed included certain amount of gas in pore space. This will increase the internal friction between soil particles.

The major difference between the present solution and Yamamoto's solution [14] is that we consider the seabed of finite thickness, which is expected to have significant contribution to the soil response. Figure 7.6 presents the amplitude of the wave-induced pore pressure ($|p|/p_o$) for various seabed thickness (h/L_o) in three different seabeds. In this example, the seabed is considered to be coarse sand, fine sand and clay. The figure clearly indicates that the relative difference of pore pressure between finite thickness (for example, $h/L_o = 0.1$) and infinite thickness reach 30 % of p_o in clay. This is greater than that in sandy seabed, which was reported to have about 20 % of p_o [5]. This implies the Coulomb-damping friction will enhance the importance of seabed thickness in softer seabed (such as clay or silt) than sandy beds.

According to Figs. 7.5 and 7.6, the influence of soil type can also be observed. Generally speaking, the maximum amplitude of the wave-induced pore pressure ($|p|/p_o$) decreases smoothly with soil depth (z/L_o) in coarse sand. However, there is a rather rapidly decrease near the seabed surface in both fine sand and clay. It is noted that a great increase in $|p|/p_o$ near the seabed bottom. This may be explained by the fact that the energy is reflected from the bottom of the seabed, leading to an increase in the amplitude of the wave-induced pore pressure.

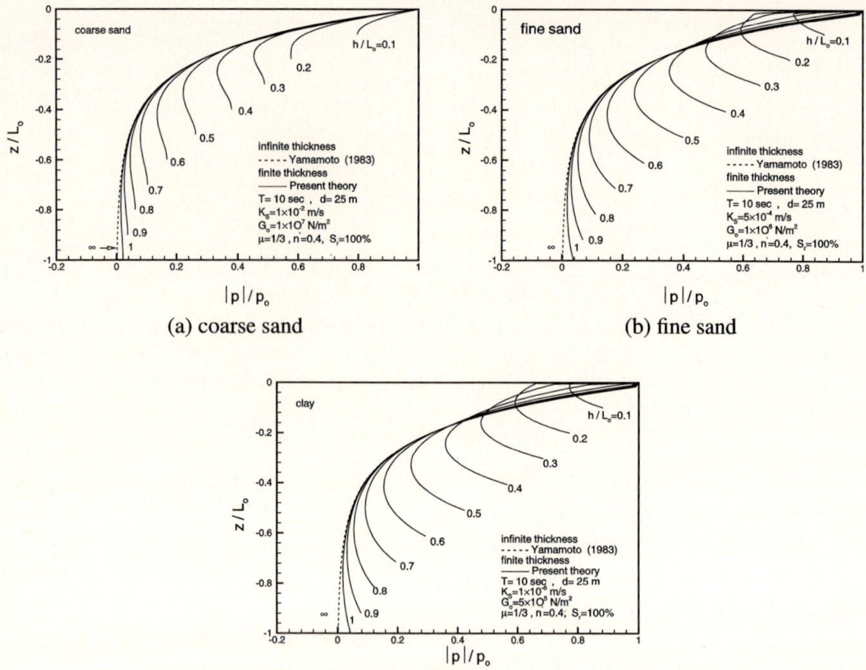

(a) coarse sand (b) fine sand

(c) clay

Fig. 7.6 Vertical distribution of the wave-induced pore pressure ($|p|/p_o$) versus soil depth (z/L_o) for various seabed thickness (h/L_o) in different seabeds

Note that the inclusion of fluid acceleration significantly contributes the decrease of pore pressure [7], compared with previous quasi-static solution [4, 5]. However, the inclusion of Coulomb-damping friction (the present solution) contributes the increase of pore pressure from the JRL model [7]. This results in the present model closes to quasi-static solution [5].

7.5.2 *Response of Seabed to Ocean Waves*

Another objective of this paper is to investigate the influence of soil properties on the wave characteristic in ocean environments. This phenomenon was firstly mentioned by Yamamoto [14]; and it has been investigated through a parametric study with the quasi-static solution [6]. However, how Coulomb-damping friction affects the wave characteristics have not been examined in the past.

7.5.2.1 Wave Damping

Figure 7.7 illustrates the wave profile in three different seabeds. The figure indicates that the damping effect in clay is more significant than sandy beds. This can

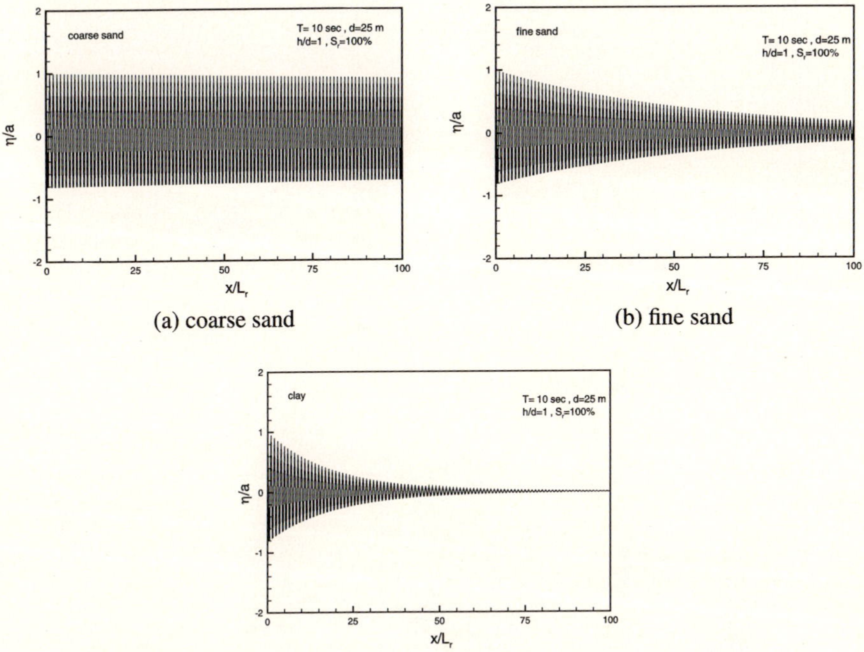

(a) coarse sand

(b) fine sand

(c) clay

Fig. 7.7 Distribution of wave profile (η/a) versus x/L_r in different seabed. ($h/d = 1.0$)

be explained by the fact that the velocities of compressional and shear waves within sediment is much greater than that of water waves in coarse sand. Thus, the phenomenon of damping is insignificant. On the other hand, the celerity of shear waves in soil is smaller than water waves in fine sand, leading to a resonance vibration in the seabed. This phenomenon is more obvious in clay.

To have a better understanding of the influence of Coulomb damping friction on the wave damping, both results for the amplitude of the wave profile ($|\eta|/a$) from the present solution (in solid lines) and the previous solution without Coulomb damping friction (in dashed lines [6]) are included in Fig. 7.8. As shown in the figure, the Coulomb damping friction definitely enhances the decaying of the wave profile, especially in fine sand (Fig. 7.8b), compared with that in coarse sand (Fig. 7.8a). It is also observed from Fig. 7.8 that the degree of saturation doesn't have significant influences on the decaying of the wave profile.

Further demonstration for the wave energy dispersion within seabed, the wave energy decaying rate, ϵ_D/EC_g, in three different seabeds are presented in Fig. 7.9. Herein, the energy decaying rate is defined by [9, 10]

$$\frac{\epsilon_D}{E_0 C_g} = \frac{\rho_f K_S \omega \lambda_r}{\eta_f} \frac{4}{2\lambda_r d + \sinh 2\lambda_r d} e^{-2\lambda_i x} \tag{7.64}$$

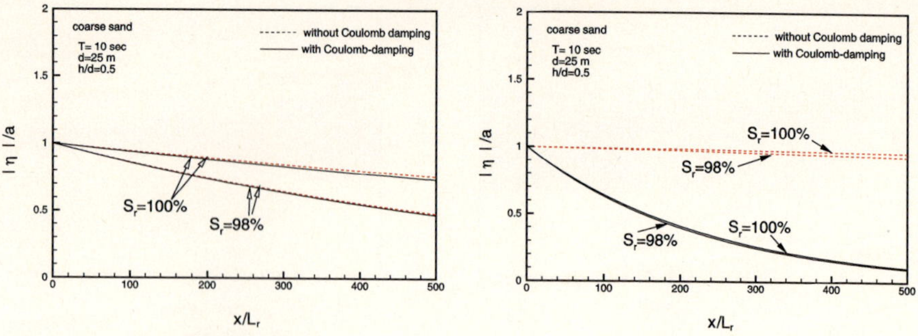

(a) coarse sand

(b) fine sand

Fig. 7.8 Distribution of the amplitude of wave profile ($|\eta|/a$) versus x/L_r in (**a**) coarse sand and (**b**) fine sand. ($h/d = 0.5$)

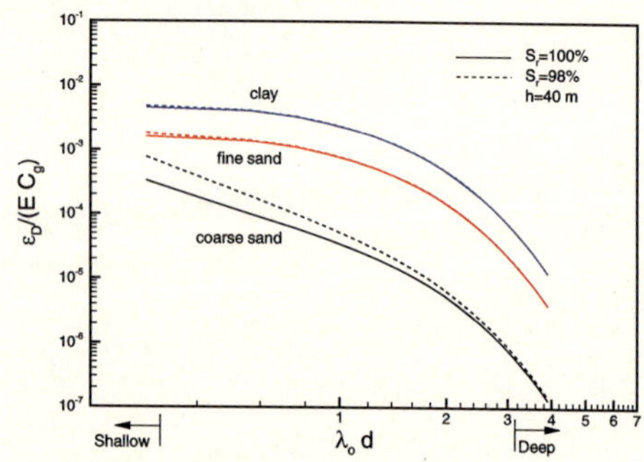

Fig. 7.9 Distribution of non-dimensional energy decaying factor (ϵ_D/EC_g) versus relative water depth ($\lambda_o d$) for different seabeds. ($h = 40$ m)

where λ_r and λ_i are the real and imaging part of the complex wave number λ, respectively.

Figure 7.9 clearly shows that the order of energy ϵ_D/EC_g in clay is about fifty times of that in coarse sand. This may be explained by the fact that the energy is much easier to transmit through the pore space in coarser material due to the high permeability. In general, the value of ϵ_D/EC_g decreases as the relative water depth ($k_o d$) increases. That is, the energy loss is more significant in shallow water. Figure 7.9 also indicates the influence of the degree of saturation on ϵ_D/EC_g, although it is insignificant.

The influence of the seabed thickness on the non-dimensional energy decaying factor (ϵ_D/EC_g) for three different seabeds are illustrated in Fig. 7.10. The energy decaying rate (ϵ_D/EC_g), in general, increases as seabed thickness increases. Furthermore, the influence of seabed thickness on ϵ_D/EC_g becomes more important in deep water, compare with that in shallow water.

7.5.2.2 Wavelength

The determination of the wavelength is particularly important in the design of wave condition near marine structures. In conventional ocean engineering, most wave the-

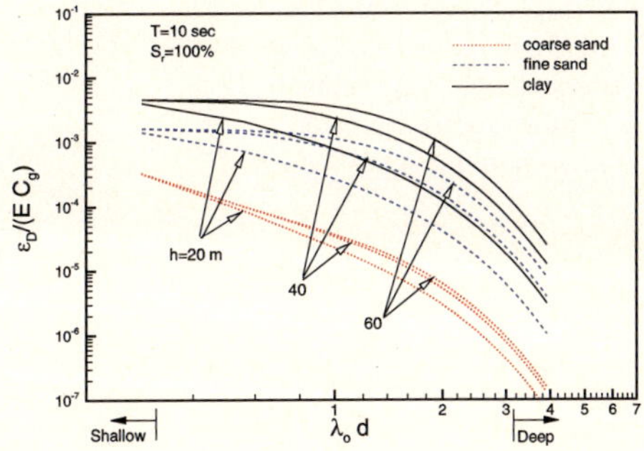

Fig. 7.10 Distribution of non-dimensional energy decaying factor (ϵ_D/EC_g) versus relative water depth ($\lambda_o d$) for various seabed thickness

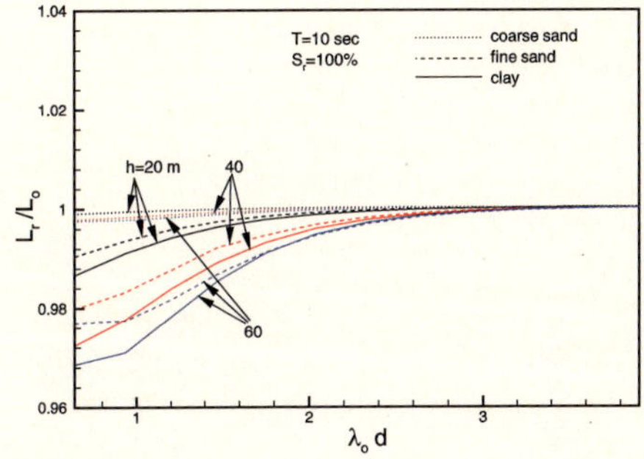

Fig. 7.11 Distribution of the wavelength ratio (L_r/L_o) versus relative water depth ($\lambda_o d$) for various seabed thickness in different seabeds

ories have been based on the assumption of wave propagating over an impermeable seabed [2]. Thus, the wavelength is a constant, which is determined by wave dispersion equation with given wave period and water depth. However, in a natural seabed, the material of the seabed is often to be porous. In such a case, the wavelength will be modified by the seabed characteristics. Therefore, the wavelength is no longer a constant. It should change as the wave cycle increases.

The effects of seabed thickness on the normalized wavelength (L_r/L_o) are presented in Fig. 7.11. It is observed that L_r/L_o decreases as the relative water depth ($\lambda_0 d = 2d/L_o$) decreases, implying the effects of soil properties on the wavelength in more important in shallow water. The figure also demonstrates the significant influence of seabed thickness on the wavelength in fine sand and clay. In general, a thick clay will reduce the wavelength more rapidly, compared with coarse sand and fine sand.

It is also interesting to examine the influence of Coulomb damping friction on the wavelength ratio (L_r/L_o). As seen in Fig. 7.12, inclusion of Coulomb damping friction will reduce the reduction of the wavelength in both saturated and unsaturated seabed.

7.5.2.3 Wave Pressure

As mentioned previously, conventional investigations didn't coupled the water wave filed and porous seabed in a model. Thus, the wave pressure at the surface of the

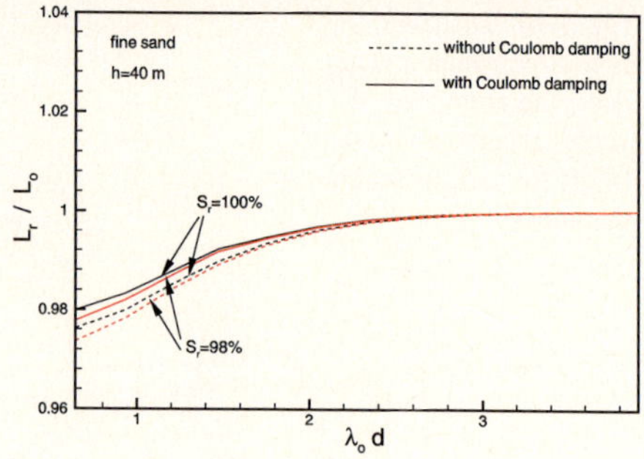

Fig. 7.12 Distribution of the wavelength ratio (L_r/L_o) versus relative water depth ($\lambda_o d$) for various degree of saturation in fine sand

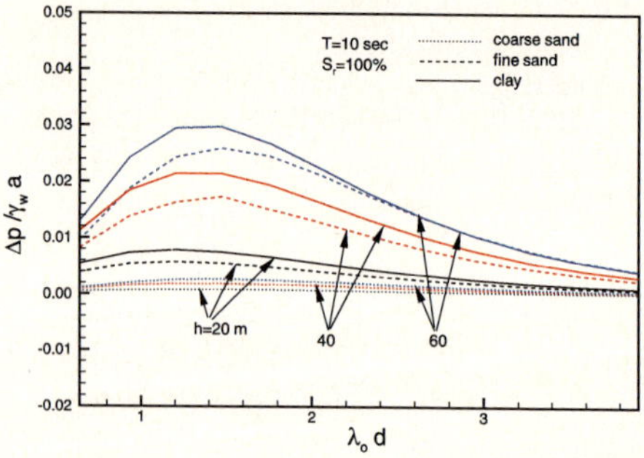

Fig. 7.13 Distribution of the relative wave pressure at the seabed surface ($\Delta p/\gamma_w a$) versus relative water depth ($\lambda_o d$) for various seabed thickness in different seabeds

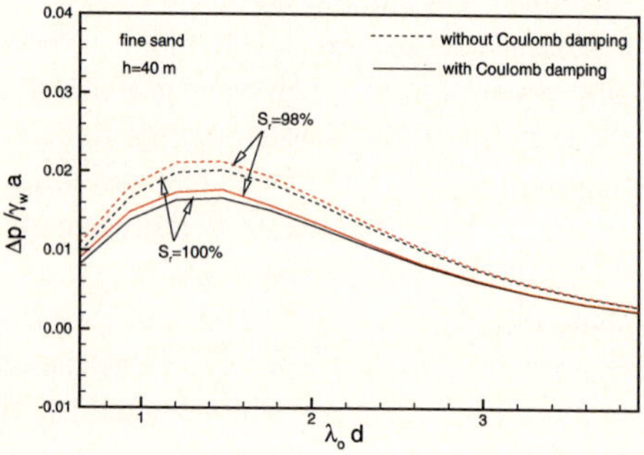

Fig. 7.14 Distribution of the relative wave pressure at the seabed surface ($\Delta p/\gamma_w a$) versus relative water depth ($\lambda_o d$) for various degree of saturation in fine sand

porous seabed is a constant. In this study, we employ a complex wave number, which allows us to couple wave field and porous flow. Therefore, the wave pressure along the surface of the seabed is affected by the soil properties.

Figure 7.13 illustrates the influences of soil properties on the relative difference of wave pressure along the seabed surface ($\Delta p/\gamma_w a$). The figure clearly indicates that the value of $\Delta p/\gamma_w a$ increases as seabed thickness increases. The maximum value of $\Delta p/\gamma_w a$ in this example can reach 3 % of $\gamma_w a$. It also observed that effects of soil properties on the wave pressure are more significant in clay and fine sand.

The influence of Coulomb damping friction on the relative difference of wave pressure along the seabed surface ($\Delta p/\gamma_w a$) is illustrated in Fig. 7.14. It is found that inclusion of Coulomb damping friction will reduced the relative difference of wave pressure.

7.6 Summary

In this chapter, a closed-form analytical solution for ocean waves propagating over a Coulomb-damping seabed is proposed. The porous seabed is considered to be finite thickness in this model. Based on the numerical results, the following conclusions can be drawn:

1. The present analytical solution overall agrees with previous experimental data in both sandy and silt beds.
2. The effect of Coulomb-damping friction on the wave-induced seabed response is important in an unsaturated seabed. The maximum difference of pore pressure from the present solution and the previous solution without Coulomb-damping friction may reach 10 % of p_o under certain combination of wave and seabed conditions.
3. The inclusion of fluid acceleration will reduce the amplitude of wave-induced pore pressure, especially in an unsaturated seabed.
4. The maximum relative difference between the present solution (finite thickness) and Yamamoto's solution (infinite thickness) reach 30 % of p_o in clay.
5. The soil properties significantly affects the wave characteristics such as the wave-length, wave profile and wave pressure. The influence of soil properties on the wave characteristics is more important in softer seabed (such as fine sand and clay).

7.7 Appendix: List of Coefficients a_1–a_6

Coefficients a_i in (7.41)–(7.44) are expressed as

$$a_i = \frac{N_i}{N_0} \quad (i = 1, 2, \ldots, 6) \tag{7.65}$$

where

$$\begin{aligned} N_i = {} & N_0^{(i)} + N_1^{(i)} e^{-2(\lambda_2+\lambda_3)h} + N_2^{(i)} e^{-2\lambda_2 h} + N_3^{(i)} e^{-2\lambda_3 h} \\ & + N_4^{(i)} e^{-(\lambda_1+\lambda_3)h} + N_5^{(i)} e^{-(\lambda_1+2\lambda_2+\lambda_3)h} + N_6^{(i)} e^{-(\lambda_2+\lambda_3)h} \\ & + N_7^{(i)} e^{-(\lambda_1+\lambda_2+2\lambda_3)h} + N_8^{(i)} e^{-2(\lambda_1+\lambda_2)h} + N_9^{(i)} e^{-2(\lambda_1+\lambda_2+\lambda_3)h} \\ & + N_{10}^{(i)} e^{-2(\lambda_1+\lambda_3)h} + N_{11}^{(i)} e^{-1(\lambda_1+\lambda_2)h} + N_{12}^{(i)} e^{-2\lambda_1 h} \\ & + N_{13}^{(i)} e^{-(2\lambda_1+\lambda_2+\lambda_3)h}. \end{aligned} \tag{7.66}$$

The N coefficients are given by

$$N_0^{(0)} = N_9^{(0)} = G\{[D_4\lambda_1 s_1^2(\lambda_2\lambda_3 - \lambda^2) + D_5\lambda_2 s_2^2(\lambda^2 - \lambda_1\lambda_3)] \times [-D_2 D_3 D_4 s_1^2 + D_1 D_3 D_5 s_2^2 + 4\lambda_3\lambda^2(D_4\lambda_2 s_1^2 - D_5\lambda_1 s_2^2)]\}, \tag{7.67}$$

$$N_1^{(0)} = N_{12}^{(0)} = G\{[D_4\lambda_1 s_1^2(\lambda_2\lambda_3 - \lambda^2) - D_5\lambda_2 s_2^2(\lambda^2 - \lambda_1\lambda_3)] \times [-D_2 D_3 D_4 s_1^2 + D_1 D_3 D_5 s_2^2 + 4\lambda_3\lambda^2(D_4\lambda_2 s_1^2 - D_5\lambda_1 s_2^2)]\}, \tag{7.68}$$

$$N_2^{(0)} = N_{10}^{(0)} = G\{[D_4\lambda_1 s_1^2(\lambda_2\lambda_3 + \lambda^2) + D_5\lambda_2 s_2^2(\lambda^2 - \lambda_1\lambda_3)] \times [-D_2 D_3 D_4 s_1^2 + D_1 D_3 D_5 s_2^2 - 4\lambda_3\lambda^2(D_4\lambda_2 s_1^2 - D_5\lambda_1 s_2^2)]\}, \tag{7.69}$$

$$N_3^{(0)} = N_{11}^{(0)} = -G\{[D_4\lambda_1 s_1^2(\lambda_2\lambda_3 - \lambda^2) - D_5\lambda_2 s_2^2(\lambda^2 - \lambda_1\lambda_3)] \times [D_2 D_3 D_4 s_1^2 - D_1 D_3 D_5 s_2^2 + 4\lambda_3\lambda^2(D_4\lambda_2 s_1^2 - D_5\lambda_1 s_2^2)]\}, \tag{7.70}$$

$$N_4^{(0)} = N_5^{(0)} = 8D_5\lambda^2\lambda_1\lambda_2\lambda_3 s_2^2 G[-D_2 D_4 s_1^2 + D_1 D_5 s_2^2 + D_3 G(D_5 s_2^2 - D_4 s_1^2)], \tag{7.71}$$

$$N_6^{(0)} = 8D_5\lambda^2\lambda_1\lambda_2\lambda_3 s_2^2 G[D_2 D_4 s_1^2 - D_1 D_5 s_2^2 + D_3 G(D_4 s_1^2 - D_5 s_2^2)], \tag{7.72}$$

$$N_7^{(0)} = N_8^{(0)} = 32 D_4 D_5 \lambda^4 \lambda_1\lambda_2\lambda_3 s_1^2 s_2^2 G, \tag{7.73}$$

$$N_0^{(1)} = N_9^{(2)} = -G[\lambda_2\lambda^2 s_3^2 - \lambda_1(\lambda^2 s_1^2 - \lambda_2\lambda_3(s_1^2 - s_3^2))] \times [-D_5 s_2^2(D_3 - D_4) - 2G\lambda^2(D_5 - 2\lambda_2\lambda_3)], \tag{7.74}$$

$$N_1^{(1)} = N_{12}^{(2)} = -G[-\lambda_2\lambda^2 s_3^2 - \lambda_1(\lambda^2 s_1^2 - \lambda_2\lambda_3(s_1^2 - s_3^2))] \times [-D_5 s_2^2(D_3 - D_4) - 2G\lambda^2(D_5 - 2\lambda_2\lambda_3)], \tag{7.75}$$

$$N_2^{(1)} = N_{10}^{(2)} = G[\lambda_2\lambda^2 s_3^2 + \lambda_1(\lambda^2 s_1^2 + \lambda_2\lambda_3(s_1^2 - s_3^2))] \times [D_5 s_2^2(D_3 - D_4) + 2G\lambda^2(D_5 - 2\lambda_2\lambda_3)], \tag{7.76}$$

$$N_3^{(1)} = N_{11}^{(2)} = G[-\lambda_2\lambda^2 s_3^2 + \lambda_1(\lambda^2 s_1^2 + \lambda_2\lambda_3(s_1^2 - s_3^2))] \times [D_5 s_2^2(D_3 - D_4) + 2G\lambda^2(D_5 - 2\lambda_2\lambda_3)], \tag{7.77}$$

$$N_4^{(1)} = 4G\lambda_2\lambda^2\lambda_3 s_3^2[\lambda_2((D_1 - D_2)s_1^2 + 2G\lambda^2) + \lambda_1((D_3 - D_4)s_2^2 + 2G\lambda^2)] = N_5^{(2)}, \tag{7.78}$$

$$N_5^{(1)} = 4G\lambda_2\lambda^2\lambda_3 s_3^2[-\lambda_2((D_1 - D_2)s_1^2 + 2G\lambda^2) + \lambda_1((D_3 - D_4)s_2^2 + 2G\lambda^2)] = N_4^{(2)}, \tag{7.79}$$

$$N_6^{(1)} = N_{13}^{(2)} = 8G\lambda^2\lambda_1\lambda_2\lambda_3\left[-s_1^2\left((D_3-D_4)s_2^2+2G\lambda^2\right)-D_5G\left(s_1^2-s_3^2\right)\right], \tag{7.80}$$

$$N_7^{(1)} = N_8^{(2)} = 2G\lambda^2\lambda_2 s_3^2\left[-4\lambda^2\lambda_1\lambda_3 G + D_5\left((D_1-D_2)s_1^2+2\lambda^2 G\right)\right], \tag{7.81}$$

$$N_8^{(1)} = N_7^{(2)} = -2G\lambda^2\lambda_2 s_3^2\left[4\lambda^2\lambda_1\lambda_3 G + D_5\left((D_1-D_2)s_1^2+2\lambda^2 G\right)\right], \tag{7.82}$$

$$\begin{aligned} N_0^{(3)} &= G\left[D_4\lambda_1 s_1^2\left(\lambda_2\lambda_3-\lambda^2\right)+D_5\lambda_2 s_2^2\left(\lambda^2-\lambda_1\lambda_3\right)\right]\left(D_1D_3-4\lambda^2\lambda_1\lambda_3 G\right) \\ &= N_{10}^{(3)}, \end{aligned} \tag{7.83}$$

$$\begin{aligned} N_3^{(3)} &= G\left[D_4\lambda_1 s_1^2\left(\lambda_2\lambda_3+\lambda^2\right)-D_5\lambda_2 s_2^2\left(\lambda^2+\lambda_1\lambda_3\right)\right]\left(D_1D_3+4\lambda^2\lambda_1\lambda_3 G\right) \\ &= N_{12}^{(3)}, \end{aligned} \tag{7.84}$$

$$N_4^{(3)} = N_5^{(4)} = 8\lambda^2\lambda_1\lambda_2\lambda_3 G\left[D_1D_5s_2^2+D_3G\left(-D_4s_1^2+D_5s_2^2\right)\right], \tag{7.85}$$

$$N_6^{(3)} = N_{13}^{(4)} = -4D_4\lambda^2\lambda_1\lambda_3 s_1^2 G(D_2\lambda_1+D_1\lambda_2), \tag{7.86}$$

$$N_7^{(3)} = N_8^{(4)} = 2D_4\lambda^2\lambda_1 s_1^2 G\left(-D_2D_3+4\lambda^2\lambda_2\lambda_3 G\right), \tag{7.87}$$

$$N_8^{(3)} = N_7^{(4)} = 2D_4\lambda^2\lambda_1 s_1^2 G\left(D_2D_3+4\lambda^2\lambda_2\lambda_3 G\right), \tag{7.88}$$

$$N_{13}^{(3)} = N_6^{(4)} = 4D_4\lambda^2\lambda_1\lambda_3 s_1^2 G(D_2\lambda_1-D_1\lambda_2), \tag{7.89}$$

$$N_0^{(5)} = N_9^{(6)} = -2i\lambda G(D_2\lambda_1-D_1\lambda_2)\left[D_4\lambda_1 s_1^2\left(\lambda_2\lambda_3-\lambda^2\right)+D_5\lambda_2\left(\lambda^2-\lambda_1\lambda_3\right)\right], \tag{7.90}$$

$$N_2^{(5)} = N_{10}^{(6)} = -2i\lambda G(D_2\lambda_1+D_1\lambda_2)\left[D_4\lambda_1 s_1^2\left(\lambda_2\lambda_3-\lambda^2\right)+D_5\lambda_2\left(\lambda^2-\lambda_1\lambda_3\right)\right], \tag{7.91}$$

$$N_4^{(5)} = N_5^{(6)} = -2i\lambda\lambda_1\lambda_2 G\left(D_4s_1^2-D_5s_2^2\right)\left(D_2D_3+4\lambda^2\lambda_2\lambda_3 G\right), \tag{7.92}$$

$$N_5^{(5)} = N_4^{(6)} = -2i\lambda\lambda_1\lambda_2 G\left(D_4s_1^2-D_5s_2^2\right)\left(D_2D_3-4\lambda^2\lambda_2\lambda_3 G\right), \tag{7.93}$$

$$N_6^{(5)} = N_{13}^{(6)} = 2i\lambda\lambda_1\lambda_2 G\left(D_4s_1^2-D_5s_2^2\right)\left(D_2D_3+4\lambda^2\lambda_2\lambda_3 G\right), \tag{7.94}$$

$$N_8^{(5)} = N_7^{(6)} = 8i\lambda^3\lambda_1\lambda_2 G\left(D_2D_4s_1^2+D_1D_5s_2^2\right), \tag{7.95}$$

$$N_{11}^{(5)} = N_3^{(6)} = 2i\lambda G(D_2\lambda_1-D_1\lambda_2)\left[D_4\lambda_1 s_1^2\left(\lambda_2\lambda_3-\lambda^2\right)-D_5\lambda_2\left(\lambda^2-\lambda_1\lambda_3\right)\right], \tag{7.96}$$

$$N_{12}^{(5)} = N_1^{(6)} = 2i\lambda G(D_2\lambda_1+D_1\lambda_2)\left[D_4\lambda_1 s_1^2\left(\lambda_2\lambda_3-\lambda^2\right)-D_5\lambda_2\left(\lambda^2-\lambda_1\lambda_3\right)\right], \tag{7.97}$$

$$N_{13}^{(5)} = N_6^{(6)} = 2i\lambda\lambda_1\lambda_2 G\left(D_4s_1^2-D_5s_2^2\right)\left[D_2D_3-4\lambda^2\lambda_2\lambda_3 G\right], \tag{7.98}$$

$$\begin{aligned} &D_1 = C+\delta_1 M, \qquad D_2 = C+\delta_2 M, \qquad D_3 = 2\lambda^2-s_3^2, \\ &D_4 = H+\delta_1 M, \qquad D_5 = H+\delta_2 M. \end{aligned} \tag{7.99}$$

References

1. Dalrymple, R., Liu, P.L.F.: Waves over soft muds: A two-layer fluid model. J. Phys. Oceanogr. **8**, 1121–1131 (1978)
2. Fenton, J.D., McKee, W.D.: On calculating the length of water waves. Coast. Eng. **14**, 499–513 (1990)
3. Hardin, B.O., Drnevich, V.P.: Shear modulus and damping in soils: Design equations and curves. J. Soil Mech. Found. Div. **98**(7), 667–692 (1972)
4. Hsu, J.R.C., Jeng, D.S.: Wave-induced soil response in an unsaturated anisotropic seabed of finite thickness. Int. J. Numer. Anal. Methods Geomech. **18**(11), 785–807 (1994)
5. Jeng, D.S.: Wave-induced seabed response in front of a breakwater. Ph.D. thesis, University of Western Australia (1997)
6. Jeng, D.S.: On calculating the length of a short-crested wave over a porous seabed. Appl. Ocean Res. **22**(2), 63–73 (2000)
7. Jeng, D.S., Rahman, M.S., Lee, T.L.: Effects of inertia forces on wave-induced seabed response. Int. J. Offshore Polar Eng. **9**(4), 307–313 (1999)
8. Lee, T.L., Tsai, C.P., Jeng, D.S.: Ocean waves propagating over a Coulomb-damped poroelastic seabed of finite thickness: An analytical solution. Comput. Geotech. **29**(2), 119–149 (2002)
9. Putnam, J.A.: Loss of wave energy due to percolation in a permeable sea bottom. Trans. Am. Geophys. Union **30**(3), 349–356 (1949)
10. Reid, R.O., Kajiura, K.: On the damping of gravity waves over a permeable sea bed. Trans. Am. Geophys. Union **38**, 662–666 (1957)
11. Stokoe, K.H., Isenhower, W.M., Hsu, J.R.: Dynamic properties of offshore silty samples. In: Proceedings 12th Annual Offshore Technology Conference, pp. 289–302 (1989)
12. Tzang, S.Y.: Unfluidized soil responses of a silty seabed to monochromatic waves. Coast. Eng. **35**(4), 283–301 (1998)
13. Yamamoto, T.: Non-linear mechanics of ocean wave interactions with sediment beds. Appl. Ocean Res. **4**(2), 99–106 (1982)
14. Yamamoto, T.: On the response of a Coulomb-damped poro-plastic bed to water waves. Mar. Geotechnol. **5**(2), 93–130 (1983)
15. Yamamoto, T., Koning, H., Sellmeijer, H., Hijum, E.V.: On the response of a poro-elastic bed to water waves. J. Fluid Mech. **87**(1), 193–206 (1978)
16. Yamamoto, T., Suchukman, B.: Experiments and theory of wave-soil interactions. J. Eng. Mech. **110**(1), 95–112 (1984)

Chapter 8
Random Wave-Induced Seabed Response

Abstract In this chapter, unlike most previous investigations for wave-induced soil response, a simple semi-analytical model for the random wave-induced soil response is established for an unsaturated seabed of finite thickness. Two different wave spectra, the B–M and JONSWAP spectra, are considered. The effects of random wave loading on the soil response is investigated by comparing with the corresponding representative regular wave results through a parametric study, which includes the effect of the degree of saturation, soil permeability, wave height, wave period and seabed thickness. The maximum liquefaction depth under the random waves is also examined. The difference on the soil response under the two random wave types, B–M and JONSWAP frequency spectra, is also discussed in the present work.

Keywords Soil response · Liquefaction · Pore pressure · Random waves · Wave spectrum

8.1 Introduction

To date, only a few studies have been carried out to consider the variations on the soil responses in the marine sediments under random wave loadings. Sumer and Fredsøe [4] performed experimental tests to investigate the effect of irregular wave on soil response, and they found the process of build-up of pore pressure in irregular waves occurs in much the same way as in the case of the regular wave. In this chapter, based on the framework proposed in Chap. 3, we will establish a semi-analytical solution for the random wave-induced soil response. Two typical frequency spectra, B–M spectrum and JONSWAP spectrum, will be utilized to generate the random waves. The present investigation is a twofold work: the soil response of infinite thickness and that of finite thickness. The objective of this study is to investigate the random wave-induced soil response in the marine sediments. First, the random wave theory in which the present research domain, random wave generation and validation, statistic features of the simulated random waves together with the representative regular wave characteristics are discussed in detail. Second, random wave-induced oscillatory pore pressure and effective stress in marine sediments will be discussed. Finally, conclusions from the present study will be drawn.

D.-S. Jeng, *Porous Models for Wave-seabed Interactions*,
DOI 10.1007/978-3-642-33593-8_8,

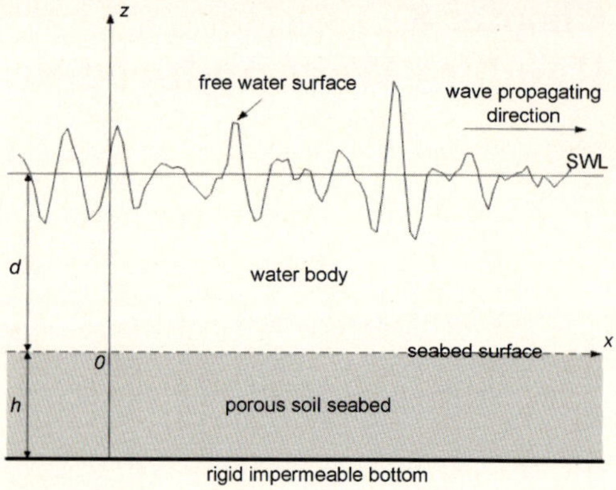

Fig. 8.1 Definition sketch of random wave propagating over a porous seabed

8.2 Random Waves

In this chapter, we consider a series of random waves propagating over a porous seafloor with an infinite thickness or a finite thickness (h) above a rigid impermeable bottom, as shown in Fig. 8.1. In realistic ocean environments, random waves instead of regular linear wave loadings always occur. Such random waves characterizing with irregular water surface elevation introduce the relevant change on the wave dynamic pressure acting at the seabed, which further induce the variation of pore water pressure and effective stresses inside the marine sediments.

Herein, we consider the flow is two-dimensional and assume the random waves with still water level (SWL) located at $z = d$ are traveling along the positive x-direction, and assume the vertical z-axis is upward from the surface of the seabed (water-soil interface, $z = 0$) as illustrated in Fig. 8.1. A porous seabed with a soil thickness h is located between the seabed surface and a rigid impermeable bottom for a finite soil depth.

8.2.1 Random Wave Generation

The profile of random sea waves can be regarded as a stationarily stochastic process, which follows the Gaussian (normal) distribution. Such a process satisfies the property of ergodicity. When we consider the profile of random sea waves, following the mathematical representations of Longuet-Higgins [3], the wave profile or the water surface elevation can be represented by,

$$\eta(x,t) = \sum_{i=1}^{\infty} a_i \cos(k_i x - 2\pi \tilde{f}_i t + \epsilon_i) \approx \sum_{i=1}^{M} a_i \cos(k_i x - 2\pi \tilde{f}_i t + \epsilon_i), \qquad (8.1)$$

with M being a sufficiently large number. In (8.1), a_i denotes the amplitude of the component wave in the ith frequency, $\tilde{f}_i$ is the ith representative frequency, which is evenly distributed in the range of (f_{i-1}, f_i), and ϵ_i is a random initial phase angle and equally distributed in the range of (0, 2 π). In (8.1), wave number of the ith component, k_i, can be determined from the dispersion relationship after knowing the corresponding representative frequency $\tilde{f}_i$ and water depth d,

$$(2\pi \tilde{f}_i)^2 = g k_i \tanh k_i d, \tag{8.2}$$

in which g denotes the gravitational acceleration. The component wave amplitude a_i is determined from a given function of the frequency spectrum $S(f)$ by,

$$a_i = \sqrt{2S(\tilde{f}_i)\Delta f_i}, \quad \tilde{f}_i = \frac{f_i + f_{i-1}}{2}, \quad \Delta f_i = f_i - f_{i-1}. \tag{8.3}$$

In this study, two commonly used frequency spectra, B–M spectrum and JONSWAP spectrum, are adopted for further discussions. Here, we summarized these two standard frequency spectral density functions as: [1]

- B–M spectrum,

$$S(f) = 0.257 H_{1/3}^2 T_{1/3}^{-4} f^{-5} \exp\left(-1.03 (T_{1/3} f)^{-4}\right), \tag{8.4}$$

where $H_{1/3}$ and $T_{1/3}$ denote the highest one-third wave (or significant wave) height and the relevant wave period respectively.
- JONSWAP spectrum,

$$S(f) = \beta_J H_{1/3}^2 T_{1/3}^{-4} f^{-5} \exp\left(-1.25 (T_p f)^{-4}\right) \gamma^{\exp(-(T_p f^{-1})^2/2\sigma^2)}, \tag{8.5}$$

in which,

$$\beta_J = \frac{0.0624(1.094 - 0.01915 \ln \gamma)}{0.230 + 0.0336\gamma - 0.185(1.9 + \gamma)^{-1}}, \tag{8.6}$$

$$T_P = \frac{T_{1/3}}{1 - 0.132(\gamma + 0.2)^{-0.559}}, \tag{8.7}$$

$$\sigma = \begin{cases} \sigma_a : & f \le f_p, \\ \sigma_b : & f \ge f_p, \end{cases} \tag{8.8}$$

where T_p represents the wave period corresponding to the frequency f_p at the spectral peak. The JONSWAP spectrum is characterized by the peak enhancement factor γ, which controls the sharpness of the spectral peak. In general, $\sigma_a = 0.07$, $\sigma_b = 0.09$ and γ = 1–7 (mean of 3.3) [1].

Two points need to be specified here,

- **Determine the Frequency Range of Random Sea Waves**: As we know, for random waves, the consisted frequency is varying from 0 Hz to $+\infty$ Hz. In the simulation, we set the frequency range between low frequency F_L and high frequency F_H. This means we neglect certain part of wave energy which is contained by the frequencies beyond this range. Assuming μ percent energy is omitted both in the range of $[0, F_L]$ and $[F\_H, +\infty]$ (i.e., $\mu = 0.2$ in the simulation), for integrable spectrums, i.e., B–M spectrum in (8.4), we have,

$$F_L = \frac{1}{T_{1/3}}\left(-\frac{1.03}{\ln \mu}\right)^{1/4} \quad \text{and} \quad F_H = \frac{1}{T_{1/3}}\left(-\frac{1.03}{\ln(1-\mu)}\right)^{1/4}. \tag{8.9}$$

For JONSWAP spectrum in (8.5), it is not so easy to be integrated. Consider the wave energy is mainly concentrated around the peak frequency, we assume,

$$F_L = 0 \quad \text{and} \quad f_H = 5 f_p, \tag{8.10}$$

where f_p can be determined from (8.7), as $f_p = 2\pi/t_p$.

- **Separate the Frequency Range** $[f_L, f_H]$: Two methods can be applied: equally separating the frequency and equally separating the energy. In the program, equally separating the frequency is utilized, which assumes the frequency is equally divided in the range $[f_L, f_H]$,

$$\Delta f = \frac{f_H - f_L}{M}. \tag{8.11}$$

Hence, the whole frequency domain is equally separated as $[f_L, f_L + \delta f], \ldots, [f_H - \delta f, f_H]$. The ith representative frequency, $\tilde{f}_i$, is evenly distributed in the range of $[f_{i-1}, f_i]$ and determined through a random number.

Using (8.3), the corresponding ith component wave amplitude can be determined. Hence, the time history of random wave surface elevations can be obtained from (8.1).

8.2.2 Random Wave Simulation

To generate the random waves with aforementioned frequency spectra, the whole range of wave frequency is evenly separated within the specified range of $[f_L, f_H]$, where most of the wave energy is located as specified in the previous context. In this study, we assume the simulated random waves consist of 100 linear wave components, i.e., $M = 100$. Goda [1] mentioned that in field observation, to keep the sampling variability of the characteristic wave heights and periods below an acceptable level, a recording length of about 20 min is employed as a compromise between the requirements of wanting low sampling variability and having a stationary, which includes about 100 consecutive waves. The data sampling interval of 1/10 to 1/20 of the significant wave period is recommended.

Figure 8.2 illustrates the time history of the water surface elevation at the fixed point. The upper panel demonstrates the result using B–M spectrum while the lower one is for the result of JONSWAP spectrum. Water surface irregularity caused by

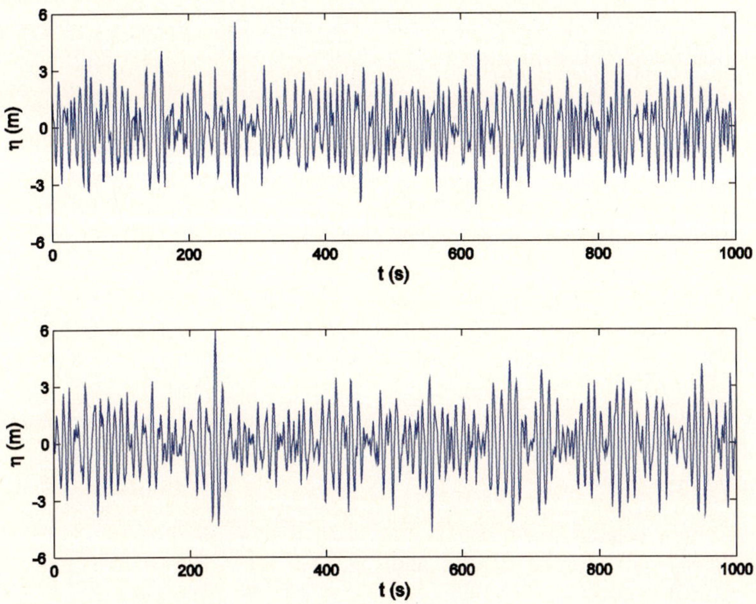

Fig. 8.2 Time history of simulated random wave profiles using different frequency spectra: B–M spectrum (*upper panel*), JONSWAP spectrum (*lower panel*)

the wave randomness can be clearly observed from this figure with water surface elevation fluctuates around the zero level. In the numerical calculation, the significant wave period $T_{1/3}$ and wave height $H_{1/3}$ are assumed to be 10 s and 6 m respectively, and the water depth is 25 m. Taking into account the statistical principle for record data length and time interval of data sampling mentioned above, time interval $\Delta t = 1$ s and a recording length $T_{total} = 1000$ seconds are adopted in the present study.

8.2.3 Random Wave Validation

After generating the random waves, validation of the simulated random wave profile is needed to assure the precision and efficiency of the numerical simulation. This is conducted by comparing the simulated or predicted frequency spectrum with the target frequency spectrum presented in (8.4) and (8.5) after considering that all the typical random wave characteristics, such as significant wave height and period, can be determined by the frequency spectrum [1].

After knowing the water surface elevation η distribution from (8.1), there are two methods which can be applied to determine the simulated (predicted) spectrum: auto-correlation method and FFT method. In this study, the auto-correlation method is utilized. Assuming the profile of random sea wave is a stationarity stochastic process, for one wave recording surface elevation data, $\eta_1, \eta_2, \ldots, \eta_N$ the auto-correlation function is estimated as,

$$R(\tau) = \frac{1}{N-\nu} \sum_{n=1}^{N-\nu} \eta_{n+\nu}\eta_n, \quad \nu = 0, 1, 2, \ldots, m. \tag{8.12}$$

Assuming the data sampling interval is Δt, then,

$$R(\tau) = R(\nu \Delta t) = \frac{1}{N-\nu} \sum_{n=1}^{N-\nu} \eta(t_n + \nu \Delta t)\eta(t_n), \quad \tau = \nu \Delta t, \nu = 0, 1, 2, \ldots, m. \tag{8.13}$$

We can have $m+1$ values of correlation function $R(\tau)$ which are equally distributed at $\tau = 0, \Delta t, 2\Delta t, \ldots, m\Delta$. The frequency spectrum $S(\omega)$ or $S(f)$ can be estimated by,

$$S(\omega) = \frac{2}{\pi} \int_0^\infty R(\tau) e^{-i\omega t}\, dt \quad \text{and} \quad S(f) = 4 \int_0^\infty R(\tau) e^{-i2\pi f t}\, dt. \tag{8.14}$$

Considering the effect of Nyquist frequency or folding frequency f_N,

$$f_N = \frac{\omega_N}{2\pi} = \frac{1}{2\Delta t}. \tag{8.15}$$

The predicted spectrum is distributed in the range of $[0, f_N]$. Assuming the equally distributed $m+1$ values of frequency $f_0 = 0, f_1, f_2, \ldots, f_m = f_N$, then for one specified frequency, f_n,

$$f_n = n\Delta f = \frac{n}{m} \frac{1}{2\Delta t}, \qquad \omega_n = 2\pi f_n. \tag{8.16}$$

What we are concerning now is to get the frequency spectrum values at these specified frequencies using (8.14). Hence, the spectrum at a certain frequency f_n, $S'(f_n)$ can be estimated as

$$S'(f_n) = \frac{2}{\pi} \sum_{\tau=0}^{m\Delta t} R(\tau) \cos(\omega_n \tau)\, d\tau = \frac{2}{\pi} \sum_{\nu=0}^{m} R(\nu \Delta t) \cos(\omega_n \nu \Delta t) \Delta t. \tag{8.17}$$

Using trapezoidal formula for integration, we have,

$$S'(f_n) = \frac{2}{\pi} \left[\frac{1}{2} R(0) + \sum_{\nu=0}^{m-1} R(\nu \Delta t) \cos(\omega_n \nu \Delta t) + \frac{1}{2} R(m\Delta t) \cos(\omega_n m \Delta t) \right] \Delta t,$$

$$n = 0, 1, 2, \ldots, m. \tag{8.18}$$

Substituting (8.16) into (8.18), we can obtain the frequency spectrum as,

$$S'(f_n) = \frac{2\Delta t}{\pi}\left[\frac{1}{2}R(0) + \sum_{\nu=0}^{m-1} R(\nu\Delta t)\cos\left(\frac{n\nu\pi}{m}\right) + \frac{1}{2}R(m\Delta t)\cos(n\pi)\right]\Delta t,$$

$$n = 0, 1, 2, \ldots, m. \tag{8.19}$$

Using the aforementioned method, the predicted spectrum $S'(f_n)$ is not completely precise, and the shape of the spectrum curve has lots of fluctuation. To improve the simulation, we introduce the weight functions. After multiplying the auto-correlation function $R(\tau)$, we can finally obtain the smoothed spectrum distribution $S(f_n)$. Such weight functions have various forms, in this study we apply the Hamming Window as,

$$S(f_n) = 0.23S'(f_{n-1}) + 0.54S'(f_n) + 0.23S'(f_{n+1}). \tag{8.20}$$

At two lateral boundaries,

$$S(f_0) = 0.54S'(f_0) + 0.46S'(f_1), \tag{8.21}$$

$$S(f_n) = 0.46S'(f_{n-1}) + 0.54S'(f_n). \tag{8.22}$$

Other frequently utilized approaches include Hanning Window, which has the following expression,

$$S(f_n) = 0.25S'(f_{n-1}) + 0.5S'(f_n) + 0.25S'(f_{n+1}). \tag{8.23}$$

At two lateral boundaries,

$$S(f_0) = 0.5S'(f_0) + 0.5S'(f_1), \tag{8.24}$$

$$S(f_n) = 0.5S'(f_{n-1}) + 0.5S'(f_n). \tag{8.25}$$

Figure 8.3 illustrates the comparison of the predicted spectrum (dashed lines) and target spectrum (solid lines) for B–M spectrum using Hamming Window (left panels) and Hanning Window (right panels) for different m values varying from 20 to 90. Considering that the sampling interval is 1 second for the simulated wave profile, the highest frequency to which the spectrum can be estimated is 0.5 Hz. From this figure, we can see that with the increasing of value, the predicted spectrum peak values increase while the fluctuation of the predicted spectrum profiles also increase, especially after the spectral peak frequency, f_p. Nevertheless, with the decreasing of value, the predicted spectrum peak values decrease while the predicted spectrum profiles become smooth (further decreasing will induce the incontinuity considering that represents the total number of the predicted frequency and relevant spectrum value). There is no clear difference on the predicted spectrum profiles between the Hamming Window and the Hanning Window.

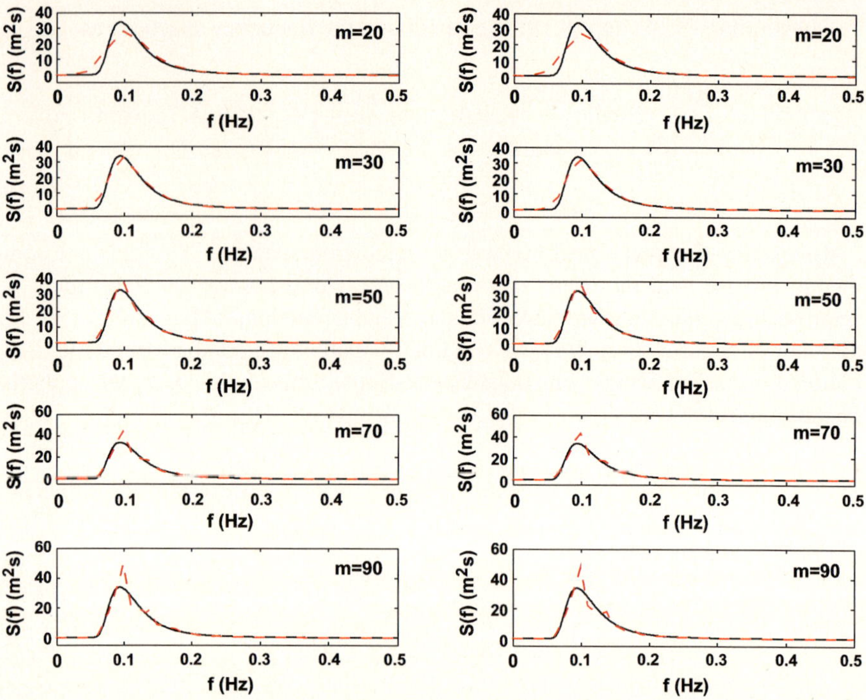

Fig. 8.3 Comparison of the predicted spectrum (*dashed lines*) and target spectrum (*solid lines*) for B–M spectrum using Hamming Window (*left panels*) and Hanning Window (*right panels*) for different values varying from 20 to 90

Figure 8.4 presents the corresponding results for JONSWAP spectrum. In this figure, only the results using Hamming Window are demonstrated. From this figure, we can see that the same phenomenon for predicted spectrum profiles can be observed: small value of underestimates the peak spectrum and large m value causes the fluctuation of the spectrum distribution. [1] suggested the value of m should be around, in which is the total number of the recorded data.

Figure 8.5 presents the comparison between the target and predicted spectra for both B–M spectrum and JONSWAP spectrum. For these two frequency spectra, the wave energy concentrates around the peak frequency f_p, which is almost equal to the frequency corresponding to the significant wave period, i.e., $f_{1/3} = 1/T_{1/3} = 0.1$ Hz (strictly speaking, it is a little bit smaller than as we can see from (8.7)). Both the profile and magnitude of the target spectrum can be captured by the present predicted spectrum, although there exists some underestimation, especially for the JONSWAP spectrum case around the maximum spectra value. As for the JONSWAP spectrum with peak enhancement factor $\gamma = 3.3$, the peak spectra value is about 2.2 times of the B–M spectrum result. As a whole, Fig. 8.5 demonstrates that the simulation of random wave is effective and can be used for the analysis of the dynamics response of seabed.

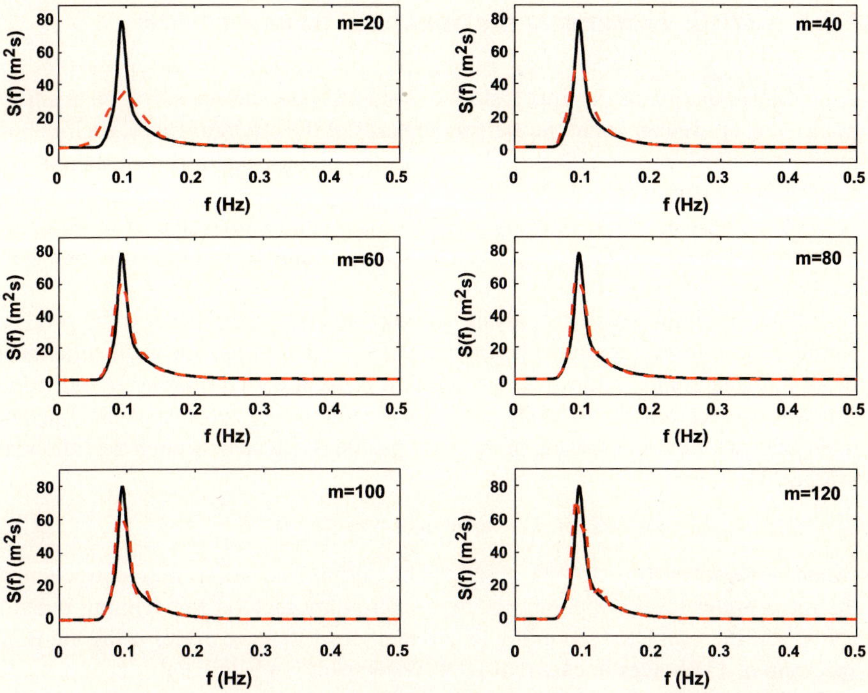

Fig. 8.4 Comparison of the predicted spectrum (*dashed lines*) and target spectrum (*solid lines*) for JONSWAP spectrum using Hamming Window for different values varying from 20 to 120

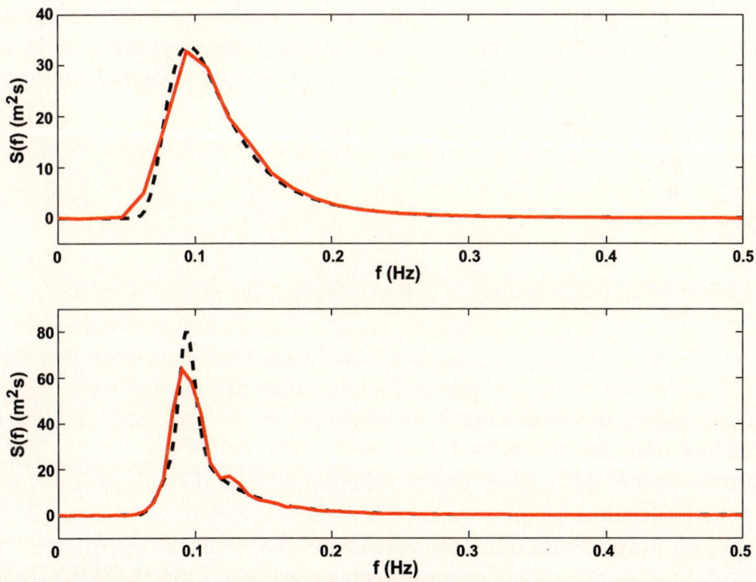

Fig. 8.5 Comparison of predicted spectrum (*solid lines*) and target spectrum (*dashed lines*): B–M spectrum (*upper panel*), JONSWAP spectrum (*lower panel*)

8.2.4 Statistic Features of the Simulated Random Waves

After the simulation of the time history of random wave surface elevation at a fixed point, $\eta(x,t)$, we can count the statistic features of the random waves, such as how many individual waves in one wave record, the highest one-nth wave height and relevant wave period ($n=1$ for mean wave value; $n=3$ for highest one-third wave value or significant wave value; $n=10$ for one-tenth wave value; $n=N_{wave}$ for highest wave value in which N_{wave} is the total individual wave number of one wave record).

Before doing the statistic analysis, we need to correct the mean water level. A simple procedure to determine the mean water level is to use the arithmetic mean of all the data points. After that, we can first count the total number of individual waves in one record. In this study, we use the definition of zero-upcrossing method to do analysis. Zero-upcrossing of the wave profile is detected through the following criteria [1],

$$\eta_i \eta_{i+1} < 0 \quad \text{and} \quad \eta_{i+1} > 0 \tag{8.26}$$

where η_i denotes the ith data point of the surface elevation after correction of the mean water level. Applying the zero-upcrossing method for defining individual waves, the simulated recording length includes 120 wave cycles using the B–M spectrum or 117 waves in case that JONSWAP spectrum is utilized.

To determine the zero-upcrossing wave height, the highest point on the surface elevation must be searched for in the time interval between two successive zero-upcrossing points. Once this point is obtained among the sampled points, say, η_i, and then the true maximum elevation $\eta_{\max}$ between two zero-upcrossing points can be estimated by fitting a parabolic curve to the three conjoint points η_{i-1}, η_i and η_{i1} in order to eliminate the underestimation. The formula for a parabolic fitting is,

$$\eta_{\max} = C - \frac{B^2}{4A}, \tag{8.27}$$

where,

$$A = \frac{1}{2}(\eta_{i-1} - 2\eta_i + \eta_{i1}), \qquad B = \frac{1}{2}(\eta_{i+1} - \eta_{i-1}), \qquad C = \eta_i. \tag{8.28}$$

The lowest surface elevation $\eta_{\min}$ is obtained by a similar process, and the wave height is calculated as the sum of the absolute values of $\eta_{\max}$ and $\eta_{\min}$.

After knowing the total number of individual waves and specifying the beginning/ending time and wave heights of every individual wave, we can obtain the statistic features of the random waves according to their definitions. Table 8.1 lists such statistic features of the simulated B–M type random waves and JONSWAP type random waves. From this table we can see, the simulated statistic features for JONSWAP spectrum present better estimation considering that the target significant wave height and period are 6 m and 10 s, respectively. As for the results of B–M spectrum, the simulated features somewhat underestimate the target values.

Table 8.1 Statistic features of the simulated random waves

N_{wave}	B–M spectrum		JONSWAP spectrum	
	120		117	
	wave height	wave period	wave height	wave period
Mean values	3.40	8.26	3.59	8.50
1/3 highest values	5.62	9.19	6.00	9.91
1/10 highest values	6.68	8.70	7.52	10.06
the highest values	9.37	7.67	10.44	9.90

8.2.5 Representative Regular Wave

As mentioned previously, most previous studies for the wave-induced soil response have focused on the linear regular wave loading, rather than more realistic random wave loading. Therefore, it is interesting to investigate the various soil responses under the random waves and try to find the difference between regular and random wave loadings.

To compare with regular wave-induced soil response, how to define the suitable representative regular wave parameters (i.e., regular wave height and wave period) in case of considered random wave characteristics is needed. As a first approximation, we may apply the significant wave height and period ($H_{1/3}$, $T_{1/3}$) as the representative regular wave parameters.Sumer and Fredsøe [4], from their irregular wave experimental measurement, found that the combination of representative wave height $H_r = H_{1/3}/\sqrt{2}$ and wave period $T_r = \overline{T}$ provides the best comparison with the regular wave results, where can be interpreted as the equivalent sinusoidal wave height of the irregular waves, and T_r represents the mean zero-upcrossing period of a random wave record, which can be determined from the frequency spectra by [1]

$$T_r = \overline{T} = \sqrt{m_0/m_2} \quad \text{and} \quad m_n = \int_0^\infty f^n S(f)\,df, \quad n = 0, 2 \tag{8.29}$$

where m_n is the ith spectral moment. Applying the random wave characteristics used in previous examples, for B–M spectrum, the corresponding representative regular wave height are $H_r = 4.24$ m and period $T_r = 7.64$ s, respectively (hence representative wavelength $L_r = 86.27$ m); for JONSWAP spectrum, these two are $H_r = 4.24$ m and period $T_r = 9.48$ s, respectively with wavelength $L_r = 102.29$ m.

Using these approach to determine the representative regular wave characteristics, both H_r and T_r are smaller than the corresponding significant values, say, $H_{1/3}$ and $T_{1/3}$. Another thing need to be pointed out is that taking Table 8.1 into account, the simulated or predicted mean zero-upcrossing wave period (8.26 s) of B–M spectrum somewhat overestimates the target value (7.64 s). Nevertheless, simulation for JONSWAP spectrum provides much better prediction after comparing with the theoretical value.

8.3 Wave-Induced Oscillatory Soil Response

8.3.1 Boundary Vale Problem

The governing equations of two-dimensional wave-induced oscillatory pore pressure in marine sediment are summarized here:

$$\frac{K_x}{K_z}\frac{\partial^2 p}{\partial x^2}+\frac{\partial^2 p}{\partial z^2}-\frac{\gamma_w n'\beta}{K_z}\frac{\partial p}{\partial t}=\frac{\gamma_w}{K_z}\frac{\partial \varepsilon}{\partial t}, \tag{8.30}$$

$$G\nabla^2 u+\frac{G}{1-2\mu}\frac{\partial \varepsilon}{\partial x}=\frac{\partial p}{\partial x}, \tag{8.31}$$

$$G\nabla^2 w+\frac{G}{1-2\mu}\frac{\partial \varepsilon}{\partial z}=\frac{\partial p}{\partial z}. \tag{8.32}$$

To solve the three coupled partial differential equations, three independent conditions are needed at each boundary.

In case of infinite soil thickness under random wave loadings, at large depth, it is obvious that no soil displacement will be experienced, and also no pore-water pressure will develop. Therefore

$$u=w=p=0 \quad \text{at } z\rightarrow -\infty. \tag{8.33}$$

In case of finite soil thickness, the soil particles rest on the impermeable rigid bottom and no soil displacement will be experienced

$$u=w=0 \quad \text{at } z\rightarrow -h. \tag{8.34}$$

At the rigid bottom, no vertical flow or velocity occurs. Considering that the flow velocity is related to the pressure gradient through Darcy's law, the following pressure condition is obtained

$$\frac{\partial p}{\partial z}=0 \quad \text{at } z=-h. \tag{8.35}$$

At the seabed surface ($z=0$), it is commonly accepted that both vertical effective stress and shear stress vanish, i.e.

$$\sigma_z'=\tau_{xz}=0 \quad \text{at } z=0 \tag{8.36}$$

and the pore pressure is equal to the wave-induced pressure at the seabed surface. Under the random wave conditions, the wave dynamic pressure along the seafloor is expressed as the superposition of a large number of linear regular wave results at the seabed

$$p(x,0,t)=p_b(x,t)=\sum_{i=1}^{M} p_b^{(i)}\cos(k_i x-2\pi \tilde{f}_i t+\varepsilon_i), \tag{8.37}$$

$$p_b^{(i)}=\frac{\gamma_w a_i}{\cosh k_i d}=\frac{\gamma_w\sqrt{2S(\hat{f}_i)\Delta f_i}}{\cosh k_i d}. \tag{8.38}$$

8.3.2 Analytical Solutions

The boundary value problem specified in above, describing the wave-seabed interaction, can be solved based on the governing equations (8.30)–(8.32) with suitable boundary conditions, i.e., (8.33)–(8.38). The analytical solution for the wave-induced pore pressure and soil displacements and can first be obtained, from which various effective stresses can be obtained.

8.3.2.1 An Infinite Seabed

Following the framework proposed in Chap. 3 for regular waves and for the present two-dimensional problem, the random wave-induced pore pressure and displacements for infinite soil thickness can be written as

$$p = \mathbf{Re}\left\{\sum_{m=1}^{M} \frac{p_b^{(m)}}{1-2\mu}\left[(1-2\mu-\lambda_m)c_1^m e^{k_m z} + \frac{\delta_m^2-k_m^2}{k_m}(1-\mu)c_2^m e^{\delta_m z}\right] e^{i(k_m x-\omega_m t+\varepsilon_m)}\right\}, \tag{8.39}$$

$$u = \mathbf{Re}\left\{\sum_{m=1}^{M} \frac{i p_b^{(m)}}{2G}\left[\left(c_0^m + c_1^m z\right)e^{k_m z} + c_2^m e^{\delta_m z}\right] e^{i(k_m x-\omega_m t+\varepsilon_m)}\right\}, \tag{8.40}$$

$$w = \mathbf{Re}\left\{\sum_{m=1}^{M} \frac{p_b^{(m)}}{1-2\mu}\left[\left(c_0^m + \frac{1+2\lambda_m}{k_m}c_1^m + c_1^m z\right)e^{k_m z} + \frac{\delta_m}{k_m}c_2^m e^{\delta_m z}\right] e^{i(k_m x-\omega_m t+\varepsilon_m)}\right\} \tag{8.41}$$

where **Re**{} represents the real part of the function, and $\omega_n = 2\pi \tilde{f}_m$ is the angular frequency of the mth component wave. Coefficients δ_m and λ_m are expressed as

$$\delta_m^2 = k_m^2 \frac{K_x}{K_z} - \frac{i\omega_m\gamma_w}{K_z}\left[n'\beta + \frac{1-2\mu}{2G(1-\mu)}\right], \tag{8.42}$$

$$\lambda_m = \frac{(1-2\mu)[k_m^2(K_z-K_x) + i\omega_m\gamma_w n'\beta]}{k_m^2(K_z-K_x) + i\omega_m\gamma_w[n'\beta + (1-2\mu)/G]}. \tag{8.43}$$

Then, the effective normal stresses, and shear stress can be determined as

$$\sigma_x' = \mathbf{Re}\left\{-\sum_{m=1}^{M} p_b^{(m)}\left[\left(k_m\left(c_0^m + c_1^m z\right) + \frac{2\mu\lambda_m}{1-2\mu}c_1^m\right)e^{k_m z} + \left(k_m - \frac{\mu(\delta_m^2-k_m^2)}{k_m(1-2\mu)}\right)c_2^m e^{\delta_m z}\right] e^{i(k_m x-\omega_m t+\varepsilon_m)}\right\}, \tag{8.44}$$

$$\sigma_z' = \mathbf{Re}\left\{\sum_{m=1}^{M} p_b^{(m)}\left[\left(k_m\left(c_0^m + c_1^m z\right) - \frac{2\lambda_m(1-\mu)}{1-2\mu}c_1^m\right)e^{k_m z}\right.\right.$$
$$\left.\left. + \frac{\delta_m^2(1-\mu) - k_m^2\mu}{k_m(1-2\mu)}c_2^m\, e^{\delta_m z}\right]e^{i(k_m x-\omega_m t+\varepsilon_m)}\right\}, \tag{8.45}$$

$$\tau_{xz} = \mathbf{Re}\left\{\sum_{m=1}^{M} i p_b^{(m)}\left[k_m c_0^m + (k_m z - \lambda_m)c_1^m\, e^{k_m z} + \delta_m c_2^m\, e^{\delta_m z}\right]e^{i(k_m x-\omega_m t+\varepsilon_m)}\right\}. \tag{8.46}$$

In the expressions of (8.39)–(8.46), the coefficients are obtained by

$$c_0^m = \frac{-\lambda_m[\mu(\delta_m - k_m)^2 - \delta_m(\delta_m - 2k_m)]}{k_m(\delta_m - k_m)[\delta_m(1-\mu) + k_m(\lambda_m + \mu)]}, \tag{8.47}$$

$$c_1^m = \frac{\delta_m - \delta_m\mu + k_m\mu}{\delta_m(1-\mu) + k_m(\lambda_m + \mu)}, \tag{8.48}$$

$$c_2^m = \frac{k_m\lambda_m}{(\delta_m - k_m)[\delta_m(1-\mu) + k_m(\lambda_m + \mu)]}. \tag{8.49}$$

8.3.2.2 A Seabed of Finite Thickness

Following the framework proposed Chap. 3 for regular waves and for the present two-dimensional problem, the random wave-induced pore pressure and displacements for finite soil thickness can be written as

$$p = \mathbf{Re}\left\{\sum_{m=1}^{M} \frac{p_b^{(m)}}{1-2\mu}\left[(1-\lambda_m-2\mu)\left(C_2^m\, e^{k_m z} - C_4^m\, e^{-k_m z}\right)\right.\right.$$
$$\left.\left. + (1-\mu)\left(\delta_m^2 - k_m^2\right)\left(C_5^m\, e^{\delta_m z} - C_6^m\, e^{-\delta_m z}\right)\right]e^{i(k_m x-\omega_m t+\varepsilon_m)}\right\}, \tag{8.50}$$

$$u = \mathbf{Re}\left\{\sum_{m=1}^{M} \frac{i p_b^{(m)}}{2Gk_m}\left[\left(C_1^m + C_2^m k_m z\right)e^{k_m z} + \left(C_3^m + C_4^m k_m z\right)e^{-k_m z}\right.\right.$$
$$\left.\left. + \left(C_5^m\, e^{\delta_m z} + C_6^m\, e^{-\delta_m z}\right)\right]e^{i(k_m x-\omega_m t+\varepsilon_m)}\right\}, \tag{8.51}$$

$$w = \mathbf{Re}\left\{\sum_{m=1}^{M} \frac{p_b^{(m)}}{2Gk_m}\left[\left(C_1^m - (1+2\lambda_m - k_m z)C_2^m\right)e^{k_m z}\right.\right.$$
$$- \left(C_3^m + (1+2\lambda_m + k_m z)C_4^m\right)e^{-k_m z}$$
$$\left.\left. + k_m\delta_m\left(C_5^m\, e^{\delta_m z} - C_6^m\, e^{-\delta_m z}\right)\right]e^{i(k_m x-\omega_m t+\varepsilon_m)}\right\}. \tag{8.52}$$

Coefficients δ_m and λ_m are expressed as (8.42)–(8.43).

Again, the effective normal stresses, and shear stress can be determined as

$$\sigma'_x = \mathbf{Re}\left\{\sum_{m=1}^{M} p_b^{(m)}\left[\left(C_1^m + C_2^m k_m z + \frac{2\mu\lambda_m}{1-2\mu}C_2^m\right)e^{k_m z}\right.\right.$$
$$+\left(C_3^m + C_4^m k_m z - \frac{2\mu\lambda_m}{1-2\mu}C_4^m\right)e^{-k_m z}$$
$$\left.\left.+\left(k_m^2 - \frac{\mu(\delta_m^2 - k_m^2)}{1-2\mu}\right)\left(C_5^m e^{\delta_m z} + C_6^m e^{-\delta_m z}\right)\right]e^{i(k_m x - \omega_m t + \varepsilon_m)}\right\}, \quad (8.53)$$

$$\sigma'_y = \mathbf{Re}\left\{\sum_{m=1}^{M} p_b^{(m)}\left[\left(C_1^m + C_2^m k_m z - \frac{2\lambda_m(1-\mu)}{1-2\mu}C_2^m\right)e^{k_m z}\right.\right.$$
$$+\left(C_3^m + C_4^m k_m z - \frac{2\lambda_m(1-\mu)}{1-2\mu}C_4^m\right)e^{-k_m z}$$
$$\left.\left.+\frac{1}{1-2\mu}\left[\delta_m^2(1-\mu) - k_m^2\mu\right]\left(C_5^m e^{\delta_m z} + C_6^m e^{-\delta_m z}\right)\right]e^{i(k_m x - \omega_m t + \varepsilon_m)}\right\}, \quad (8.54)$$

$$\tau_{xz} = \mathbf{Re}\left\{\sum_{m=1}^{M} i p_b^{(m)}\left[\left(C_1^m + (k_m z - \lambda_m)C_2^m\right)e^{k_m z}\right.\right.$$
$$-\left(C_3^m + (k_m z + \lambda_m)C_4^m\right)e^{-k_m z}$$
$$\left.\left.+k_m\delta_m\left(C_5^m e^{\delta_m z} - C_6^m e^{-\delta_m z}\right)\right]e^{i(k_m x - \omega_m t + \varepsilon_m)}\right\}. \quad (8.55)$$

In the expressions of (8.50)–(8.55), there are six resulting coefficients, to for each mth component regular wave, whose values are systematically presented in [2] for hydraulically anisotropic and unsaturated soil conditions.

8.4 Numerical Results

Based on the general solutions presented above, it is possible to investigate the effect of wave randomness on the wave-induced soil response. All the physical variables to be presented are non-dimensionalized with respect to the hydrostatic water pressure at the seabed surface ($\gamma_w d$). The present numerical simulation is performed for a hydraulically isotropic coarse sand. Input data of the numerical examples are tabulated in Table 8.2, which includes significant wave height $H_{1/3}$, significant wave period $T_{1/3}$, water depth d, finite soil depth h, permeability K_x and K_z, soil porosity n', Poisson's ratio μ, shear modulus G, degree of saturation S_r and unit weight of submerged solid grain γ_s. All the calculations are based on these parameters unless specified otherwise.

8.4.1 Comparison Between Regular and Random Wave-Induced Soil Responses

Based on the input data tabulated in Table 8.2, a comparison between regular and random wave-induced soil responses can be conducted for both infinite and finite soil thickness. Figures 8.6 and 8.7 illustrate the time-varying normalized pore pressure (p), effective normal stresses (σ_x' and σ_z') and shear stress (τ_{xz}) variations in one random record for an infinite soil thickness by using B–M spectrum at two different elevations, say, $z = -0.25L_r$ and $z = -0.50L_r$, respectively. Figures 8.8 and 8.9 are the corresponding results using JONSWAP spectrum. In these figures, two horizontal dashed lines represent the corresponding soil responses range under the representative regular wave loading. No doubt, for random wave loading, at some certain duration, i.e., $t = 150$ seconds for B–M spectrum or $t = 250$ seconds for JONSWAP spectrum, the wave-induced soil response is greater than the results of the regular waves at the same time due to wave irregularity owing to the wave randomness.

Figure 8.10 demonstrates the time-varying normalized pore pressure distribution ($p/\gamma_w d$) of a seabed of finite soil thickness within one wave record by using B–M spectrum. The upper panel illustrates the results at $z = -0.5h$, while the lower panel

Table 8.2 Input data of numerical example

Wave characteristics			Soil characteristics						
$H_{1/3}$	$T_{1/3}$	d	h	$K_x = K_z$	n'	μ	G	S_r	γ_s
6	10	25	12.5	0.01	0.3	1/3	10^7	0.975	2.0

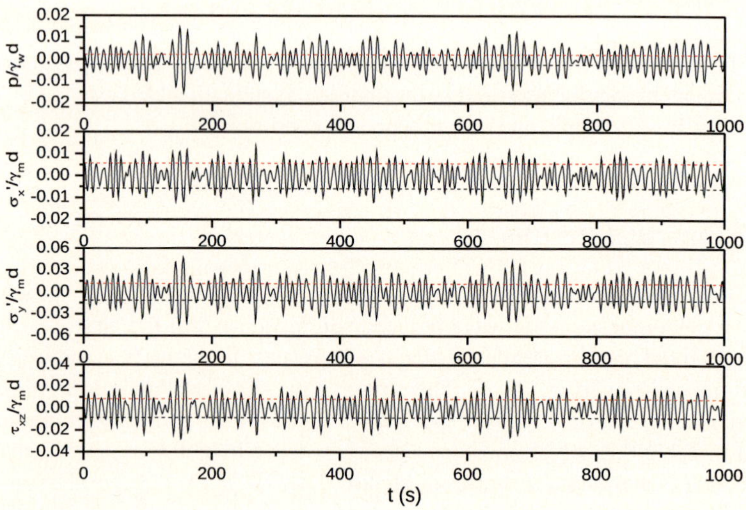

Fig. 8.6 Time-varying normalized pressures, effective normal stress and shear stress variations at level $z = -0.25L_r$ for infinite soil thickness using B–M spectrum

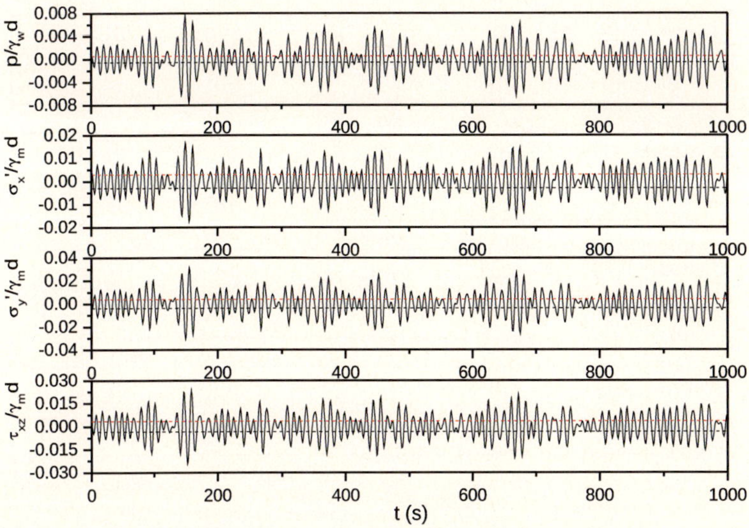

Fig. 8.7 Time-varying normalized pressures, effective normal stress and shear stress variations at level $z = -0.5L_r$ for infinite soil thickness using B–M spectrum

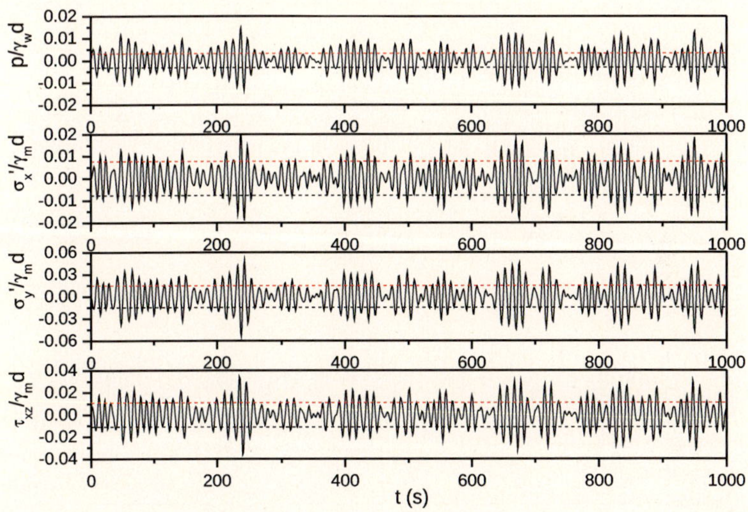

Fig. 8.8 Time-varying normalized pressures, effective normal stress and shear stress variations at level $z = -0.25L_r$ for infinite soil thickness using JONSWAP spectrum

is for the soil response at the rigid impermeable bottom, i.e., $z = -h$. In the figure, two horizontal dashed lines represent the soil response range under the representative regular wave loading at the relevant elevations. The corresponding results using JONSWAP spectrum is shown in Fig. 8.11. From these two figures, the irregularity

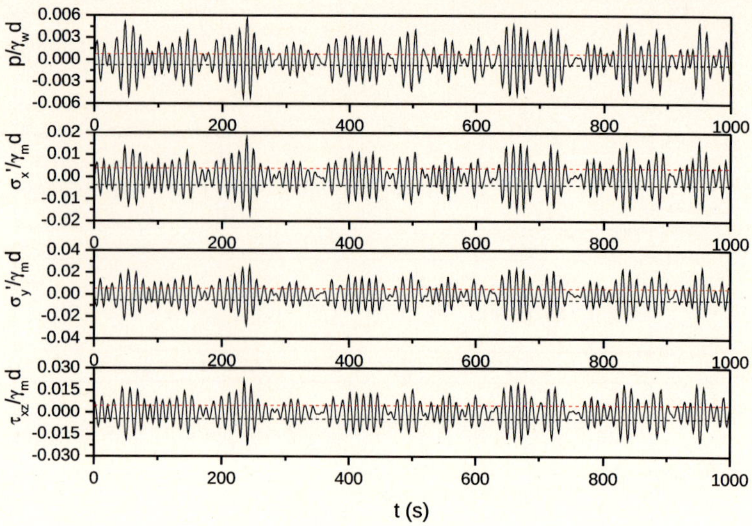

Fig. 8.9 Time-varying normalized pressures, effective normal stress and shear stress variations at level $z = -0.5L_r$ for infinite soil thickness using JONSWAP spectrum

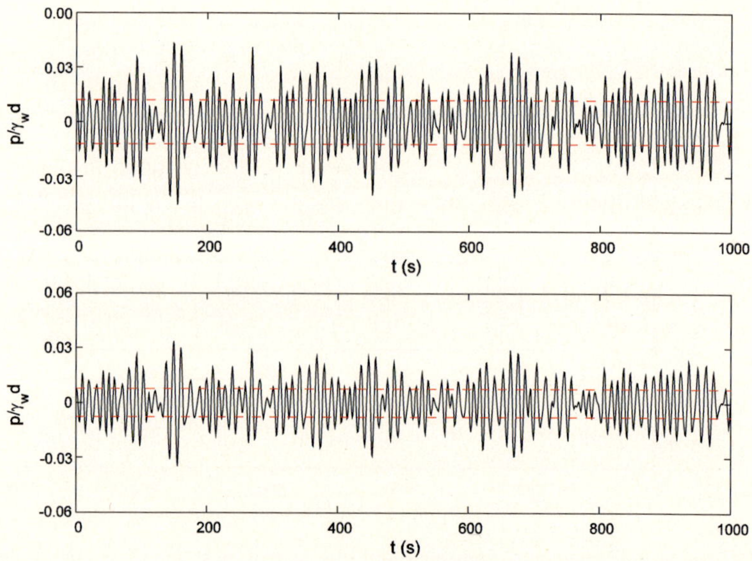

Fig. 8.10 Time-varying normalized pore pressure distribution of a finite soil thickness using B–M spectrum: $z = -0.5h$ (*upper panel*) and $z = -h$ (*lower panel*)

of pore pressure variation caused by the wave randomness can be clearly observed. For random wave loading, at some specified time, the wave induced pore pressure is greater than the results of regular waves at the same time. Comparing the results at

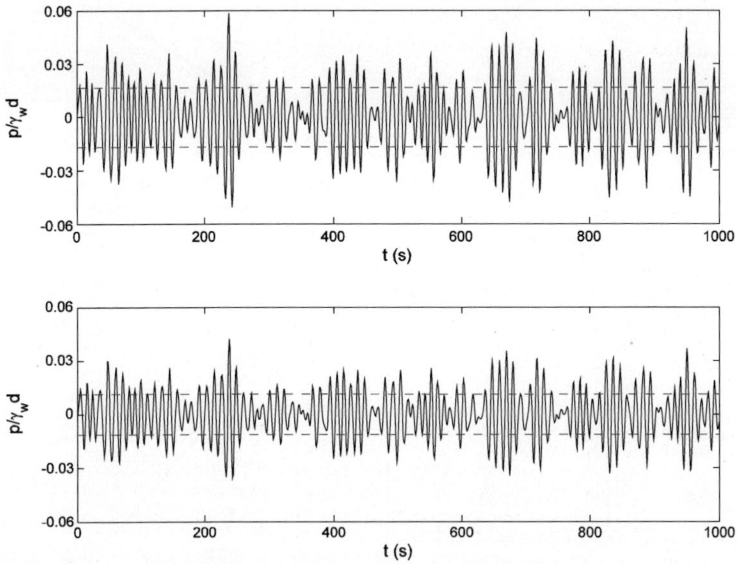

Fig. 8.11 Time-varying normalized pore pressure distribution of a finite soil thickness using JONSWAP spectrum: $z = -0.5h$ (*upper panel*) and $z = -h$ (*lower panel*)

$z = -0.5h$ and $z = -h$, we see that the pore pressure values are decreasing with the increase of soil depth $|z/h|$, which shows that the spatial distribution of the random wave-induced pore pressure occurs in much the same way as in the case of regular waves since the dashed line range also becomes smaller as $|z/h|$ increases.

Considering the difference between B–M and JONSWAP spectra results in Figs. 8.10 and 8.11, we can see the irregularity of the temporal pore pressure variation is more significant for JONSWAP spectrum case, which can be confirmed through the standard deviation of the simulated results. The mean values of the pore pressure variation at different depths are almost zero. For B–M spectrum, the standard deviation of normalized pore pressure distribution at $z = -0.5h$ is 0.016, and 0.012 at $z = -h$. Both are smaller than the corresponding values for JONSWAP spectrum, which are 0.018 and 0.013, respectively. The temporal pore pressure variation under the JONSWAP spectrum is more active with large fluctuations along the time history.

For the engineering practice, it is important to investigate the vertical distribution of the maximum soil responses. Figure 8.12 presents such results under the random and representative regular wave loadings, which include the pore pressure, the horizontal and vertical effective normal stresses and shear stresses. Due to the wave randomness, the maximum random wave-induced soil response is larger than the corresponding representative regular wave results. However, the vertical distribution tendency is the same for these two wave loadings. i.e., maximum pore pressure $(p/\gamma_w d)$ is attenuating with the increase of soil depth $|z/h|$. Looking at the results of representative regular waves, we can see that the soil responses for the "JON-

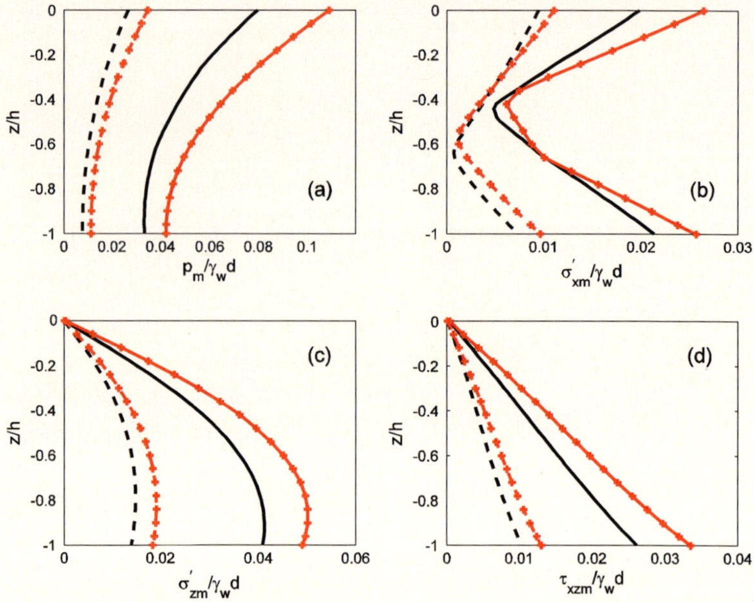

Fig. 8.12 Vertical distribution of the maximum soil responses under the random waves ('+' for JONSWAP spectrum and '−−' for B–M spectrum) and representative regular wave ('+' for JONSWAP spectrum and '−−' for B–M spectrum) loadings. (**a**) Pore pressure p_m, (**b**) horizontal effective normal stress σ'_x, (**c**) vertical effective normal stress σ'_z and (**d**) shear stress τ_{xz}

SWAP" regular wave (JRW) are larger than that of "B–M" regular wave (BRW). This is because that the longer JRW period, as mentioned in Sect. 8.2.1, induces a smaller wave number due to a longer wavelength. Hence, the amplitude of pore pressure at the seabed is larger for JRW case.

Further investigating the results of random waves, in general, the maximum soil responses under the JONSWAP spectrum are still greater than the B–M spectrum cases although the significant wave height and period are the same for these two different spectra. Taking Fig. 8.5 into account, the maximum frequency spectra, which is related to the maximum component wave energy or wave height, is much larger for JONSWAP spectrum than for B–M spectrum. Considering the significant wave period, which is almost equal to the maximum component wave period [1], is same for these two spectra, the maximum pore pressure at the seabed is larger for JONSWAP spectrum due to aforementioned larger wave height. This means under the same wave and soil characteristics, the JONSWAP type random waves introduce more significant soil responses within the marine sediments.

Figure 8.13 illustrates vertical distribution of relative maximum soil responses between random and representative regular wave loadings ($\Delta x_m = x_{mi} - x_{mr}$, where x_{mi} and x_{mr} denote the irregular/random and regular wave-induced soil response respectively) in case of B–M and JONSWAP spectra. As shown in the figure, the difference between random and regular wave results is decreasing with increasing

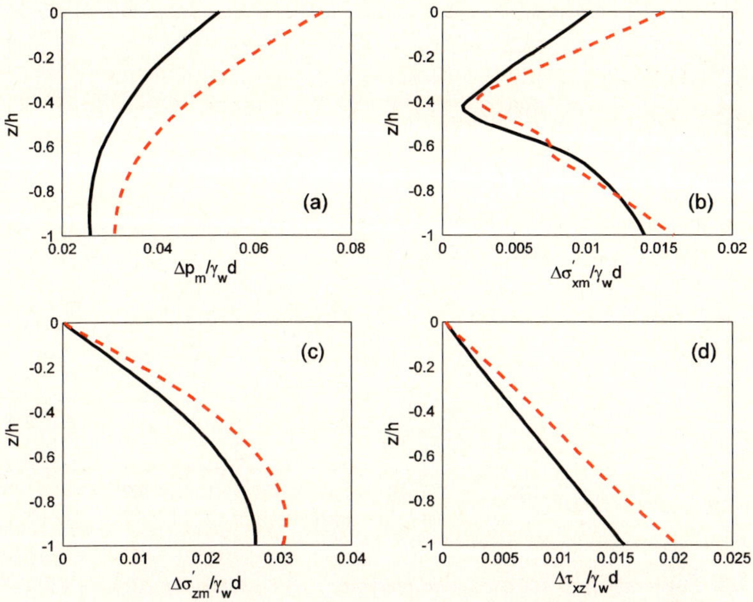

Fig. 8.13 Vertical distributions of relative maximum soil responses between random and representative regular wave loadings (*solid lines* for B–M spectrum and *dashed lines* for JONSWAP spectrum). (**a**) Relative pore pressure Δp_m, (**b**) relative horizontal effective normal stress $\Delta\sigma'_{xm}$, (**c**) relative vertical effective normal stress $\Delta\sigma'_{zm}$ and (**d**) relative shear stress $\Delta\tau_{xzm}$

$|z/h|$, while for shear stress, it is increasing with $|z/h|$ in the present simulated soil depth. Considering the results between B–M and JONSWAP spectra, the pore pressure difference is decreasing as $|z/h|$ increases; the shear stress difference is increasing with $|z/h|$. However, with the increase of soil depth, the relative soil response difference (between the random and regular waves or different type random waves) will decrease because various soil responses go to zero at the infinite depth, which will be further discussed in the following section.

8.4.2 Effect of Soil Parameters on Random Wave-Induced Soil Response

In this section, the influence of two key soil parameters, i.e., the degree of saturation (S_r) and vertical soil permeability (K_z), on the random wave-induced soil response will be investigated. The numerical calculation is performed under the same wave and soil characteristics listed in the previous section unless specified otherwise. For simplicity, the discussion is only for the random wave results under the B–M frequency spectrum.

For Figs. 8.14 and 8.15, the numerical calculation is performed for an infinite soil thickness at the level, above which is the region likely to be liquefied [2]. Fig-

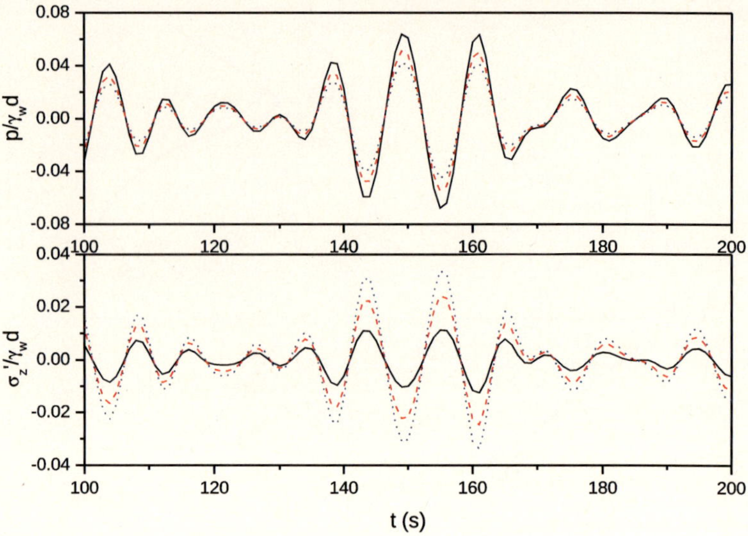

Fig. 8.14 Time-varying normalized pore pressure (*upper panel*) and effective vertical normal stress (*lower panel*) distributions for three different degrees of soil saturation. '—' for $S_r = 1.0$; '– –' for $S_r = 0.975$; '· · ·' for $S_r = 0.95$ (infinite seabed)

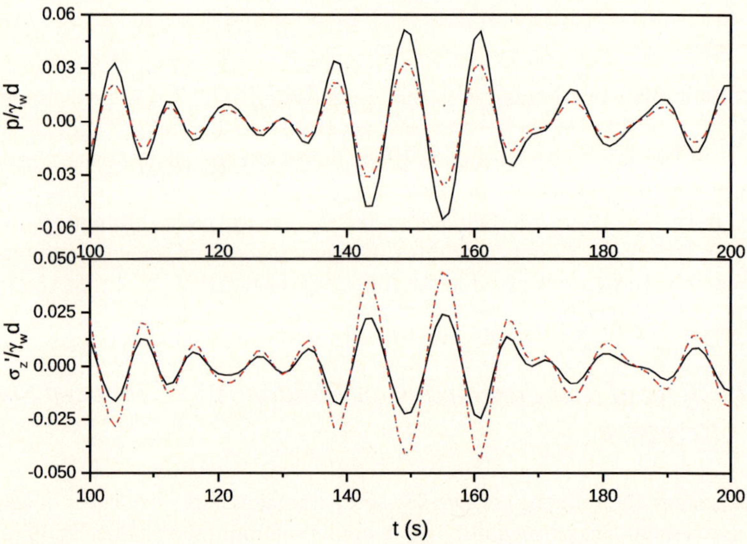

Fig. 8.15 Time-varying normalized pore pressure (*upper panel*) and effective vertical normal stress (*lower panel*) distributions for three different vertical soil permeabilities. '—' for $K_z = 10^{-2}$ m/s; '– –' for $K_z = 10^{-4}$ m/s; '· · ·' for $K_z = 10^{-6}$ m/s (infinite seabed)

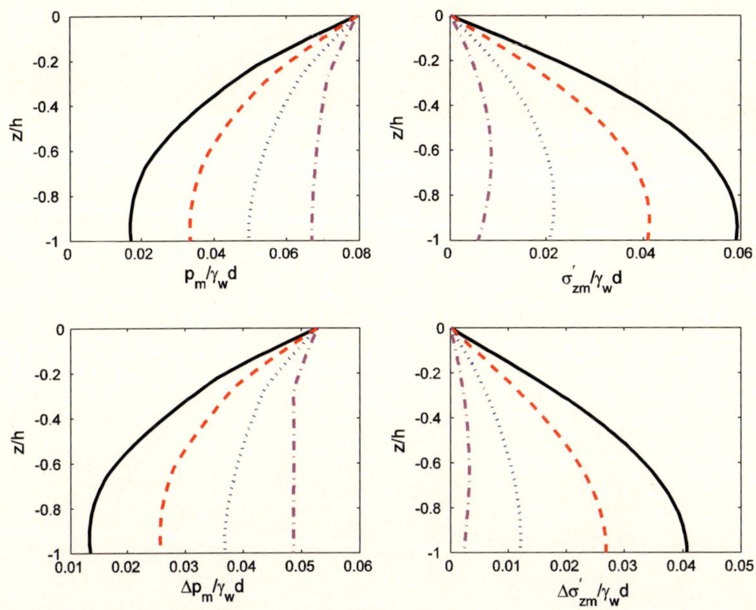

Fig. 8.16 Time-varying normalized pore pressure (*upper panel*) and effective vertical normal stress (*lower panel*) distributions for three different degrees of soil saturation. '—' for $S_r = 1.0$; '−−' for $S_r = 0.975$; '···' for $S_r = 0.95$ (finite thickness)

ure 8.14 illustrates the time-varying normalized pore pressure (p) and effective vertical normal stress (σ'_z) distributions for three different soil saturations, i.e., $S_r = 1$, $S_r = 0.975$ and $S_r = 0.95$. General speaking, the wave-induced pore pressure decreases as the degree of saturation decreases, while an opposite trend is observed for the vertical effective normal stress.

Soil permeability is another important parameters, which directly affect the wave-induced soil response. Figure 8.15 presents the distribution of the pore pressure and vertical effective normal stress for three different soil permeabilities, i.e., $K_z = 10^{-2}$ m/s, $K_z = 10^{-4}$ m/s and $K_z = 10^{-6}$ m/s. As shown in the figure, there is a significant difference between the results of $K_z = 10^{-2}$ m/s and $K_z = 10^{-4}$ m/s with pore pressure being larger and vertical effective stress being smaller for higher soil permeability, while the difference between $K_z = 10^{-4}$ m/s and $K_z = 10^{-6}$ m/s is ambiguous.

In Figs. 8.16 and 8.17, the numerical calculation is performed for a finite soil thickness. As mentioned previously, it is important to investigate the vertical profile of the maximum soil response. In the following analysis, we will only focus on two kinds of soil responses, maximum pore pressure (p_m) and maximum vertical effective normal stresses (σ'_{zm}).

Figure 8.16 demonstrates the effect of degree of saturation on the maximum random wave-induced soil response as well as the relative difference between random and regular waves. Four degrees of saturation are considered here, e.g., $S_r = 0.975$,

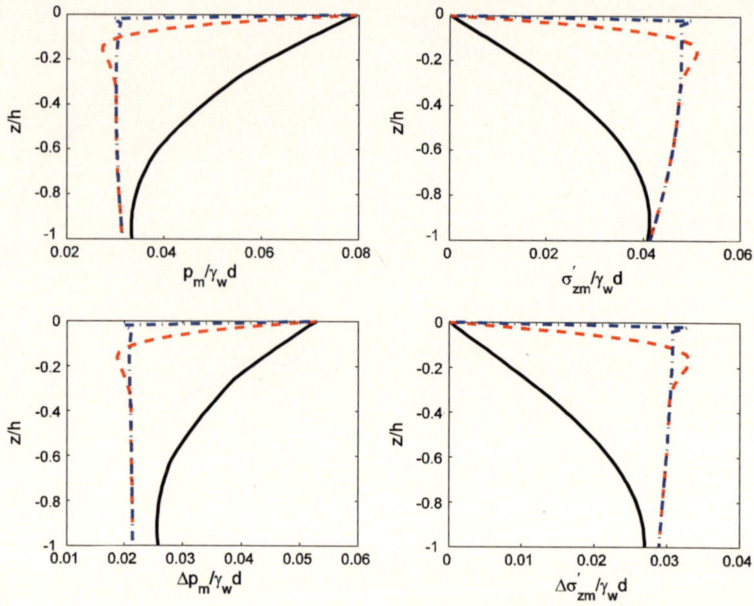

Fig. 8.17 Time-varying normalized pore pressure (*upper panel*) and effective vertical normal stress (*lower panel*) distributions for three different vertical soil permeabilities. '—' for $K_z = 10^{-2}$ m/s; '−−' for $K_z = 10^{-4}$ m/s; '···' for $K_z = 10^{-6}$ m/s (finite thickness)

0.99 and 1.0. With the increase of the degree of saturation, p_m increases and σ'_{zm} decreases along the vertical depth. Considering the relative difference between random and regular wave loadings, the same conclusions can be obtained because the regular wave-induced soil responses present the same tendency with rather small values.

Figure 8.17 presents the influence of soil permeability on the maximum random wave-induced soil responses together with the relative difference between random and regular wave-induced results. Three different soil permeabilities are utilized, i.e., $K_z = 10^{-2}$ m/s for coarse sand, $K_z = 10^{-4}$ m/s for fine sand and $K_z = 10^{-6}$ m/s for silt. With the decrease of soil permeability, p_m decreases and σ'_{zm} increases. The difference between fine sand and silt is negligible except at the vicinity of seabed where there is a sudden decrease for p_m and increase for σ'_{zm}. For coarse sand, the spatial distribution of soil response is gentler. Similar results occur for Δp_m and $\Delta\sigma'_{zm}$, which indicate the random wave-induced soil response is in the much same way as the regular wave cases.

8.4.3 Effect of Wave Characteristics on Random Wave-Induced Soil Response

In this section, the influence of two random wave parameters, i.e., significant wave height ($H_{1/3}$) and wave period ($T_{1/3}$), on the random wave-induced soil response of a finite soil thickness will be investigated.

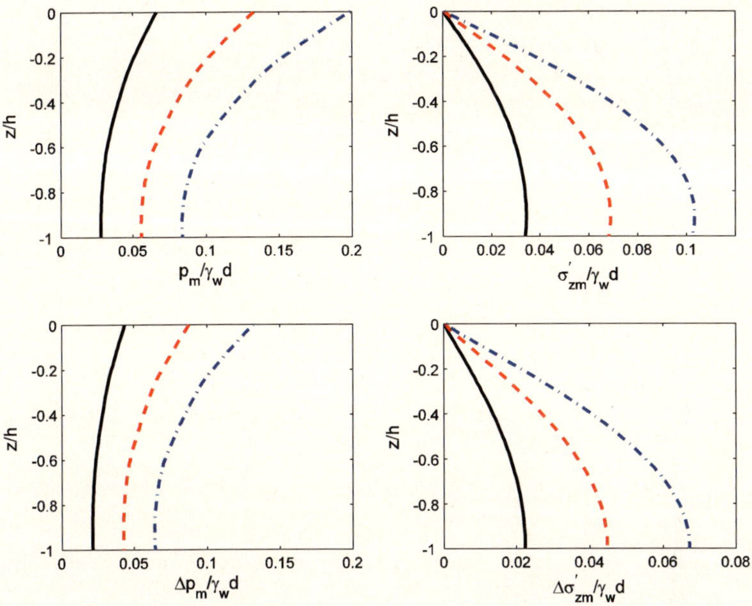

Fig. 8.18 Vertical distribution of p_m (*left panels*) and σ'_{zm} (*right panels*) for random wave results (*upper panels*) and relative results (*lower panels*) under different significant wave heights. *Solid lines* for $H_{1/3} = 5$ m, *dashed lines* for $H_{1/3} = 10$ m and *dashed-dotted lines* for $H_{1/3} = 15$ m (finite thickness)

Figure 8.18 shows the influence of significant wave height on maximum random wave-induced soil response. Three typical wave heights (before wave breaking) are utilized here, i.e., $H_{1/3} = 0.2d$, $0.4d$ and $0.6d$. Figure 8.19 presents the influence of significant wave period. Three wave periods are applied, i.e., $T_{1/3} = 5$, 10 and 20 seconds. From these two figures, we can see: with the increase of $H_{1/3}$ and $T_{1/3}$, both p_m and σ'_{zm} increase due to the increase of pore pressure at the seabed. Looking at the relative difference between the random and regular waves, similar scenario can be observed, which further proves that the effect of wave characteristics on the maximum random wave-induced soil response are similar to the regular wave cases.

8.4.4 Effect of Seabed Thickness on Random Wave-Induced Soil Response

The influence of seabed thickness on the random wave-induced maximum soil responses is investigated and illustrated in Fig. 8.20. Different finite soil thicknesses together with the results under infinite soil depth (dashed lines) are presented in the figure.

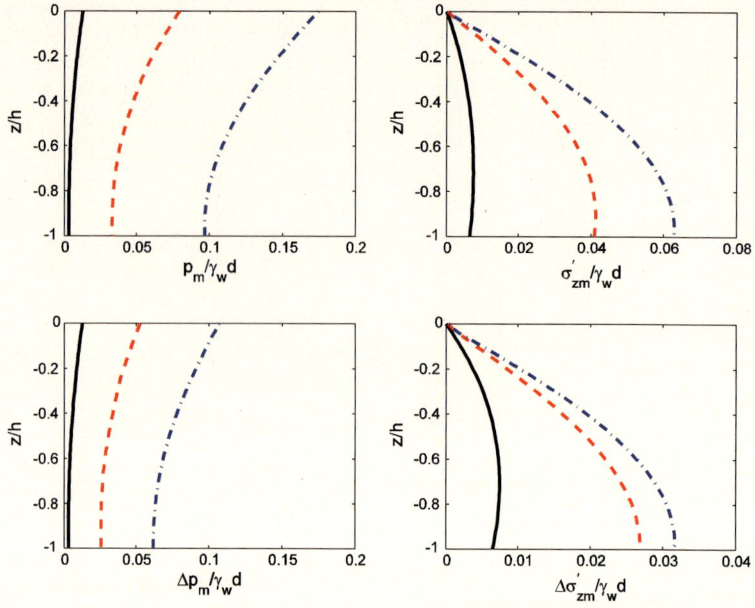

Fig. 8.19 Vertical distribution of p_m (*left panels*) and σ'_{zm} (*right panels*) for random wave results (*upper panels*) and relative results (*lower panels*) under different significant wave periods. *Solid lines* for $T_{1/3} = 5$ s, *dashed lines* for $T_{1/3} = 10$ s and *dashed-dotted lines* for $T_{1/3} = 20$ s (finite thickness)

In the vicinity of seabed, the soil responses for different soil depths are almost uniform. Different seabed thicknesses affect the soil response around the rigid impermeable bottom. Comparing with the infinite soil depth results, for finite soil depth, the maximum pore pressure around the rigid bottom increases. With the increase of soil depth, the vertical pore pressure distribution is approaching the results under the infinite soil depth. As for vertical effective normal stress σ'_{zm}, with the increase of soil thickness, its value first increases from the case $h/d = 0.25$ to $h/d = 2$; while σ'_{zm} decreases somewhat from $h/d = 2$ to $h/d = 4$. Further increase the soil thickness, the σ'_{zm} value decreases and its profile approaches the results for infinite soil depth. For small seabed thickness ($h/d = 0/25$ and 0.5), the σ'_{zm} profile still cannot completely developed before reaching the rigid bottom. After the σ'_{zm} profile entirely grows ($h/d = 1$), the influence of rigid bottom becomes significant, which "compresses" the vertical σ'_{zm} distribution and makes the profile more forwardly protuberant. This makes the maximum value of σ'_{zm} for case $h/d = 2$ around level $z/d = 0.7$ larger than the infinite result. Further increase the soil thickness, such "compression" due to finite seabed thickness releases gradually, and soil can hardly "feel" the existence of rigid impermeable bottom. Hence, the maximum value of σ'_{zm} decreases, and finally σ'_{zm} profile returns back to the infinite case.

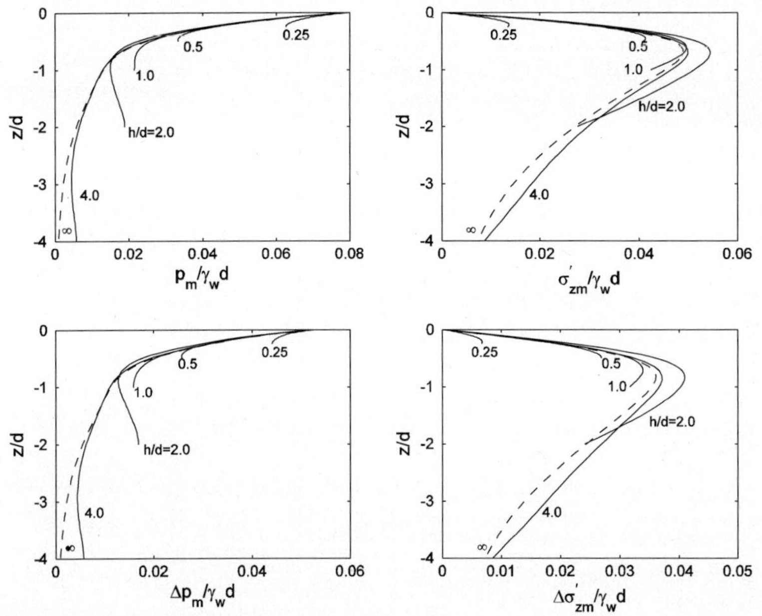

Fig. 8.20 Vertical distribution of p_m (*left panels*) and σ'_{zm} (*right panels*) for random wave results (*upper panels*) and relative results (*lower panels*) under different soil thickness. *Dashed lines* represent the results of infinite thickness

8.5 Summary

In this chapter, an analytical solution for ocean wave-induced oscillatory pore pressure and effective stresses in a porous seabed was presented. The new model provides coastal engineers a better understanding of the wave-seabed interaction system in realistic ocean environments. Based on new models, we perform a series of parametric study to explore the effects of wave and soil characteristics on wave-induced soil response in marine sediments. the numerical examples demonstrated the significant influence of rand wave loadings.

References

1. Goda, Y.: Random Seas and Design of Marine Structures. World Scientific, Singapore (2000)
2. Hsu, J.R.C., Jeng, D.S., Tsai, C.P.: Short-crested wave-induced soil response in a porous seabed of infinite thickness. Int. J. Numer. Anal. Methods Geomech. **17**(8), 553–576 (1993)
3. Longuet-Higgins, M.S.: The statistical analysis of a random, moving surface. Philos. Trans. R. Soc. **249**, 321–387 (1957)
4. Sumer, B.M., Fredsøe, J.: The Mechanism of Scour in the Marine Environment. World Scientific, New Jersey (2002)

Chapter 9
Wave-Induced Pore Pressure Accumulation in Marine Sediments

Abstract The residual mechanism of wave-induced pore water pressure accumulation in marine sediments is re-examined. In this chapter, an analytical approximation is derived using a linear relation for pore pressure generation in cyclic loading, and mistakes in previous solutions are corrected. A numerical scheme is then employed to solve the case with a non-linear relation for pore pressure generation. Both analytical and numerical solutions are verified with experimental data, and provide a better prediction of pore pressure accumulation than the previous solution. The parametric study concludes that the pore pressure accumulation and use of full non-linear relation of pore pressure become more important under the following conditions: (1) large wave amplitude, (2) longer wave period, (3) shallow water, (4) shallow soil and (5) softer soils with a low consolidation coefficient.

Keywords Pore pressure accumulation · Shear stress · Cyclic loading · Liquefaction

9.1 Introduction

In general, two mechanisms for wave-induced pore pressure have been observed in field measurements and laboratory experiments, as shown in Fig. 1.1. The first results from the transient or oscillatory excess pore pressure and is accompanied by attenuation of the amplitude and phase lag in the pore pressure changes [8, 20]. The second mechanism is termed the residual pore pressure, which is the build-up of excess pore pressure caused by contraction of the soil under the action of cyclic loading [17, 19].

Numerous studies for wave-induced momentary liquefaction, caused by oscillatory pore pressure, were completed in the past few decades. Among these, Madsen [8] and Yamamoto *et al.* [20] considered an infinite seabed with a similar framework. Mei and Foda [10] propose a boundary-layer approximation to derive a rather simplified formulation for the wave-induced transient pore pressure, valid for coarse sand [5]. Okusa [12] further demonstrated the significant influence of the degree of saturation on the pore pressure and proposed the criteria of liquefaction. Jeng [6] derived a series of analytical solutions for the oscillatory pore pressure within marine sediments, which has been included in Chap. 3.

D.-S. Jeng, *Porous Models for Wave-seabed Interactions*,
DOI 10.1007/978-3-642-33593-8_9,

The mechanism of pore pressure build-up due to ocean waves has also been considered. Seed and Rahman *et al.* [17] established a simple one-dimensional finite element model by taking into account the distribution of cyclic shear stresses in the soil profile, and pore- pressure dissipation. Sekiguchi *et al.* [18] proposed an elasto-plastic model for the standing wave-induced liquefaction using a Laplace transformation. Later, some numerical models for post-liquefaction and progressive liquefaction and densification in marine sediments were developed [11, 13–15].

In addition to numerical modeling, McDougal *et al.* [9] proposed a set of analytical solutions for wave-induced pore pressure build-up in a uniform layer of soil, based on the assumption of an incompressible soil. In their approach, the source term in the modified Biot's consolidation equation is derived using a linear relationship between pore pressure ratio (u_g/σ_0') and cyclic ratio (N/N_ℓ) [3]. To provide a convenient practical result for engineers, McDougal *et al.* [9] presented their solutions for the cases of shallow, finite and deep soil depths. These analytical solutions are useful for both engineers and researchers, as they can be used for either the investigation of qualitative behavior of complicated engineering problems or the validation of numerical methods. Recently, using a similar approach, Cheng *et al.* [3] re-examined the analytical solution of [9] and proposed a numerical model to investigate the same problem. As pointed out by Cheng *et al.* [3], the analytical solution proposed by McDougal *et al.* [9] revealed some errors in the formulations. However, after a close examination of both solutions [3, 9], Jeng *et al.* [7] found errors in both publications. Here, we list the original formulae of both solutions [3, 9]:

Original formulae in McDougal *et al.* [9] [(28), (33), (36) and (37), pp. 6 and 7]:

$$F = \frac{d^2 f}{c u_*}, \tag{M-28}$$

$$\frac{\partial U}{\partial T} = \frac{\partial^2 U}{\partial Z^2} + Z\big[F(Z)\big]^{-1/\beta_1^*}, \tag{M-33}$$

$$U(T,Z) = \sum_{n=0}^{\infty} G_n\big(1 - e^{-k_n^2 t}\big)\sin(K_n Z), \tag{M-36}$$

$$G_n = \frac{2}{K_n^2}\int_0^1 s\big[F(s)\big]^{-1/\beta_1^*}\,ds. \tag{M-37}$$

Original formulae in Cheng *et al.* [3] [p. 886]:

$$\frac{\partial U}{\partial T} = \frac{\partial^2 U}{\partial z^2} + \left(\frac{1+2k_0}{3}\right)^{1+1/\beta_1^*} Z\big[F(Z)\big]^{-1/\beta_1^*}, \tag{C-33}$$

$$G_n = \left(\frac{1+2k_0}{3}\right)^{1+1/\beta_1^*} \frac{2}{K_n^2}\int_0^1 s\big[F(s)\big]^{-1/\beta_1^*}\,ds. \tag{C-37}$$

The correct forms of these equations should read as follows [7]:

$$\frac{\partial U}{\partial T} = \frac{\partial^2 U}{\partial Z^2} + Z\big[F(Z)\big]^{-1/\beta_1^*}, \tag{9.1}$$

$$U(T, Z) = \sum_{n=0}^{\infty} G_n\big(1 - e^{-k_n^2 T}\big)\sin(K_n Z), \tag{9.2}$$

$$G_n = \frac{2}{K_n^2}\int_0^1 s\big[F(s)\big]^{-1/\beta_1^*}\,ds, \tag{9.3}$$

$$F = \frac{d^2}{cu_*}\left[\frac{\tau_{xz}(Z)}{ZP_0\lambda}\right]. \tag{9.4}$$

The definitions of the symbols used above are the same as those used in the original publications of [3, 9], except β_1^* representing β_1 in the original paper. As shown above, errors in (M-28), (M-36) and (M-37) occurred in the original paper [9], while errors of (C-33) and (C-37) occurred in [3].

All the aforementioned investigations are based on the linear relation of pore pressure generation to cyclic loading proposed in [16, 19]. From previous experiments [1], this relation is non-linear, rather than linear. However, to date, no analytical solution for the non-linear relations of pore pressure generation is available so numerical methods must be used to get numerical approximation, for example, [17] adopted the finite element method.

In this chapter, analytical and numerical techniques for the wave-induced pore water pressure accumulation within marine sediment will be proposed. We derive an analytical solution for the linear relation of pore pressure generation to cyclic loading. Then, we investigate the problem with a non-linear relation using a numerical scheme. Both analytical approximation and numerical model are verified with the previous experimental data. A parametric study then examines the influence of wave and soil characteristics on the pore pressure accumulation in a marine sediment. The effect of a non-linear relation on pore pressure generation is also examined.

9.2 Boundary Value Problem

For the problem of wave-seabed interaction, the Biot consolidation equations [2] have been generally adopted to model the dynamic response of marine sediments for various applications. In general, the wave-induced pore pressure within marine sediments consists of two components: oscillatory ($\tilde{p}$) and residual (u) mechanisms, which can be expressed as

$$p = \tilde{p} + u \tag{9.5}$$

where p is the pore water pressure, $\tilde{p}$ represents the oscillatory pore pressure that leads to momentary liquefaction, while u represents the period-averaged pore pressure that leads to residual liquefaction, and is defined by

$$u = \frac{1}{T}\int_{t}^{t+T} p\,dt \tag{9.6}$$

where T is the wave period and t is the time.

A series of analytical solutions for the oscillatory pore pressure within marine sediments have been developed since the 1970's [6, 8, 20]. The amplitude of the oscillatory pore pressure ($\tilde{p}$) and shear stress (τ_0) for a saturated seabed are given for a seabed of finite thickness by:

$$\tilde{p}_0 = \frac{p_b}{1-2\mu}\left[(1-2\mu)\left(C_2 e^{-kz} - C_4 e^{kz}\right) + (1-\mu)\left(\delta^2 - k^2\right)\left(C_5 e^{-\delta z} - C_6 e^{\delta z}\right)\right], \tag{9.7}$$

$$\tau_0 = p_b\left[(C_1 C_2 kz)\, e^{-kz} - (C_3 - C_4 kz)\, e^{kz} + k\delta\left(C_5 e^{-\delta z} - C_6 e^{\delta z}\right)\right] \tag{9.8}$$

where $P_b = \gamma_w H/2\cosh kd$ is the amplitude of dynamic wave pressure, μ is the Poisson's ratio, k is the wave number and the C_i $(i = 1, \ldots, 6)$ coefficients and the parameter δ can be found in [6].

For an infinite seabed,

$$p_\infty = P_b\, e^{-kz} \quad \text{and} \quad \tau_\infty = P_b kz\, e^{-kz}. \tag{9.9}$$

As shown in (9.9), the wave-induced pore pressure and shear stress in an infinite seabed are independent of soil characteristic and are identical to the solution for an incompressible soil and pore fluid [9]. However, (9.7) and (9.8) are more general than the solution derived from Laplace's equations [9].

The residual pore pressure (u) in a homogeneous, isotropic soil can be derived from the one-dimensional Biot's consolidation equation [19]

$$\frac{\partial u}{\partial t} = c_v \frac{\partial^2 u}{\partial z^2} + f \tag{9.10}$$

where f is the mean accumulation pore pressure source term associated with the surface water waves [9]. A detailed derivation of (9.10) can be found in [19]. In (9.10), c_v is the coefficient of consolidation, given by

$$c_v = \frac{2Gk_z(1-\mu)}{\gamma_w(1-2\mu)}. \tag{9.11}$$

To solve (9.10), the following boundary and initial conditions are required:

$$u(z,0) = u(0,t) = 0 \tag{9.12}$$

and

$$\frac{\partial u(h,t)}{\partial z} = 0 \quad \text{or} \quad u(\infty,t) = 0. \tag{9.13}$$

9.3 Source Term

We now investigate the "*source term*" of the pore pressure generation (f), and consider both linear and nonlinear mechanisms of pore pressure generation.

9.3.1 Nonlinear Mechanism of Pore Pressure Generation

The laboratory results of deAlba [1] relate the development of pore water pressure to the number of load cycles in simple shear. Their non-linear relationship is given by

$$\frac{u_g}{\sigma_0'} = \frac{1}{2} + \frac{1}{\pi} \sin^{-1}\left[2\left(\frac{N}{N_\ell}\right)^{1/\theta} - 1\right] \tag{9.14}$$

or

$$\sin\left(\frac{\pi u_g}{2\sigma_0'}\right) = \left(\frac{N}{N_\ell}\right)^{\frac{1}{2\theta}} \tag{9.15}$$

where u_g is the pore pressure generation due to cyclic loading, σ_0' is the effective over burden, N is the number of cyclic loading, N_ℓ is the number of cycles to liquefaction, and θ is the shape factor suggested to be 0.7 [16].

The pore pressure source term in (9.10) is given by [16] as

$$f = \frac{\partial}{\partial t}\left(\sigma_0' \frac{N}{N_\ell}\right). \tag{9.16}$$

9.3.2 Linear Mechanism of Pore Pressure Generation

To simplify the problem, a linear mechanism of pore pressure generation was proposed [16, 19]

$$\frac{u_g}{\sigma_0'} = \frac{N}{N_\ell} \tag{9.17}$$

from which the source term of pore pressure generation can be expressed as

$$f = \frac{\partial}{\partial t}\left(\sigma_0' \frac{N}{N_\ell}\right) = \frac{\partial}{\partial t}\left(\sigma_0' \frac{t/T}{N_\ell}\right) = \frac{\sigma_0'}{T N_\ell} \tag{9.18}$$

where T is the wave period. In (9.18), N_ℓ is the number of cycles to liquefaction, which is a function of the cyclic shear stress ratio [17, 19],

$$N_\ell = \left[\frac{\tau_0}{\alpha \sigma_0'}\right]^{-1/\beta} \tag{9.19}$$

where τ_0 is the amplitude of wave-induced shear stress, and α and β are the functions of the soil type and relative density.

Substituting (9.19) into (9.18), we have

$$f = \frac{\sigma_0'}{T}\left[\frac{\tau_0}{\alpha\sigma_0'}\right]^{-1/\beta} \tag{9.20}$$

which is a generalized definition of the source term.

It is noted that the linear mechanism of pore pressure generation was first applied to the wave-induced pore pressure build-up in marine sediment by [17]. Since then, this relation has been commonly used in various approaches [3, 9, 19]. We will consider the full non-linear relation of pore pressure generation and examine its influence.

9.4 Theoretical Models

9.4.1 Analytical Approximation for Linear Mechanism

The shear stress in the source term depends on both wave and soil characteristics. Since the expression for the shear stress in a seabed of finite thickness, i.e., (9.8), is complicated, similar to [9], we consider two simplified cases, shallow soil depth and deep depth in addition to the finite soil model. We only outlined the analytical solutions in this section, as detailed derivations are available in Sect. 9.8.

9.4.1.1 Finite Soil Model

For the finite soil depth, the complete function, (9.18) is required for the source term. Using a Fourier series expansion, the residual pore pressure can be expressed as

$$u = \sum_{n=1}^{\infty} a_n\left(1 - e^{-c_v\kappa_n^2 t/h^2}\right)\sin\left(\frac{\kappa_n z}{h}\right), \tag{9.21}$$

$$a_n = \frac{2h}{c_v k_n^2}\int_0^h f(r)\sin\left(\frac{\kappa_n r}{h}\right)dr \tag{9.22}$$

where f is given in (9.20), and $\kappa_n = \frac{(2n-1)\pi}{2}$.

9.4.1.2 Shallow Soil Model

For relatively shallow soil depth, $h/L \leq 0.1$, we consider the shallow depth approximation, in which the shear stress can be expressed as

$$\tau_0 = m P_b z \tag{9.23}$$

where the value of m can be determined by equating (9.8) and (9.23), and integrated over the entire soil depth, yielding

$$m = \frac{2}{kh^2}\{-(C_1 e^{-kh} + C_3 e^{kh}) + (C_2 e^{-kh} - C_4 e^{kh}) + C_1 - C_2 + C_3 + C_4 + kh(C_2 e^{-kh} + C_4 e^{kh}) - k^2(C_5 e^{-\delta h} + C_6 e^{\delta h}) + k^2(C_5 + C_6)\}. \quad (9.24)$$

Then, the source term for shallow soil depth can be further simplified as

$$f = az, \quad (9.25)$$

$$a = \frac{(1+2K_0)\gamma'}{3T}\left[\frac{3mP_b}{\alpha(1+2K_0)\gamma'}\right]^{1/\beta}. \quad (9.26)$$

The residual pore pressure can again be given by a Fourier series as

$$u = \frac{a}{2c_v}\left[\left(h^2 z - \frac{z^3}{3}\right) - \sum_{n=1}^{\infty} a_n e^{-c_v \kappa_n^2 t/h^2} \sin\left(\frac{\kappa_n z}{h}\right)\right], \quad (9.27)$$

$$a_n = \frac{2}{h}\int_0^h \left(h^2 r - \frac{r^3}{3}\right)\sin\left(\frac{\kappa_n r}{h}\right) dr. \quad (9.28)$$

9.4.1.3 Deep Soil Model

For soil depth in the range $h/L \geq 0.3$, we consider the deep (infinite) depth approximation, in which the shear stress is given by (9.9). The source term for deep soil can be expressed as

$$f = bz e^{-\lambda z}, \quad (9.29)$$

$$\lambda = \frac{k}{\beta} \quad \text{and} \quad b = a\left(\frac{k}{m}\right)^{1/\beta} \quad (9.30)$$

where a is defined in (9.26).

Then, the residual pore pressure is calculated using a Laplace transformation as

$$u = \frac{2b}{c_v \lambda^3}\left[1 - \left(\frac{\lambda z}{2} + 1\right)e^{-\lambda z} - \frac{1}{\pi}\int_0^{\infty} \frac{e^{-r c_v \lambda^2 t}}{r(1+r)^2}\sin(\sqrt{r}\lambda z)\, dr\right]. \quad (9.31)$$

The solution (9.31) is similar to the one proposed by McDougal *et al.* [9] with corrections pointed out previously. It is noted that the shear stress used in the [9] solution was based on the assumption of an incompressible soil.

Sumer and Fredsøe [19] also obtained an analytical solution, but in a different form:

$$u = \frac{2}{\pi}\int_0^t \int_0^{\infty} \int_0^{\infty} e^{-cv\xi^2(t-t')} \sin(\xi z) f(z') \sin(\xi z')\, dz'\, d\xi\, dt', \quad (9.32)$$

which is in their notation [19]. To compare (9.32) to (9.31), let $f(z') = Az' e^{-\lambda z}$ and $\xi = \sqrt{r}\lambda$, then we have

$$\frac{1}{\pi}\int_0^{\infty} \frac{\sin(\sqrt{r}\lambda z}{r(1+r^2)} = 1 - \left(\frac{\lambda z}{2} + 1\right) e^{-\lambda z}. \tag{9.33}$$

The present solution provides a much simplified form by using a different approach.

9.4.2 Numerical Scheme

Equation (9.10) and the Neumann boundary condition are discretized by the following second-order accuracy formulations

$$\begin{aligned}\frac{u(z,t+\Delta t)-u(z,t)}{\Delta t} &= \frac{1}{2}c_v\left[\frac{u(z+\Delta z,t+\Delta t)-2u(z,t+\Delta t)+u(z-\Delta t+\Delta t)}{\Delta z^2}\right.\\ &\quad \left.+\frac{u(z+\Delta z,t)-2u(z,t)+u(z-\Delta z,t)}{\Delta z^2}\right]\\ &\quad +\frac{1}{2}\left[f(z,t+\Delta t)+f(z,t)\right]\end{aligned} \tag{9.34}$$

and

$$\frac{3u(h,t+\Delta t)-4u(h-\Delta z,t+\Delta t)+u(h-2\Delta z,t+\Delta t)}{2\delta z} = 0. \tag{9.35}$$

It is easy and fast to numerically solve (9.34) and (9.35) with the initial and boundary conditions (9.12) and (9.13), if we algebraically deal with last two equations to rewrite the system as a tri-diagonal system. In our computations, we use a time transformation so that we can use a small time step size and finite computational times to obtain the numerical solution for large times.

9.4.3 Comparisons

Here the experimental data [4] is compared with the present solutions. Several tests were conducted in a small wave tank [4], and numerous sets of wave-induced liquefaction observations were presented in their report. A comparison of the calculated and measured pore pressure accumulation for the two soils is shown in Fig. 9.1. The input data of the experiments [4] and the numerical calculations are tabulated in Table 9.1. The results of the previous analytical solutions [9] (dashed lines), the present linear approximation (solid lines) and nonlinear numerical solutions (solid lines with symbols) are included in the comparison. The relative soil depth for the data is in the range of $0.2 \leq h/L \leq 0.3$, hence the soil depth is intermediate. As

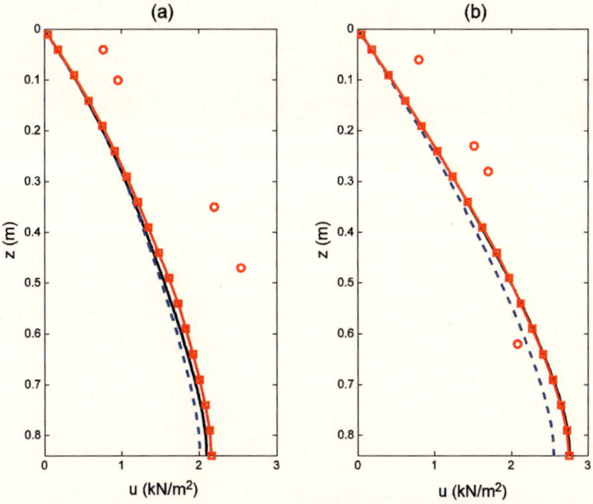

Fig. 9.1 Comparison of the previous model [9], the present analytical and numerical models and experimental data [4]. Notations: *dashed lines* = the previous model [9], *solid lines* = the present analytical solution, *solid lines with symbols* = the present numerical model, and ∘ = experimental data

Table 9.1 Input data of experiments [4]

Wave characteristics		
	Case (a)	Case (b)
wave period (T)	1.76 s	2.02 s
water depth (d)	0.3 m	0.3 m
wavelength (L)	3.473 m	4.103 m
wave height (H)	0.22 m	0.21 m

Soil characteristics	
seabed thickness (h)	0.84 m
Poisson's ratio (μ)	0.40
soil porosity (n_e)	0.46 m
shear modulus (G)	5.6×10^5 N/m^2
soil permeability (k)	4×10^{-8} m/s
unit weight of soil (γ_s)	$1.8668\gamma_w$
unit weight of pore fluid (γ_w)	1980 N/m^3
degree of saturation (S)	1.0
coefficient of earth pressure (K_0)	0.4
coefficient of consolidation (c_v)	0.000165
residual parameter (α)	0.246
residual parameter (β_1)	0.165

shown in the figure, the present solutions (both analytical and numerical) are approaching to the experimental data, and provide a slightly better prediction than that of [9].

Table 9.2 Input data of examples

Soil characteristics	Soil A	Soil B
Poisson's ratio (μ)	0.35	0.49
soil porosity (n_e)	0.3	0.46
shear modulus (G)	5×10^6 N/m^2	10^6 N/m^2
soil permeability (k)	10^{-6} m/s	10^{-7} m/s
unit weight of soil (γ_s)	$2.65\gamma_w$	$2.65\gamma_w$
unit weight of pore fluid (γ_w)	9806 N/m^3	9806 N/m^3
degree of saturation (S)	1.0	1.0
coefficient of earth pressure (K_0)	0.5	0.5
coefficient of consolidation (c_v)	0.00261	0.00052
residual parameter (α)	0.246	0.246
residual parameter (β_1)	0.165	0.165

9.5 Parametric Study

The primary objective of this paper is to investigate the effects of the non-linear relation of pore pressure generation relation on the wave-induced pore water pressure accumulation in marine sediments. In this section, a parametric study is carried out to examine the influence of five important parameters: wave height, wave period, water depth, seabed thickness and soil type. The input data for the parametric study is given in Table 9.2. Wave height is one important wave characteristics in the design of coastal structures, as it directly affects the wave forces, pressure and energy acting on the structures. It also significantly affects the wave-induced pore pressure and effective stresses in marine sediments [6]. Here we consider the case of non-breaking waves, in which the maximum wave steepness satisfies:

$$\left(\frac{H}{L}\right)_{\max} \le 0.142 \tanh\left(\frac{2\pi d}{L}\right) \quad \text{or} \quad \left(\frac{H}{L}\right)_{\max} \le 0.78. \tag{9.36}$$

Figure 9.2 illustrates the distribution of pore pressure accumulation (u/σ_0') versus time (t/T) for two different soils for three values of H/L. In the figure, solid lines represent the results using the non-linear relation (numerical model) and dashed lines denote the traditional linear relation (analytical approximation). As shown in Fig. 9.2, the pore pressure accumulates faster under large wave loading (Fig. 9.2b and c), while it will take longer to build-up with small waves (Fig. 9.2a). Also, the influence of non-linear relations on the pore pressure accumulation becomes more important as wave steepness increases, comparing Fig. 9.2a–c. It is noted that the line at $u/\sigma_0' = 1$ is the criterion of residual liquefaction.

Wave period and water depth are two parameters that are used to determine the wavelength. In general the wave period varies from 5 s to 15 s for ocean waves. The pore pressure accumulations for three wave periods are plotted in Fig. 9.3. The figure clearly indicates that the pore pressure builds up faster for short period waves because the pore pressure and energy are more difficult to drain out in a short duration. However, the accumulated pore pressure increases as wave period increases because the higher wave energy can accumulate in a long period wave. It is noted

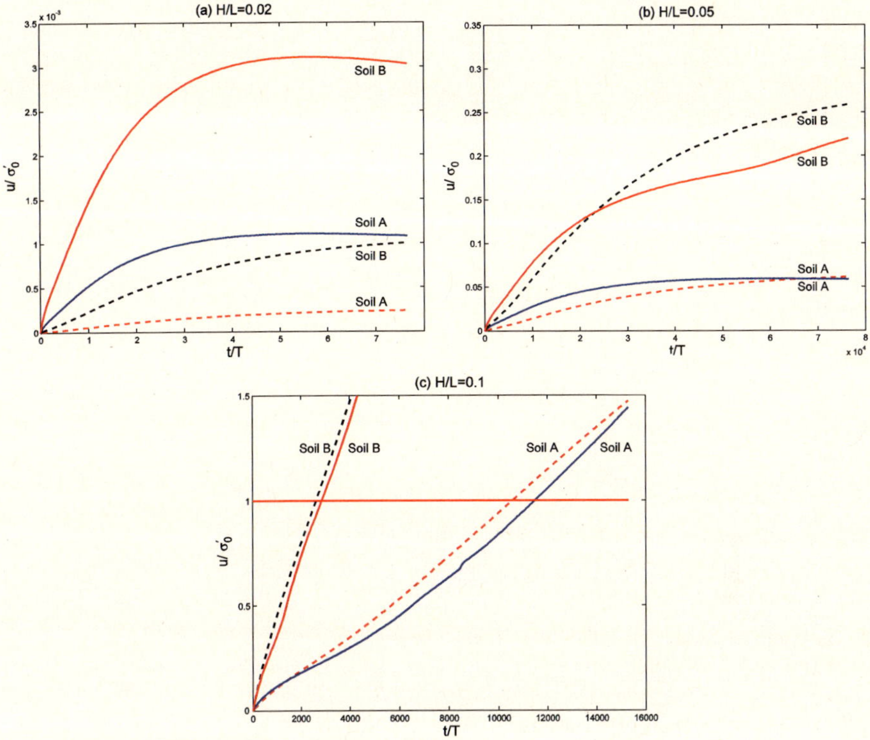

Fig. 9.2 Time development of the pore pressure at $z = 2$ m ($T = 10$ s, $d = 16$ m, $h = 20$ m). Notation: *dashed lines* = linear relation, *solid lines* = non-linear relation

that the results presented in Fig. 9.3 are based on the same wave height for various wave periods. For a short period wave, such as $T = 5$ s, the pore pressure can be built up quickly and become larger under large wave loading.

The relative water depth (d/L) is an index to classify the wave field. For example, we consider water to be shallow if $d/L \leq 0.05$, and deep if $d/L \geq 0.5$, and intermediate water between them. The effect of relative water depth on the pore pressure accumulation is illustrated in Fig. 9.4. The results clearly indicate that the accumulated pore pressure is larger in shallow water than that in deep water, and that the effect of using the nonlinear relation is more significant in shallow water. Note that the vertical scales in the sub-figures in Fig. 9.4 are different.

Based on the results presented in Figs. 9.2–9.4, it can be concluded that the mechanism of pore pressure build-up and the nonlinear relation for pore pressure generation are more important under the conditions of: (1) larger amplitude waves, (2) longer wave periods, and (3) shallow water.

Besides wave characteristics, soil properties are also important for the wave-induced pore pressure in marine sediment [6]. In this section, we examine two important soil parameters, seabed thickness and soil types.

Figure 9.5 illustrates the effects of seabed thickness on the pore pressure accumulation in marine sediments. Here the pore pressure can build up faster and accu-

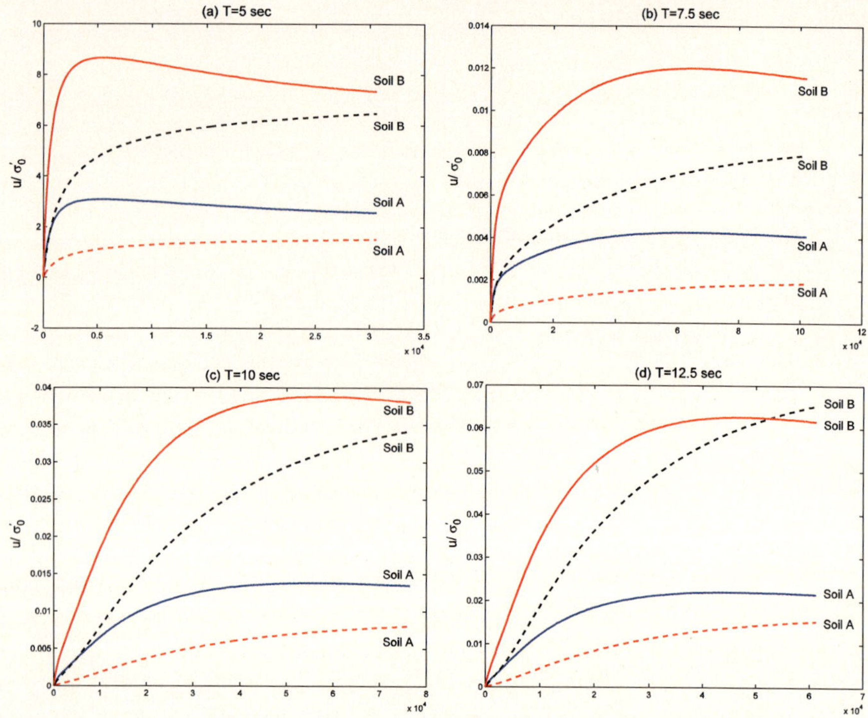

Fig. 9.3 Time development of the pore pressure at $z = 2$ m for various wave period. (**a**) $T = 5$ s, (**b**) $T = 7.5$ s, (**c**) $T = 10$ s and (**d**) $T = 12.5$ s ($H = 4$ m, $d = 16$ m, $h = 20$ m). Notation: *dashed lines* = linear relation, *solid lines* = non-linear relation

mulate to a large value in shallow soil (Fig. 9.5a). The numerical solution for the soil B in Fig. 9.5a shows the singularity which is caused by a singular point at some time in the source term with the nonlinear relation.

Referring to Figs. 9.2–9.5, two different soils are considered in the examples. Soil B with a low consolidation coefficient ($c_v = 0.0052$) can accumulate the pore pressure to a large value, which implies that residual liquefaction is more likely to occur.

9.6 A Simplified Approximation for an Infinite Seabed

9.6.1 Scaling Analysis

The first objective of this study is to clarify the applicable ranges of the two mechanisms, We compare the amplitude of the oscillatory pore pressure ($\tilde{p}$) and the equilibrium residual pore pressure ($u(z, \infty)$):

$$P_{\text{osc}} = |\tilde{p}| = P_b \exp(-kz) \tag{9.37}$$

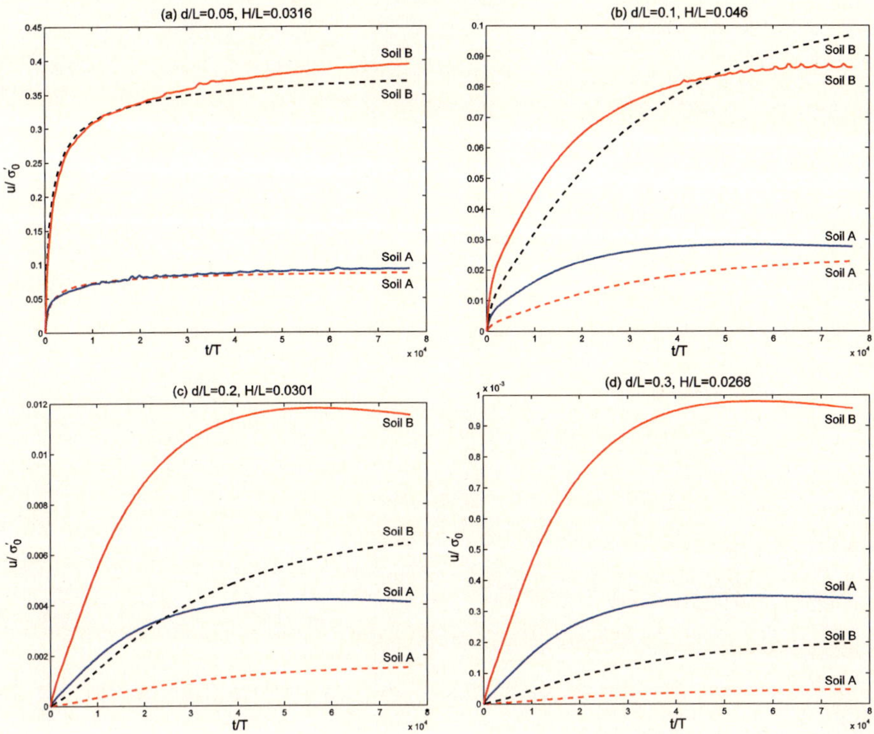

Fig. 9.4 Time development of the pore pressure at $z = 2$ m for various relative water depth. (**a**) $d/L = 0.05$, $H/L = 0.0316$; (**b**) $d/L = 0.1$, $H/L = 0.046$; (**c**) $d/L = 0.2$, $H/L = 0.0301$; and (**d**) $d/L = 0.3$, $H/L = 0.0268$ ($T = 10$ s, $h = 20$ m). Notation: *dashed lines* = linear relation, *solid lines* = non-linear relation

and

$$P_{\text{res}} = u(z, \infty) = \frac{2A}{c_v}\lambda^3\left[1 - \left(\frac{\lambda z}{2} + 1\right)\exp(-\lambda z)\right]. \tag{9.38}$$

To examine the role of each mechanism, we define the amplitude ratio as

$$\epsilon(z) = \frac{P_{\text{res}}}{P_{\text{osc}}}. \tag{9.39}$$

Distributions of the scaling factor (ϵ) versus soil depth (z/L) for various wave conditions are illustrated in Fig. 9.6. Three different wave steepness and four relative water depths are considered. Figure 9.6 clearly indicates that the residual mechanism becomes more important as the scaling factor (ϵ) increases, the wave steepness (H/L) increases, or the relative water depth (d/L) decreases.

To further clarify the range of residual mechanism, we plot the critical line of $\epsilon = 1$ with wave steepness and relative water depth at $z = 5$ m in Fig. 9.7. The lines represent the critical relationship of H/L and d/L with given a consolidation coefficient (c_v). The region below the curve denotes conditions when the transient

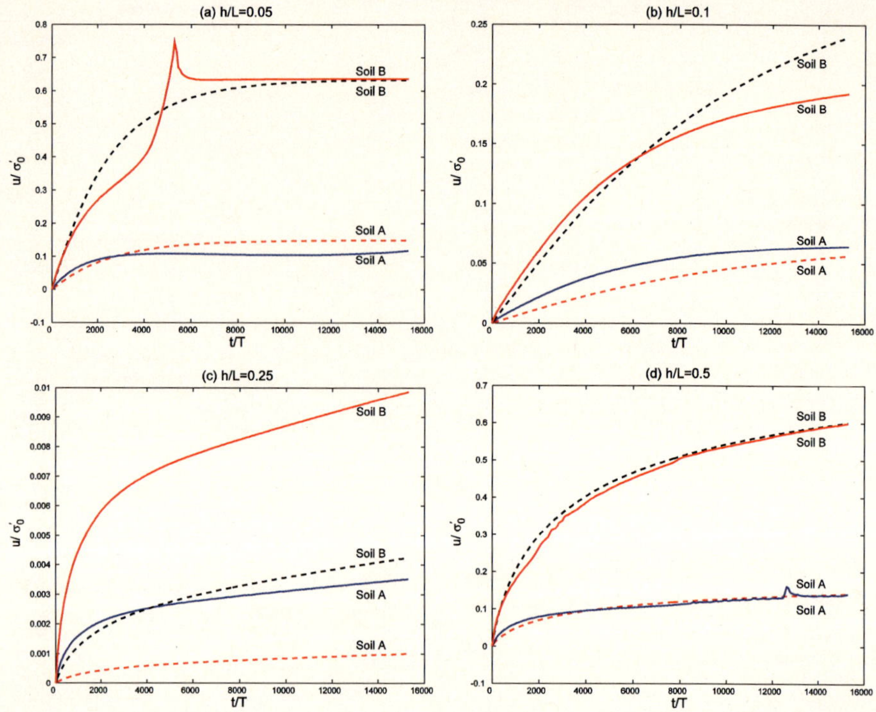

Fig. 9.5 Time development of the pore pressure at $z = 2$ m for various relative seabed thickness. (**a**) $h/L = 0.05$, (**b**) $h/L = 0.1$; (**c**) $h/L = 0.25$, and (**d**) $h/L = 0.5$ ($T = 10$ s, $H = 4$ m and $d = 16$ m). Notation: *dashed lines* = linear relation, *solid lines* = non-linear relation

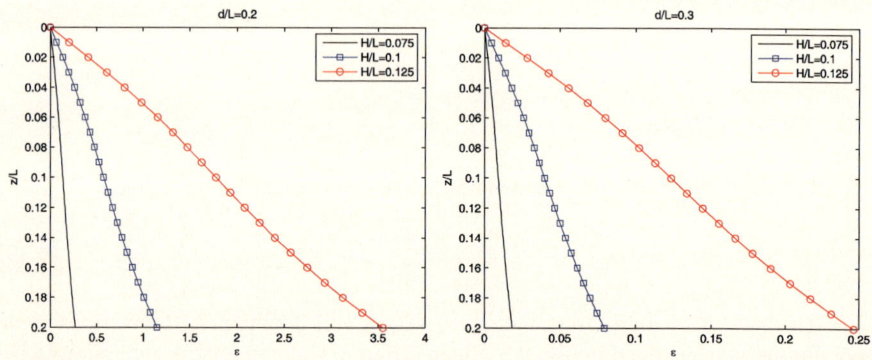

Fig. 9.6 Distribution of the maximum scaling factor (ϵ) versus soil depth (z/L) for various wave steepness (H/L). (**a**) $d/L = 0.2$ and (**b**) $d/L = 0.3$. [Input data: $T = 10$ s, $n_e = 0.33$, $\mu = 0.33$, $\alpha = 0.246$, $c_v = 0.022$, $\eta = 0.25$]

(oscillating) mechanism dominates, while the region above the curve denotes conditions when the residual mechanism dominates. The critical curve will move up as the consolidation coefficient (c_v) increases.

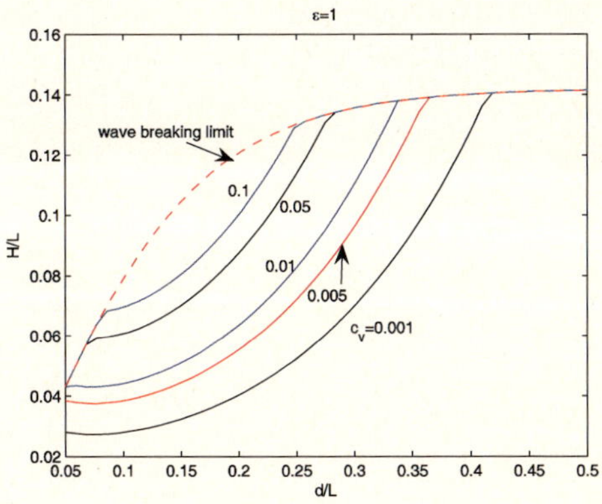

Fig. 9.7 Distribution of the maximum scaling factor (ϵ) versus soil depth (z/L) for various wave steepness (H/L). (**a**) $d/L = 0.2$ and (**b**) $d/L = 0.3$. [Input data: $T = 10$ s, $n_e = 0.33$, $\mu = 0.33$, $\alpha = 0.246$, $c_v = 0.022$, $\eta = 0.25$]

9.6.2 A Simplified Approximation for Wave-Induced Liquefaction

For engineers, the most important task is to examine where liquefaction will occur and how deep it is. The criterion of residual liquefaction is

$$\frac{P_{\text{res}}}{\sigma_0'} = 1, \tag{9.40}$$

which leads to

$$P_{\text{res}} = \frac{2A}{c_v\lambda^3}\left[1 - \left(\frac{\lambda z_L}{2} + 1\right)\exp(-\lambda z_L)\right] = \sigma_0' = \frac{(1+2K_0)}{3}\gamma' z_L, \tag{9.41}$$

where z_L is the liquefaction depth.

Let

$$B = \frac{(1+2K_0)\gamma' c_v \lambda^2}{6b}, \tag{9.42}$$

then, the relationship of the maximum liquefied depth (z_L) and the parameter B can be plotted (see Fig. 9.8). For engineering applications, given wave and soil conditions, we can determine the parameter B from (9.42). The maximum liquefied depth can then be easily determined from Fig. 9.8. It is noted that the relation presented in Fig. 9.8 is universal, applicable to all engineering conditions.

Herein, we further consider a special case: assuming λz_L is small. With this assumption, we further expend $\exp(-\lambda z_L)$ with Taylor's expansion, and then use the first three terms, we have the solution of (9.41) as $z = 0$ or

$$z_L = \frac{2}{\lambda}\sqrt{\frac{1}{2} - B}. \tag{9.43}$$

Note that (9.43) is only valid under the condition of $B \leq 1/2$. It is noted that $B > 1/2$ will provide an unrealistic solution. Under such condition, $z = 0$ will be the solution (i.e., no liquefaction occurs).

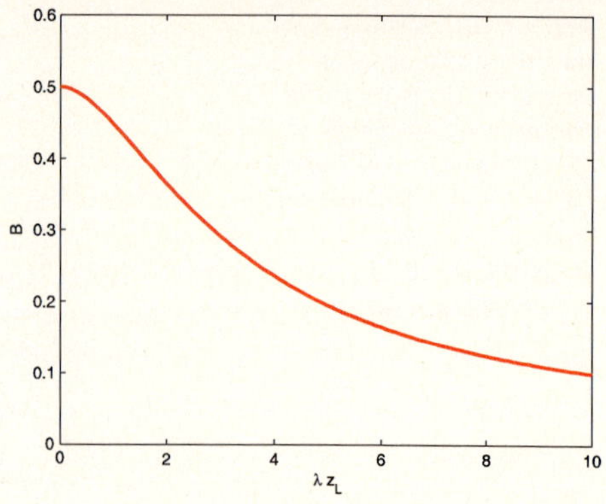

Fig. 9.8 Distribution of the maximum liquefied depth (z_L) and parameter B

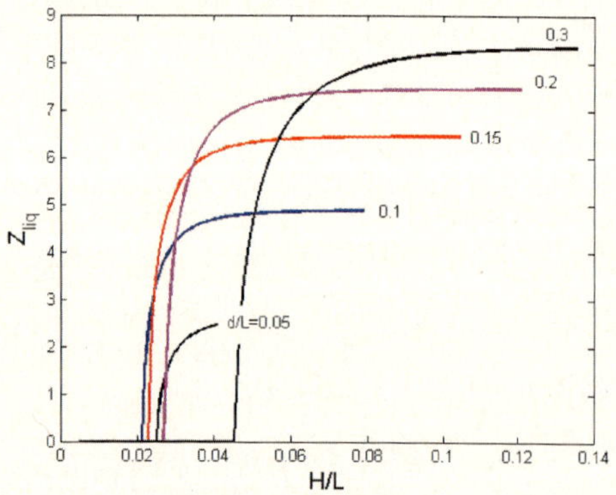

Fig. 9.9 Distribution of maximum liquefaction depth (z_L) versus wave steepness (H/L) for various relative water depth (d/L). [Input data: $n_e = 0.33$, $\mu = 0.33$, $k_z = 10^{-6}$ m/s, $G = 5 \times 10^6$ N/m^2, $\alpha = 0.246$, $\beta = 0.25$, $c_v = 0.0022$ and $T = 10$ s]

Based on (9.43), a numerical example for the prediction of maximum liquefaction depth (z_L) is presented in Fig. 9.9. In the example, the relative water depth (d/L) varies from shallow water ($d/L = 0.05$) to intermediate water ($d/L = 0.3$). As the water depth increases, the maximum wave steepness for inducing soil liquefaction will increase, which will enhance the liquefaction potential. The figure also indicates that the maximum residual liquefaction depth will increase as the wave steepness increases. This finding is also consistent with the result presented in Fig. 9.7 that the residual mechanism is important for the large wave conditions.

9.7 Summary

In this chapter, the pore water pressure accumulation in marine sediments due to ocean waves is investigated. First, an analytical approximation is proposed for the linear relation of pore pressure generation. Then, a numerical scheme is employed for the case of a nonlinear relation. Based on the numerical results presented, the following conclusions can be drawn:

1. A comprehensive comparison between the present analytical approximation and numerical model against the experimental data [4] was performed. The comparison indicates that the present models provide a better prediction of pore pressure accumulation than the previous analytical solution [9].
2. An investigation of the effects of wave characteristics on the pore pressure accumulation was presented in Figs. 9.2–9.4. The results indicate that the mechanism of pore pressure build-up and the nonlinear relation of pore pressure generation are more important under the conditions of (1) larger wave, (2) longer wave period, and (3) shallow water.
3. For soil characteristics, the pore pressure accumulates faster in shallow water and Soil A with a low consolidation coefficient.
4. A simplified approximation is proposed for the case of infinite seabed. Numerical results indicate that the residual mechanism is particularly important for large wave loading, while the oscillatory mechanism dominates the pore pressure under small wave loading. Figure 9.7 clearly indicates the range of residual mechanisms for various values of c_v.
5. Based on the new solution, a simplified approximation for the prediction of the maximum liquefaction depth is proposed for engineering practice. The universal relationship of the parameter B and the maximum liquefied depth (z_L) provides coastal geotechnical engineers a first-hand effective method for the evaluation of the residual liquefaction potential.

9.8 Appendix: Mathematical Derivation of Analytical Solutions

The boundary value problem for a seabed of finite depth are outlined here:

$$\frac{\partial u}{\partial t} = c_v \frac{\partial^2 u}{\partial z^2} + f, \tag{9.44}$$

$$u(z,0) = u(0,t) = 0 \tag{9.45}$$

and

$$\frac{\partial u(h,t)}{\partial z} = 0. \tag{9.46}$$

Since the source term is a time-independent function, the residual pore pressure (u) can be rewritten as

$$u = u_1(z) + u_2(z,t). \tag{9.47}$$

Then, the boundary value problem can be rewritten as

$$c_v \frac{\partial^2 u_1}{\partial z^2} = -f, \qquad \frac{\partial u_2}{\partial t} = c_v \frac{\partial^2 u_2}{\partial z^2}, \tag{9.48}$$

$$u_1(0) = 0, \qquad \frac{\partial u_1(h,t)}{\partial z} = 0 \tag{9.49}$$

and

$$u_2(z,0)=0, \qquad u_2(0,t)=-u_1(0), \qquad \frac{\partial u_2(h,t)}{\partial z}=0. \tag{9.50}$$

9.8.1 Finite Soil Model

The solution of $u_1(z)$ can be obtained through a Fourier series expansion and written as

$$u_1=\sum_{n=1}^{\infty} a_n \sin\left(\frac{\kappa_n z}{h}\right), \qquad a_n=\frac{2h}{c_v k_n^2}\int_0^h f(r)\sin\left(\frac{\kappa_n r}{h}\right)dr. \tag{9.51}$$

Then, we can have the solution of $u_2(z,t)$ as

$$u_2=-\sum_{n=1}^{\infty} a_n\, e^{-c_v\kappa_n^2 t/h^2}\sin\left(\frac{\kappa_n z}{h}\right). \tag{9.52}$$

The final solution of pore pressure will be identical to (9.21).

9.8.2 Shallow Soil Model

The solution for shallow soil model can be obtained with a similar framework as finite soil model. Thus, we can directly integrate (9.25) to obtain u_1 (z) as

$$u_1=\frac{a}{2c_v}\left(h^2 z-\frac{z^3}{3}\right). \tag{9.53}$$

Again, we can have u_2 through a Fourier series expansion as

$$u_2=-\frac{a}{2c_v}\sum_{n=1}^{\infty} a_n\, e^{-c_v\kappa_n^2 t/h^2}\sin\left(\frac{\kappa_n z}{h}\right), \qquad a_n=\frac{2}{h}\int_0^h\left(h^2 r-\frac{r^3}{3}\right)\sin\left(\frac{\kappa_n r}{h}\right)dr. \tag{9.54}$$

Then, we have the solution for shallow soil model as (9.27).

9.8.3 Deep Soil Model

The boundary value problem for residual pore pressure in an infinite (deep) seabed can be summarized as:

$$\frac{\partial u}{\partial t}=c_v\frac{\partial^2 u}{\partial z^2}+bz\exp(-\lambda z), \tag{9.55}$$

and

$$u(z,0) = u(0,t) = 0, \quad \text{and} \quad u(\infty,t) = 0. \tag{9.56}$$

Since it is an infinite domain, this can be easy solved by a Laplace transformation.

Now, we scale the problem as follows:

$$\chi = c_v \lambda^2 t, \qquad U = \frac{c_v \lambda^3 u}{b}, \quad \text{and} \quad y = \lambda z. \tag{9.57}$$

Then, (9.55) and (9.56) become

$$U_\chi = U_{yy} + y e^{-y} \quad \text{and} \quad U(0,\chi) = U(y,0) = 0. \tag{9.58}$$

Take Laplace transformation in χ:

$$L\big(U(\chi,y)\big) = Q(s,y) = \int_0^\infty e^{-s\chi} U(\chi,y)\, d\chi. \tag{9.59}$$

Then

$$sQ = Q_{yy} + \frac{1}{s} y e^{-y} \quad \text{and} \quad Q(s,y) = 0, \; Q(s,\infty) = 0. \tag{9.60}$$

The solution of (9.60) is

$$Q(s,y) = \frac{ys - y - 2}{s(s-1)^2} e^{-y} + \frac{2}{s(s-1)^2} e^{-\sqrt{s}y}. \tag{9.61}$$

We need to invert (9.61). Let's split into two parts:

$$R(s,y) = \frac{ys - y - 2}{s(s-1)^2} e^{-y} \quad \text{and} \quad V(s,y) = \frac{2}{s(s-1)^2} e^{-\sqrt{s}y}. \tag{9.62}$$

For $R(s,y)$ the inversion is straight forward: it has a simple pole at $s = 0$ and a double pole at $s = 1$. $V(s,y)$ has a branch point at $s = 0$ and a double pole at $s = 1$. This inversion entails a careful integration in the complex s-plane around a '*keyhole*' contour. The combined inversion gives:

$$U(y,\chi) = 2\left[1 - \left(\frac{y}{2} + 1\right)\right] e^{-y} - \frac{1}{\pi} \int_0^\infty \frac{e^{-r\chi}}{r(r+1)^2} \sin(\sqrt{r}y)\, dr. \tag{9.63}$$

Now back in dimensional variables:

$$u = \frac{2b}{c_v \lambda^3}\left[1 - \left(\frac{\lambda z}{2} + 1\right) e^{-\lambda z} - \frac{1}{\pi} \int_0^\infty \frac{e^{-r c_v \lambda^2 t}}{r(1+r)^2} \sin(\sqrt{r}\lambda z)\, dr\right] \tag{9.64}$$

which is identical to (9.31).

References

1. de Alba, P., Seed, H.B., Chan, C.K.: Sand liquefaction in large-scale simple shear tests. J. Geotech. Div. **102**, 909–928 (1976)
2. Biot, M.A.: General theory of three-dimensional consolidation. J. Appl. Phys. **26**(2), 155–164 (1941)
3. Cheng, L., Sumer, B.M., Fredsøe, J.: Solution of pore pressure build up due to progressive waves. Int. J. Numer. Anal. Methods Geomech. **25**, 885–907 (2001)
4. Clukey, E.C., Kulhawy, F.H., Liu, P.L.F.: Laboratory and field investigation of wave-sediment interaction. Tech. rep. report 83-1, Joseph H. Defrees Hydraulics Laboratory, School of Civil and Environmental Engineering, Cornell University, Ithaca (1983)
5. Hsu, J.R.C., Jeng, D.S.: Wave-induced soil response in an unsaturated anisotropic seabed of finite thickness. Int. J. Numer. Anal. Methods Geomech. **18**(11), 785–807 (1994)
6. Jeng, D.S.: Wave-induced seabed response in front of a breakwater. Ph.D. thesis, University of Western Australia (1997)
7. Jeng, D.S., Seymour, B.R., Li, J.: A new approximation for pore pressure accumulation in marine sediment due to water wave. Int. J. Numer. Anal. Methods Geomech. **31**(1), 53–69 (2007)
8. Madsen, O.S.: Wave-induced pore pressures and effective stresses in a porous bed. Géotechnique **28**(4), 377–393 (1978)
9. McDougal, W.G., Tsai, Y.T., Liu, P.L.F., Clukey, E.C.: Wave-induced pore water pressure accumulation in marine soils. J. Offshore Mech. Arct. Eng. **111**(1), 1–11 (1989)
10. Mei, C.C., Foda, M.A.: Wave-induced response in a fluid-filled poro-elastic solid with a free surface-a boundary layer theory. Geophys. J. R. Astron. Soc. **66**, 597–631 (1981)
11. Miyamoto, J., Sassa, S., Sekiguchi, H.: Progressive solidification of a liquefied sand layer during continued wave loading. Géotechnique **54**(10), 617–629 (2004)
12. Okusa, S.: Wave-induced stress in unsaturated submarine sediments. Géotechnique **35**(4), 517–532 (1985)
13. Sassa, S., Sekiguchi, H.: Wave-induced liquefaction of beds of sand in a centrifuge. Géotechnique **49**(5), 621–638 (1999)
14. Sassa, S., Sekiguchi, H.: Analysis of wave-induced liquefaction of sand beds. Géotechnique **51**(2), 115–126 (2001)
15. Sassa, S., Sekiguchi, H., Miyamamot, J.: Analysis of progressive liquefaction as moving-boundary problem. Géotechnique **51**(10), 847–857 (2001)
16. Seed, H.B., Martin, P.O., Lysmer, J.: The generation and dissipation of pore water pressure during soil liquefaction. Tech. rep., College of Engineering, University of California, Berkeley, California (1975)
17. Seed, H.B., Rahman, M.S.: Wave-induced pore pressure in relation to ocean floor stability of cohesionless soils. Mar. Geotechnol. **3**(2), 123–150 (1978)
18. Sekiguchi, H., Kita, K., Okamoto, O.: Response of poro-elastoplastic beds to standing waves. Soil Found. **35**(3), 31–42 (1995)
19. Sumer, B.M., Fredsøe, J.: The Mechanism of Scour in the Marine Environment. World Scientific, New Jersey (2002)
20. Yamamoto, T., Koning, H., Sellmeijer, H., Hijum, E.V.: On the response of a poro-elastic bed to water waves. J. Fluid Mech. **87**(1), 193–206 (1978)

Chapter 10
Wave-Induced Progressive Liquefaction in a Porous Seabed

Abstract In this chapter, the prediction model proposed by Sassa *et al.* (in Géotechnique 51(10), 847–857 (2001)) for the wave-induced progressive liquefaction in marine sediment, based on two-layered inviscid fluid system, is re-examined. An alternative approach with a similar framework (Sassa *et al.* in Géotechnique 51(10), 847–857 (2001)) is developed to correct the inappropriate mechanism of wave components used. Then, a two-layered wave model which includes viscous effects is established and applied to describe the progressive nature of wave-induced liquefaction. A comprehensive comparison shows that the previous model (Sassa *et al.* in Géotechnique 51(10), 847–857 (2001)) overestimates the maximum liquefaction depth. It is found that the viscosity of liquefied soil cannot be ignored and the solution for an infinite seabed is not suitable for liquefaction analysis of shallow seabed. A parametric study demonstrates the significant influence of numerous wave and soil characteristics on the liquefaction depth.

Keywords Progressive liquefaction · Wave loading · Seabed · Viscous effect · Pore pressure

10.1 Introduction

Recently, a series of centrifuge tests were conducted by Sassa *et al.* [5]. Based on those tests, the importance of the cyclic plasticity of soil was emphasized and a new model was proposed for progressive liquefaction [6]. In their approach, the liquefied soil is treated as a heavy fluid and the potential model of a two-layered wave system proposed by Lamb [4] is used to describe the problem after the occurrence of liquefaction. However, the viscosity of a heavy fluid is not negligible. It has been reported that viscosity did exist for liquefied sand in experiments [3, 7]. Thus, consideration of viscosity is necessary in the analysis of wave-induced progressive liquefaction. Another issue in their study is that the solution for an infinite seabed is used to calculate the distribution of shear stress in a porous seabed of finite thickness. It has been reported that the relative difference between the wave-induced pore pressure determined by the solutions for infinite and finite thickness may reach 40 % [2]. Therefore, the applicability of the infinite solution needs to be checked.

D.-S. Jeng, *Porous Models for Wave-seabed Interactions*,
DOI 10.1007/978-3-642-33593-8_10,

The aim of this study is to investigate the viscous effects of a liquefied layer on the wave-induced progressive liquefaction potential. Sassa's model [6] will firstly be re-examined with an appropriate procedure. Then, the progressive liquefaction model is re-derived to include viscous effects in the liquefied layer, which was treated as an ideal fluid in Sassa's model. Another new feature of the present model is that the cyclic shear stresses are determined by the solution for a bed of finite thickness [2]. Finally, a parametric study will be conducted to examine the effects of wave and soil characteristics on the liquefaction potential.

10.2 Two-Layered Fluid System

Considering a progressive wave with a wave period (T) and wave length (L) propagating over the seabed, as depicted in Fig. 10.1, the liquefaction zone may develop with time for certain cases. In this situation, it can be assumed that the onset of liquefaction occurs at $t = 0$, and the liquefaction front advances to $z = z_L$ after a time interval, t. This phenomenon can be described by a two-layered fluid system propagating on the sub-liquefied soil (Fig. 10.1). The entire system consists of three layers, i.e. the original seawater with density ρ_1 ($0 \le z \le h$), the liquefied soil modeled as a fluid with density ρ_2 ($-z_L \le z \le 0$) and the sub-liquefied soil ($-D \le z \le -z_L$). The system consists of a surface wave $m_0 \exp(i(kx - \omega t))$ and an internal wave $a_0 \exp(i(kx - \omega t))$ at the interface between the water and the liquefied zone. The wave pressure at the original mudline ($z = 0$) is $P_0 \exp(i(kx - \omega t))$, and the wave pressure at the surface of sub-liquefied soil ($z = -z_L$) is $P_L \exp(i(kx - \omega t))$. The frequency ω and wave number k of the surface wave and internal wave are assumed to be the same.

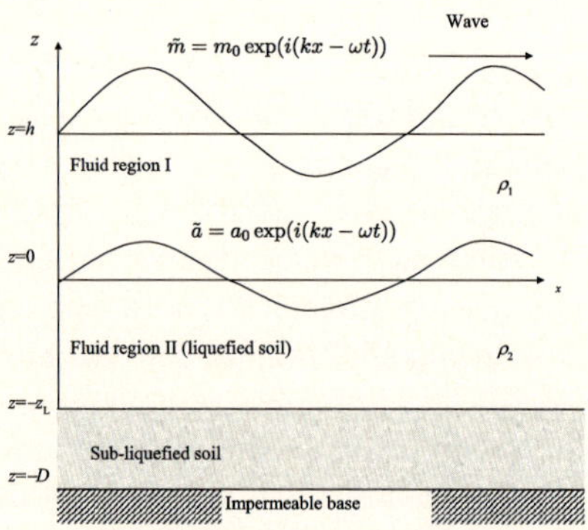

(a) two-layered wave system

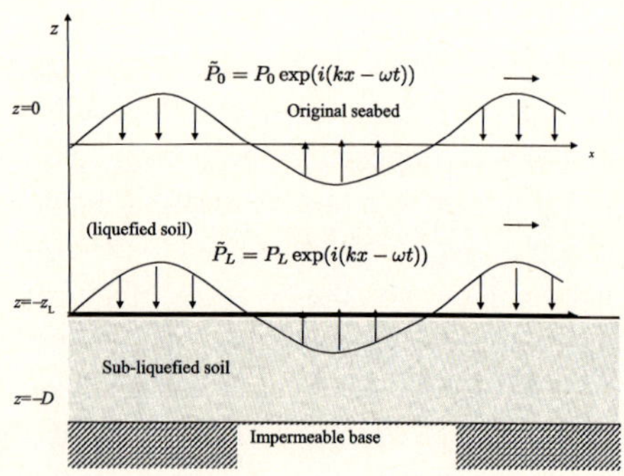

(b) corresponding wave pressure

Fig. 10.1 Definition of the progressive liquefaction problem

10.2.1 Two-Layered Inviscid Fluid Model

The basic equations in the two-layered inviscid model used by Sassa *et al.* [6] are outlined here. The dispersion relationship for a two-layered fluid system can be extended from the classical theory [4] as

$$\left[1-\left(\frac{Ngk}{\omega^2}\right)^2\right]\tanh(kh)\tanh(-kz_L)=\frac{\rho_2}{\rho_1}\frac{[1-\frac{Ngk}{\omega^2}\tanh(kh)]}{[1+\frac{Ngk}{\omega^2}\tanh(-kz_L)]}, \tag{10.1}$$

where $N =$ scale factor (only for centrifuge models), $g =$ gravity acceleration, $k =$ wave number, $\omega =$ frequency of the wave system, $h =$ water depth, $-z_l =$ the location of liquefaction front. The value of z_L will be updated with the advance of liquefaction front. For the case $-z_l = 0$ and $N = 1$, (10.1) will degenerate to the familiar dispersion relationship for a progressive wave on a single fluid layer.

The relationship between the wave pressure P_0 and the amplitude of the internal wave a_0 at $z = 0$ is

$$P_0=-\frac{\rho_2\omega^2+(\rho_2-\rho_1)Ngk\tanh(-kz_l)}{k\tanh(-kz_L)}a_0. \tag{10.2}$$

And that between the amplitude of surface wave m_0 and the amplitude of internal wave a_0 is

$$m_0=\frac{\omega^2}{\cosh(kh)[\omega^2-Ngk\tanh(kh)]}a_0. \tag{10.3}$$

The pore pressure at a generic point in the liquefied soil is

$$P=\frac{\rho_2\omega^2[\cosh(kz)-\sinh(kz)\tanh(kz_L)]}{\rho_2\omega^2+(\rho_2-\rho_1)Ngk\tanh(-kz_L)}P_0 \quad (-z_L\le z\le 0). \tag{10.4}$$

The amplitude of fluid pressure at $z=-z_L$ is

$$P_L=\frac{\rho_2\omega^2}{\rho_2\omega^2\cosh(-kz_L)+(\rho_2-\rho_1)Ngk\sinh(-kz_L)}P_0. \tag{10.5}$$

It is noticed that the location of liquefaction front in Fig. 10.1 is $-z_l$, which is different from Sassa's definition [6]. For more details of the model and a derivation of the equations above, readers are referred to [6].

10.2.2 Two-Layered Viscid Fluid Model

In the two-layered inviscid fluid model, the liquefied soil is considered as inviscid dense fluid. Such a treatment is only valid when the viscosity of liquefied soil is negligible. However, it is found that the viscosity of liquefied soil is much higher than that of the pore water which is normally treated as an inviscid fluid [3, 7]. Considerable viscosity exists for dynamic problems because of the friction between grains in the liquefied soil.

In the shaking table experiments [7], a steel model pipe was embedded in loose and water-saturated Toyoura sand. When the sand was liquefied by the shaking, the drag force needed to pull the pipe laterally was monitored. Then the viscosity of liquefied sand was back calculated from the monitored drag force together with the displacement of the pipe. In their experiments, the viscosity in the liquefied soil reached 2.16×10^6 Pa s [7]. Considering the viscosity of pure water at 20 °C is 1.002×10^{-3} Pa s, the values obtained in shaking table experiments are all at the order of 108 times the viscosity of pure water. Thus the inviscid wave model is inappropriate for the analysis of the progressive liquefaction problem and a two-layered viscid wave model is proposed.

Following the model developed by Dalrymple and Liu [1] for ocean waves over mud, both seawater and liquefied soil are treated as viscid fluids and the laminar Navier-Stokes equations are used to describe the motion of the two-layered wave system. The coordinates and the definitions are the same as those shown in Fig. 10.1. The free surface displacement is $m_0 \exp(i(kx-\omega t))$ and the interface displacement is $a_0 \exp(i(kx-\omega t))$. The frequency ω and wave number k of the surface wave and internal wave are assumed to be the same. In fluid region I ($h \geq z \geq 0$) and fluid region II ($0 \geq z \geq z_L$):

$$\frac{\partial \tilde{u}_i}{\partial t} = -\frac{1}{\rho_i}\frac{\partial \tilde{P}_i}{\partial x} + \nu_i\left(\frac{\partial^2 \tilde{u}_i}{\partial x^2} + \frac{\partial^2 \tilde{u}_i}{\partial z^2}\right), \tag{10.6}$$

$$\frac{\partial \tilde{v}_i}{\partial t} = -\frac{1}{\rho_i}\frac{\partial \tilde{P}_i}{\partial z} + \nu_i\left(\frac{\partial^2 \tilde{v}_i}{\partial x^2} + \frac{\partial^2 \tilde{x}_i}{\partial z^2}\right), \tag{10.7}$$

where $\tilde{u}$ and $\tilde{v}$ are the horizontal and vertical velocities, respectively; $\tilde{P}$ is the dynamic pressure; ρ is the density of fluid; and the coefficient $\nu = \mu/\rho$, is kinematic viscosity with the subscripts $i = 1, 2$ indicating the upper and lower layers respectively.

The equation for mass conservation is:

$$\frac{\partial \tilde{u}_i}{\partial x} + \frac{\partial \tilde{v}_i}{\partial z} = 0. \tag{10.8}$$

The solutions for those two equations are assumed to be:

$$\begin{cases} \tilde{u} = u(z)\exp(i(kx-\omega t)), \\ \tilde{v} = v(z)\exp(i(kx-\omega t)), \\ \tilde{p} = p(z)\exp(i(kx-\omega t)) \end{cases} \tag{10.9}$$

where k is wave number and ω is frequency of the wave system.

Introducing (10.9) into the equation of conservation, (10.8), yields

$$u_i(z) = i\frac{v_i'(z)}{k}. \tag{10.10}$$

Substituting (10.10) into (10.6) yields

$$P_i(z) = \frac{\rho_i \nu_i}{k^2}\left[v_i''' - v_i'\left(k^2 - \frac{i\omega}{\nu_i}\right)\right]. \tag{10.11}$$

Then, substituting $P_i(z)$ into (10.7) yields the governing equation

$$v_i^{(iv)} - \left(k^2 + \lambda_i^2\right)v_i'' + k^2\lambda_i^2 v_i = 0, \tag{10.12}$$

where

$$\lambda^2 = k^2 - \frac{i\omega}{\nu_i}. \tag{10.13}$$

For most fluids, the value of μ is much smaller than the wave number k and the value of λ is governed by ν and is much larger than k [1]. However, for liquefied soil, ν is larger than k, and λ is of the same order as k. If ν is much larger, λ will be almost the same value of k.

The general solution of (10.12) can be assumed as

$$v_1(z) = A\,e^{kz} + B\,e^{-kz} + C\,e^{\lambda_1 z} + D\,e^{-\lambda_1 z}, \tag{10.14}$$

$$v_2(z) = E\,e^{k(z+z_L)} + F\,e^{-k(z+z_L)} + G\,e^{\lambda_2(z+z_L)} + H\,e^{-\lambda_2(z+z_L)}. \tag{10.15}$$

The coefficients A, B, C, D, E, F, G, H in (10.14)–(10.15) and the amplitude of the interface wave (a_0) are unknowns to be solved with the following boundary conditions.

- Boundary conditions at the bottom ($z = -z_L$):

$$\begin{cases} \tilde{u}_2|_{z=-z_L} = 0, \\ \tilde{v}_2|_{z=-z_L} = 0. \end{cases} \tag{10.16}$$

- Boundary conditions at the free surface ($z = h$):

$$\frac{\partial \tilde{m}_0}{\partial t} = \tilde{v}_1, \tag{10.17}$$

$$\tilde{P}_1 = 2\rho_1\nu_1\frac{\partial \tilde{v}_1}{\partial z} - \rho_1 g\tilde{m}_0 = 0, \tag{10.18}$$

$$\rho_1\nu_1\left(\frac{\partial \tilde{u}_1}{\partial z} + \frac{\partial \tilde{v}_1}{\partial x}\right) = 0. \tag{10.19}$$

- Continuous conditions at the interface ($z = 0$):

$$\frac{\partial \tilde{a}_0}{\partial t} = \tilde{v}_1, \qquad \tilde{v}_1 = \tilde{v}_2, \qquad \tilde{u}_1 = \tilde{u}_2, \tag{10.20}$$

$$\tilde{P}_1 - 2\rho_1\nu_1\frac{\partial \tilde{v}_1}{\partial z} - \rho_1 g\tilde{a}_0 = \tilde{P}_2 - 2\rho_1\nu_2\frac{\partial \tilde{v}_2}{\partial z} - \rho_2 g\tilde{a}_0, \tag{10.21}$$

$$\rho_1\nu_1\left(\frac{\partial \tilde{u}_1}{\partial z} + \frac{\partial \tilde{v}_1}{\partial x}\right) = \rho_2\nu_2\left(\frac{\partial \tilde{u}_2}{\partial z} + \frac{\partial \tilde{v}_2}{\partial x}\right). \tag{10.22}$$

Introducing (10.10)–(10.11) and (10.14)–(10.15) into (10.16)–(10.22), yields the relationships between the coefficients for the assumed solutions:

$$k(E-F)+\lambda_2(G-H)=0, \tag{10.23}$$

$$E+F+G+H=0, \tag{10.24}$$

$$A+B+C+D=-i\omega a_0, \tag{10.25}$$

$$A+B+C+D=E\,e^{kz_L}+F\,e^{-kz_L}+G\,e^{\lambda_2 z_L}+H\,e^{-\lambda_2 z_L}, \tag{10.26}$$

$$kA-kB+\lambda_1 C-\lambda_1 D=kE\,e^{kz_L}-kF\,e^{-kz_L}+\lambda_2 G\,e^{\lambda_2 z_L}-\lambda_2 H\,e^{-\lambda_2 z_L}, \tag{10.27}$$

$$\begin{aligned}&\rho_1\nu_1\big[2k^2(A+B)+\big(\lambda_1^2+k^2\big)(C+D)\big]\\&\quad=\rho_2\nu_2\big[2k^2\big(E\,e^{kz_L}+F\,e^{-kz_L}\big)+\big(\lambda_2^2+k^2\big)\big(G\,e^{\lambda_2 z_L}+H\,e^{-\lambda_2 z_L}\big)\big],\end{aligned} \tag{10.28}$$

$$\begin{aligned}&\left(\frac{i\rho_1\omega}{k}-2\rho_1\nu_1 k\right)(A-B)-2\rho_1\nu_1\lambda_1(C-D)\\&\quad=\left(\frac{i\rho_2\omega}{k}-2\rho_2\nu_2 k\right)\big(E\,e^{kz_L}-F\,e^{-kz_L}\big)\\&\qquad-2\rho_2\nu_2\lambda_2\big(G\,e^{\lambda_2 z_L}-H\,e^{-\lambda_2 z_L}\big)-(\rho_2-\rho_1)ga_0,\end{aligned} \tag{10.29}$$

$$A\,e^{kh}+B\,e^{-kh}+C\,e^{\lambda_1 h}+D\,e^{-\lambda_1 h}=-i\omega m_0, \tag{10.30}$$

$$\left(\frac{i\rho_1\omega}{k}-2\rho_1\nu_1 k\right)\big(A\,e^{kh}-B\,e^{-kh}\big)-\rho_1\nu_1\lambda_1\big(C\,e^{\lambda_1 h}-D\,e^{-\lambda_1 h}\big)=\rho_1 g m_0, \tag{10.31}$$

$$2k^2\big(A\,e^{kh}+B\,e^{-kh}\big)+\big(\lambda_1^2+k^2\big)\big(C\,e^{\lambda_1 h}+D\,e^{-\lambda_1 h}\big)=0. \tag{10.32}$$

The coefficients A–H, the amplitude of interface wave a_0 and wave number k can be solved from (10.23)–(10.32) by substitution and iteration methods. Then the amplitude of fluid pressure at the bottom of fluid region can be derived from (10.11) as

$$P_2|_{z=-z_L}=\frac{i\omega\rho_2}{k}(E-F). \tag{10.33}$$

10.3 Poro-Elastoplastic Soil Model

10.3.1 Boundary Value Problem

In this study, the poro-elastoplastic model proposed by Sassa *et al.* [6] is adopted to simulate the generation and dissipation of residual pore pressure in the sub-liquefied soil layer with the aforementioned two-layer fluid system. The storage equation used for the poro-elastoplastic soil with contractancy [5] is

$$\frac{\partial p_r}{\partial \xi} = \Phi \frac{\partial^2 p_r}{\partial (k_0 z)^2} + M \frac{\partial v^p}{\partial \xi} \quad \text{for } -D \le z \le -z_L, \tag{10.34}$$

in which $\xi = \omega t/(2\pi)$, M is the constrained modulus of the soil skeleton and Φ is the partial drainage factor.

$$\Phi = 2\pi \frac{k_z M}{\gamma_f \omega} k_0^2, \tag{10.35}$$

where k_z is the coefficient of permeability and γ_f is the unit weight of the fluid.

The plastic volumetric strain at any time is defined as $v^p(\xi, \chi)$ and the following constitutive relationship is assumed for general loading cases:

$$\frac{\partial v^p}{\partial \xi} = \beta \big[v^p_\infty(\chi) - v^p(\xi, \chi) \big], \tag{10.36}$$

in which v^p_∞ is the final plastic volumetric strain

$$v^p_\infty(\chi) = R\big[\exp(\alpha\chi) - 1\big] \tag{10.37}$$

where α, β and R are material parameters and $\chi = \tau/\sigma'_0$ is the cyclic shear stress ratio.

The constrained modulus M is assumed to increase linearly with effective confining pressure

$$M = -\frac{z + z_L}{D} M_D \quad \text{for } -D \le z \le -z_L, \tag{10.38}$$

in which M_D is the constrained modulus at the bottom $z = -D$.

And the boundary conditions for this problem are

$$p_r = \gamma' z_L \quad \text{on } z = -z_L \le 0, \tag{10.39}$$

$$\frac{\partial p_r}{\partial z} = 0 \quad \text{on } z = -D. \tag{10.40}$$

The cyclic shear stress ratio used to define the plastic volumetric strain is the key factor in the model. Based on different assumptions, there are different solutions. Two elastic solutions are cited here.

10.3.2 Cyclic Shear Stress in an Infinite Seabed

Before liquefaction occurs, the amplitudes of the wave-induced shear stress and oscillatory pore pressure are calculated using poro-elastic solutions for an infinite seabed [8] as

$$\tau = -k_0 P_0 z \exp(k_0 z), \qquad p = P_0 \exp(k_0 z) \quad \text{for } z \le 0 \tag{10.41}$$

where k_0 is the wave number before liquefaction, and P_0 is the wave pressure before liquefaction

$$P_0 = \frac{\gamma_w H}{2\cosh(k_0 d)}. \tag{10.42}$$

The vertical normal effective stress is

$$\sigma'_{v0} = \gamma' z. \tag{10.43}$$

Thus, the cyclic shear stress ratio is

$$\frac{\tau}{\sigma'_{v0}} = \frac{k_0 P_0}{\gamma'} \exp(k_0 z) \quad \text{for } z \le 0. \tag{10.44}$$

Once liquefaction occurs and the liquefaction front advances to $z = -z_L$, the surface of the sub-liquefied soil layer can be treated as the position where fluid pressure is exerted as the distributed force. After coordinate transformation, the corresponding equations are

$$\tau = -kP_L(z+z_L)\exp\big(k(z+z_L)\big), \quad \text{for } z \le -z_L, \tag{10.45}$$

$$p = P_L \exp\big(k(z+z_L)\big), \quad \text{for } z \le -z_L, \tag{10.46}$$

$$\frac{\tau}{\sigma'_{vi}} = \frac{kP_L}{\gamma} \exp\big(k(z+z_L)\big), \quad \text{for } z \le -z_L. \tag{10.47}$$

Note that σ'_{vi} is calculated by

$$\sigma'_{vi} = -\gamma'(z+z_L) \quad \text{for } z \le -z_L. \tag{10.48}$$

Equations (10.45)–(10.47) were used to determine shear stress in the centrifuge models [6]. However, the soil thicknesses in their centrifuge tests were 0.08–0.1 m, which represents a maximum value of 5 m in the prototype. As reported in [2], the relative difference between the solutions for infinite and finite bed thickness is significant. Thus, the infinite seabed model is inappropriate for these tests.

10.3.3 Cyclic Shear Stress in a Seabed of Finite Thickness

A closed form solution for a seabed with finite thickness was given Sect. 3.4.3. If a progressive wave is propagating over a fully saturated poro-elastic seabed with finite depth, the shear stress and wave-induced pore pressure at a generic point are

$$\begin{aligned} \tau = ip_o\big\{&(C_1 + C_2 k_0 z)\, e^{k_0 z} - (C_3 + C_4 k_0 z)\, e^{-k_0 z} \\ &+ k_0\delta\big(C_5\, e^{\delta z} - C_6\, e^{-\delta z}\big)\big\}\exp\big(i(k_0 x - \omega t)\big), \end{aligned} \tag{10.49}$$

$$P = \frac{p_o}{1-2\mu}\{(1-\lambda-2\mu)(C_2 e^{k_0 z} - C_4 e^{-k_0 z})$$
$$+ (1-\mu)(\delta^2 - k_0^2)(C_5 e^{\delta z} + C_6 e^{-\delta z})\} \exp(i(k_0 x - \omega t)), \quad (10.50)$$

where

$$\delta^2 = k_0^2 - \frac{i\omega\gamma_w(1-2\mu)}{2Gk_z(1-\mu)} \quad (10.51)$$

where k_z is soil permeability, G is shear modulus, μ is Poisson's ratio, ω is wave frequency, k_0 is the wave number before liquefaction. Coefficients C_1–C_6 are available in [2].

If liquefaction occurs, the shear stress and pore pressure in the seabed can be calculated using (10.49) and (10.50). A minor change is that P_0 and k_0 should be P_L and k_0 which should be updated from the two-layered wave model. The vertical normal effective stress can be calculated using (10.43) or (10.48) and then the shear stress ratio $\chi = \tau/\sigma'_{v0}$ can be further calculated.

10.3.4 Numerical Scheme and Procedure

By adopting a finite difference method, the governing equation (10.34) can be discretized as

$$\frac{P_r(Z, \xi + \Delta\xi) - p_r(Z, \xi)}{\Delta\xi} - M_q$$
$$= \frac{p_r(Z + \Delta Z, \xi + \Delta\xi) - 2p_r(Z, \xi + \Delta\xi) + P_r(Z - \Delta Z, \xi + \Delta\xi)}{(\Delta Z)^2}, \quad (10.52)$$

in which $q = \frac{\partial v^p}{\partial \xi}|_\xi$, $Z = k_0 z$, $\Delta\xi =$ the increment of wave cycle, $\Delta Z =$ the increment of general depth.

The value of q can be determined by (10.36), which can be calculated using the backward Euler scheme. The left of (10.36) can be discretized as

$$\frac{dv^p}{d\xi} = \frac{v^p(\xi, \chi) - v^p(\xi - \Delta\xi, \chi)}{\Delta\xi}. \quad (10.53)$$

Thus $v^p(\xi, \chi)$ can be determined based on $(\xi - \Delta\xi, \chi)$

$$\frac{v^p(\xi, \chi) - v^p(\xi - \Delta\xi, \chi)}{\Delta\xi} = \beta(v^p_\infty(\chi) - v^p(\xi, \chi)), \quad (10.54)$$

$$v^p(\xi, \chi) = \frac{\beta v^p_\infty(\chi)\delta\xi + v^p(\xi - \Delta\xi, \chi)}{1 + \beta\delta\xi}. \quad (10.55)$$

With the initial condition that the plastic volumetric strain is zero, the value of $v^p(\xi, \chi)$ can be determined and then q can be calculated for each step.

Considering the boundary conditions at the fluid-soil interface and assuming an impermeable bottom, dividing the seabed thickness D into a mesh of $n \times \Delta z$, (10.52) can be transformed into

$$\begin{bmatrix} 1 & 0 & \cdots & \cdots & \cdots & \cdots & 0 \\ A & B & A & 0 & \cdots & \cdots & \vdots \\ 0 & A & B & A & 0 & \cdots & \vdots \\ \vdots & \ddots & \ddots & \ddots & \ddots & & \vdots \\ \vdots & & \ddots & \ddots & \ddots & \ddots & \vdots \\ \vdots & & & 0 & A & B & A \\ 0 & \cdots & \cdots & \cdots & 0 & A & B^* \end{bmatrix} \begin{bmatrix} p_{r1} \\ p_{r2} \\ p_{r3} \\ \vdots \\ \vdots \\ p_{rn} \\ p_{rn+1} \end{bmatrix}_{\xi+\Delta\xi} = \begin{bmatrix} 0 \\ p_{r2} \\ p_{r3} \\ \vdots \\ \vdots \\ p_{rn} \\ p_{rn+1} \end{bmatrix} + \begin{bmatrix} 0 \\ Mq \\ Mq \\ \vdots \\ \vdots \\ Mq \\ Mq \end{bmatrix} \Delta\xi \tag{10.56}$$

where

$$A = -\Phi \frac{\Delta\xi}{(\Delta Z)^2}, \qquad B = 1 + 2\Phi \frac{\Delta\xi}{(\Delta Z)^2}, \qquad B^* = 1 + \Phi \frac{\Delta\xi}{(\Delta Z)^2}. \tag{10.57}$$

Based on above model, the detailed procedure of the analysis is as follows:

1. Start analysis from $k = k_0$, $z_L = 0$, $\xi = 0$;
2. Before liquefaction, the wave pressure is calculated by the common formula for a progressive wave: $P_0 = \gamma_f H/(2\cosh k_0 d)$;
3. Solve (10.56) to get the pressure at each point for $\xi = \xi + \Delta\xi$;
4. Repeat step 2 and step 3 until p_{r2} increases to the value $\gamma' \Delta z$;
5. In the following steps, the two-layered wave model is used to calculate k and the amplitude of fluid pressure over sub-liquefied soil P_L;
6. At the same time, the amplitude of vertical displacement of soil at $z = 0$ is updated;
7. Following that, the shear stress ratio is calculated to determine $\frac{\partial v^p}{\partial \xi}$;
8. Solve (10.56) and judge whether liquefaction front will advance or not to update z_L;
9. Repeat step 5–step 8 until the end of the analysis.

10.4 Results and Discussions

By combining the two-layered fluid wave model and poro-elastoplastic soil model, the progressive nature of liquefaction can be simulated by solving (10.56). Based on different assumptions, five models are listed in Table 10.1. The objectives of this study are to: (1) re-examine Sassa's model [6] with an alternative approach; (2) investigate the effects of the viscosity of the liquefied layer on liquefaction depth; (3) investigate the influence of different solutions for shear stress on liquefaction depth.

10.4.1 Comparison with Sassa's Model [6]

As shown in Table 10.1, Model S is Sassa's model [6], in which the inviscid two-layered model is used and the wave pressure P_0 at the original mudline ($z = 0$) is assumed constant. To compare other models with model S, the author firstly made a program based on model S and reproduced the case analysis for Keisa No. 7 sand in [6]. As shown in Fig. 10.2, this program works well.

The assumption of $P_0 =$ constant in [6] is questionable, even though it is more convenient for the control of experimental environments. The amplitude of the surface wave is back calculated using (10.3). The case used for comparison with the results obtained in the centrifuge test [6] was: centrifuge acceleration 50 g, fluid depth $h = 0.09$ m, $f = 11$ Hz, $k_0 = 12.2\ \mathrm{m}^{-1}$, $P_0 = 5$ kPa, soil thickness $D = 0.1$ m, $\rho_1 = 980\ \mathrm{kg/m^3}$, $\rho_2 = 1840\ \mathrm{kg/m^3}$. The calculated wave height H ($H = 2m_0$) is shown in Fig. 10.3. The figure clearly shows that the wave height increases with the advance of the liquefaction front. Meanwhile, the wave number decreases from $12.2\ \mathrm{m}^{-1}$ to $10.4\ \mathrm{m}^{-1}$, this implies that the wavelength increases with the advance of the liquefaction front. With constant frequency, this leads to the conclusion that the energy in one wave period increases with the advance of the liquefaction front. This is physically inconsistent, because the liquefied sand absorbs a part of the wave

Table 10.1 Supplementary assumptions

	Type of 2-layered fluid model	Solution for shear stress in seabed	Supplement assumptions
Model S	Inviscid	Infinite solution	$P_0 =$ constant
Model A	Inviscid	Infinite solution	$m_0 =$ constant
Model B	Inviscid	Finite solution	$m_0 =$ constant
Model C	Viscid	Infinite solution	$m_0 =$ constant
Model D	Viscid	Finite solution	$m_0 =$ constant

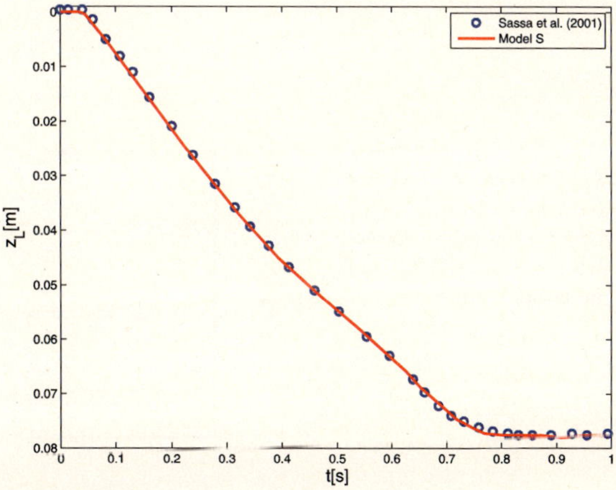

Fig. 10.2 Reproduction of Sassa's procedure

energy, causing an effect similar to damping. Thus, the wave height should decrease with the advancement of the liquefaction front to satisfy the conservation of energy.

If the liquefaction depth is not extremely large, it is acceptable to assume the height of the surface wave is constant. With this assumption, Models A–D are proposed. A comparison of the amplitude of the interface wave a_0 and the wave pressures between Models S and A are illustrated in Figs. 10.4 and 10.5. It is found

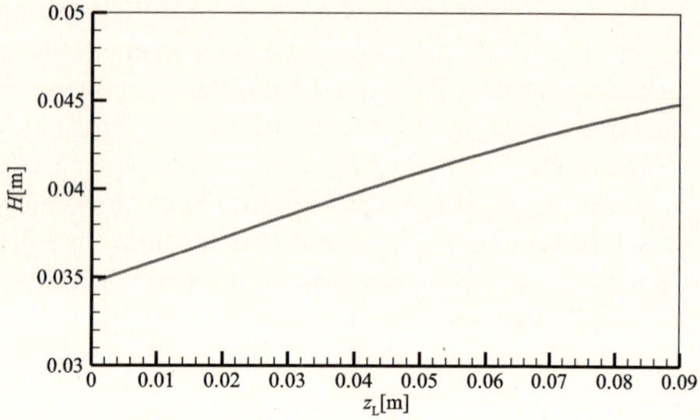

Fig. 10.3 The variation of H with assumption of $P_0 =$ constant ($P_0 = 5$ kPa)

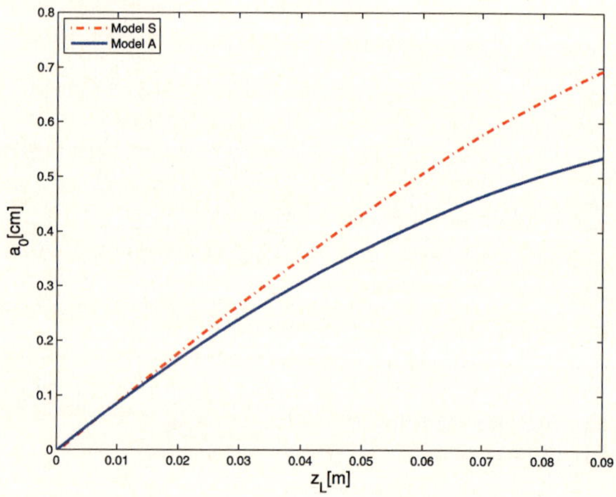

Fig. 10.4 Comparison of the amplitude of the interface wave with different assumptions ($P_0 = 5$ kPa in model S, $m_0 = 0.0347$ m in model A)

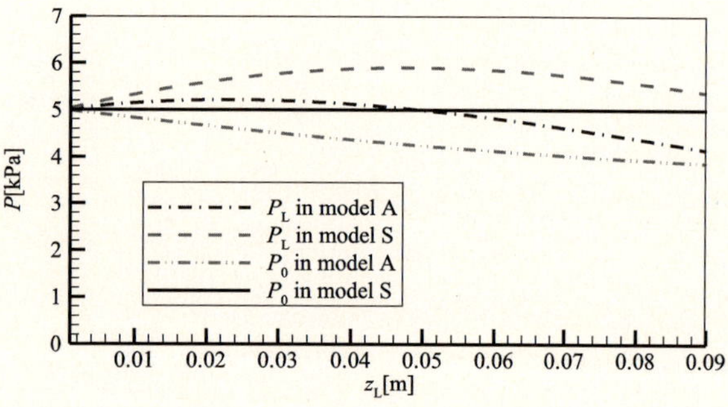

Fig. 10.5 The variation of wave pressure at $z = 0$ and $z = -z_L$ with different assumptions ($P_0 = 5$ kPa in model S, $m_0 = 0.0347$ m in model A)

that model S overestimated the soil displacement by nearly 30 %. When the liquefied depth is 0.09 m, the amplitude of fluid pressure P_0 at $z = 0$ will decrease to 3.865 kPa in model A, which is only 77 % of that used as constant in model S. This means the wave pressure at $z = 0$ cannot be assumed as constant without significant error. Two curves of P_L calculated under different assumptions are shown in Fig. 10.5. The value of P_L obtained from model S is 20 % greater than that from model A. The source term is related to the shear stress ratio which is directly determined by P_L. Thus the residual pressure in model S is larger than that in model A. This shows the assumption $P_0 =$ constant will cause an overestimation of liquefaction depth.

10.4.2 Viscous Effect and the Influence of Shear Stress

As mentioned previously, the liquefied soil behaves like a viscid fluid rather than an inviscid fluid. Furthermore, there are differences in the wave-induced pore pressure and stresses in the soil calculated under finite and infinite assumptions. A comparison of the different methods is shown in Fig. 10.6 to examine those influences. The input data are tabulated in Tables 10.2. It is shown that model S will lead to the deepest final liquefaction depth, which is 4.26 m. If the wave height is set as constant, the final liquefaction depth is 3.48 m, 82 % of the valued obtained under the assumption of constant wave pressure. In [6], the predicted theoretical values agree

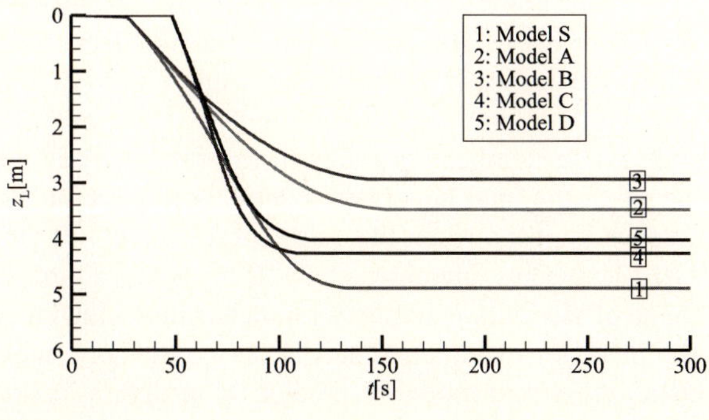

Fig. 10.6 Temporal changes of the location of the liquefaction front z_L with different models ($H = -6.5$ m, $T = 10$ s, $h = 20$ m, $D/L = 0.05$ and $k_z = 1.5 \times 10^{-4}$ m/s)

Table 10.2 Input data of numerical examples

Soil parameters

1-8 D/L	k_z (m/s)	ρ_s (kg/m^3)	M_D (kPa)	R	α	β	N
0.05	1.5×10^{-4}	1800	3.12×10^4	1.8×10^5	55	0.04	1

Wave/fluid parameters

1-8 h(m)	T (s)	L (m)	H (m)	ν_1 m^2/s	ν_2 m^2/s
20	10	121.2	6.5	10^{-6}	10

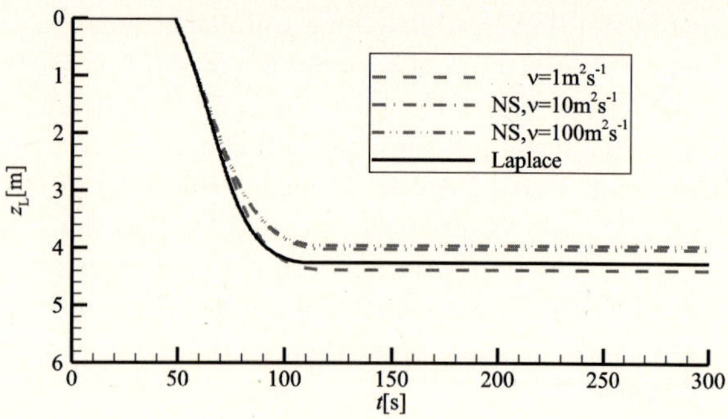

Fig. 10.7 The influence of viscosity on liquefaction history ($H = -6.5$ m, $T = 10$ s, $h = 20$ m, $D/L = 0.05$ and $k_z = 1.5 \times 10^{-4}$ m/s)

with their measured values because the experiment was conducted under constant P_0 by increasing the input energy from the wave-maker. However, in real conditions, the conservation of energy must be considered and the proposed procedure is more applicable. If the viscid two-layered wave model is used to calculate the fluid pressure and the infinite solution is used to calculate shear stress, the final liquefaction depth will be 2.94 m, 69 % of the result obtained by Sassa's procedure. This means the viscous effects will play a negative role, preventing liquefaction progress. Lines 4 and 5 are the results obtained by adopting the finite solution to calculate shear stress. Those results are deeper than the corresponding results obtained using the infinite solution to calculate shear stress. For a shallow seabed, the error due to the assumption of an infinite bed is significant and therefore it will underestimate the final liquefaction depth.

The viscous effect is further studied by varying the viscosity of the liquefied soil. As shown in Fig. 10.7, the final liquefaction depth for $\nu = 1$ m^2/s shows no remarkable difference with that obtained using the inviscid two-layered wave model. It is obvious that if the viscosity of liquefied soil is small, the two-layered viscid solution will not be significantly different from the inviscid model. If the value increases to 10 m^2/s, the final liquefaction depth will decrease by 8 %. But no remarkable difference happens when the value increases to 100 m^2/s. This can be explained by (10.13); when ν increases from 10 m^2/s to 100 m^2/s, λ only changes a little. The value of viscosity is important to determine which model should be used. However, the behavior of liquefied soil is not completely understood. The precondition for using the viscid model to predict the progress of liquefaction is to determine the viscosity correctly.

10.4.3 Effect of Parameters α, β and R in the Build-Up Pattern

The pore pressure build-up pattern controls the generation rate of residual pore pressure. It is necessary to carry out a parametric study on the three parameters α, β and R in the build-up pattern. As shown in Figs. 10.8, 10.9, 10.10, it is found that the three parameters are very sensitive. The decrease of α from 55 to 53 will cause the final liquefaction depth to more than half. If the decrease of α continues to less than

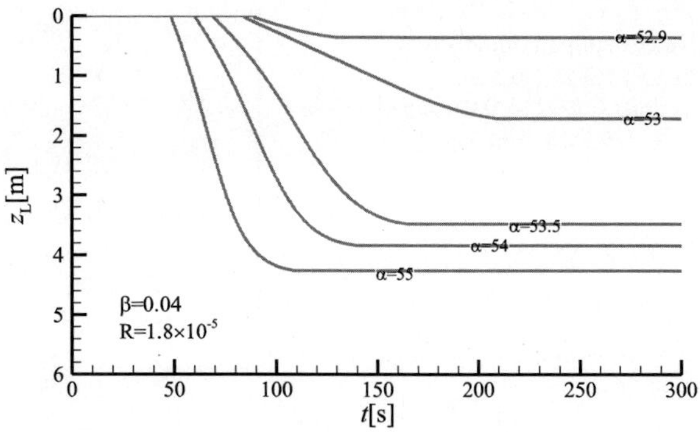

Fig. 10.8 The influence of α on liquefaction history ($H = -6.5$ m, $T = 10$ s, $h = 20$ m, $D/L = 0.05$ and $k_z = 1.5 \times 10^{-4}$ m/s)

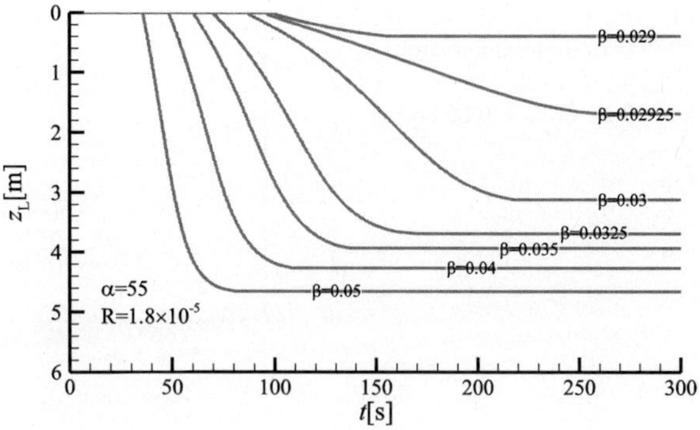

Fig. 10.9 The influence of α on liquefaction history ($H = -6.5$ m, $T = 10$ s, $h = 20$ m, $D/L = 0.05$ and $k_z = 1.5 \times 10^{-4}$ m/s)

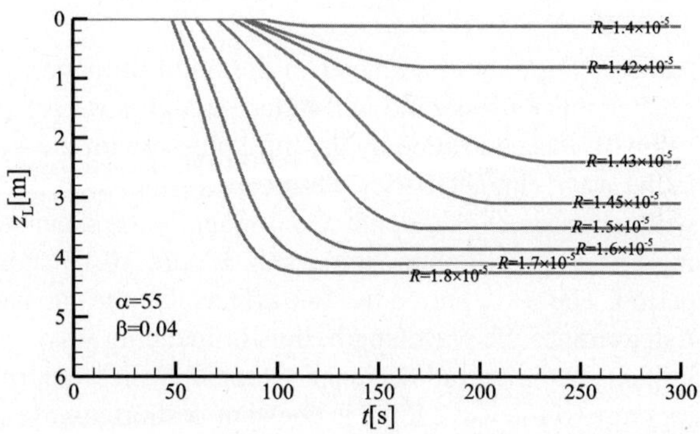

Fig. 10.10 The influence of α on liquefaction history ($H = -6.5$ m, $T = 10$ s, $h = 20$ m, $D/L = 0.05$ and $k_z = 1.5 \times 10^{-4}$ m/s)

52.9, no liquefaction occurs. A similar thing happens when β or R decreases. These changes will not only affect the final depth; the shape of the time history curves will also be influenced.

Comparatively, the variations of $\alpha$and R will have a more remarkable influence on the curves and final liquefaction depth than that of β. From (10.37), it can be found that the ultimate value of the plastic volumetric strain only depends on α and R. The ultimate value of the plastic volumetric strain is linearly related to R, while exponentially related to α. Thus a small change of α will affect the result significantly. Much attention should be paid to the calibration of the parameters.

Fig. 10.11 The influence of water depth on liquefaction history ($H = -6.5$ m, $T = 10$ s, $D/L = 0.05$ and $k_z = 1.5 \times 10^{-4}$ m/s)

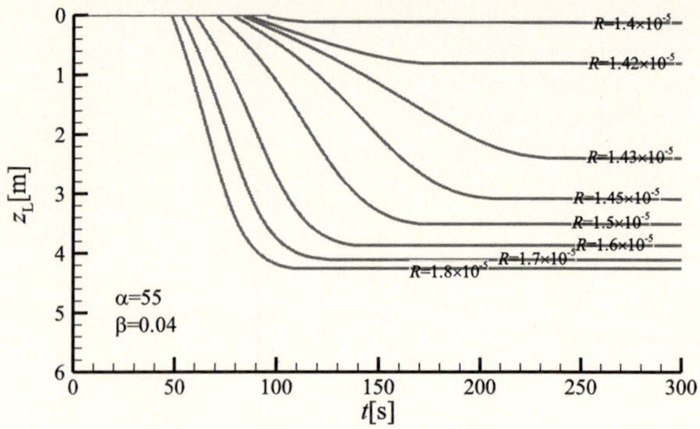

Fig. 10.12 The influence of wave period on liquefaction history ($H = -6.5$ m, $h = 20$ m, $D/L = 0.05$ and $k_z = 1.5 \times 10^{-4}$ m/s)

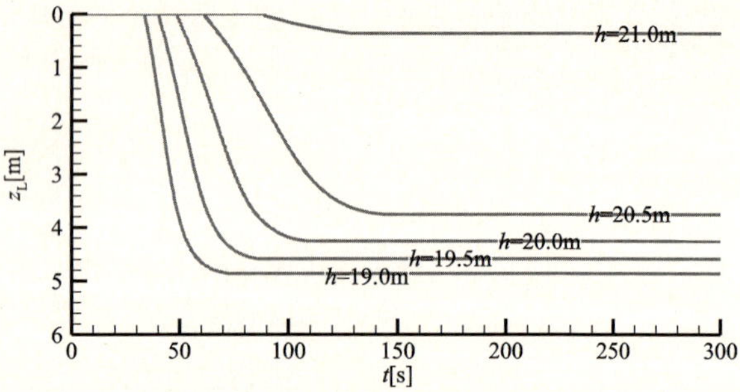

10.4.4 Effect of Wave and Soil Characteristics

Another important parameter in the build-up pattern is the shear stress ratio χ. A greater shear stress ratio will induce a higher rate of pore pressure generation. The value of χ is controlled by the fluid pressure on the surface of soil, which is related to the wave characteristics. Figure 10.11 indicates that the wave pressure decreases with increasing water depth. Consequently the shear stress ratio will decrease, causing a shallower liquefaction depth. Figure 10.12 shows the influence of the wave period. The wave period has two effects. On the one hand, together with water depth it determines the wavelength, thus influencing wave pressure. On the other hand, it influences the partial drainage factor Φ which controls the dissipation of residual pressure. If the wave force is constant, a shorter wave period is more likely to cause liquefaction because residual pressure has less time to dissipate in one period.

Besides the generation of pore pressure, the dissipation of pore pressure is another important factor in the boundary problem. In the case of a high value of the partial drainage factor, the residual pore pressure cannot build up to the initial effective vertical normal stress and no liquefaction occurs. In (10.35), it is found that the value of the partial drainage factor is linearly related to k_z. The influence of permeability is shown in Fig. 10.13. When the permeability decreases from 1.04×10^{-4} m/s to 2.0×10^{-4} m/s, the final liquefaction depth will decrease by 40 %. However, if the permeability is larger than 2.1×10^{-4} m/s, no liquefaction will occur. As the permeability is obtained from soil tests, much attention is required to avoid errors.

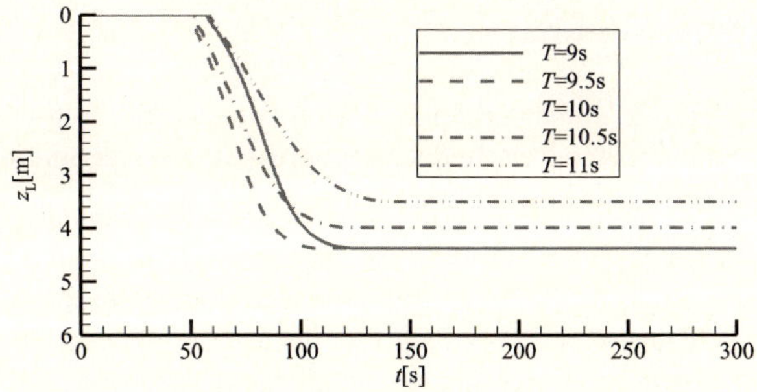

Fig. 10.13 The influence of permeability on liquefaction history ($H = -6.5$ m, $T = 10$ s, $h = 20$ m and $D/L = 0.05$)

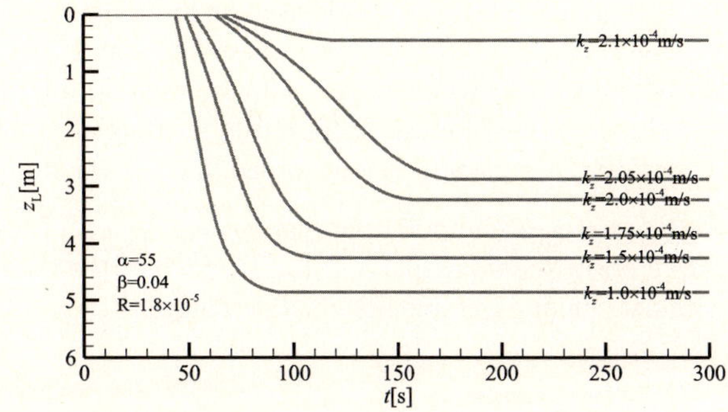

Fig. 10.14 Residual pore pressure development history at different depth in the seabed

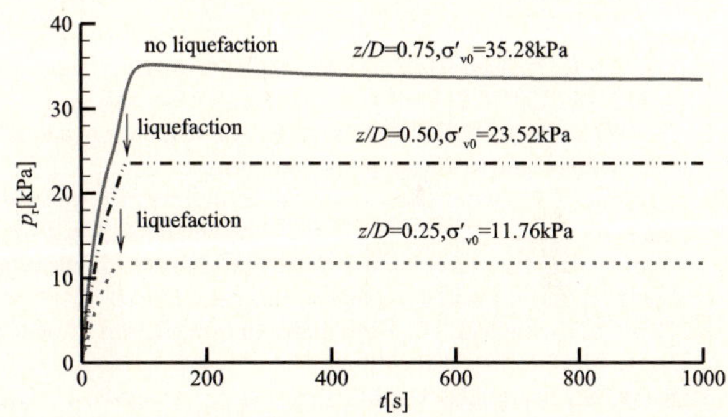

Fig. 10.15 Total pore pressure development history at the depth $z/D = 0.25$

10.4.5 Pore Pressure History

Figures 10.14 and 10.15 show the residual pressure development and the total pore pressure development in the seabed obtained by method B. These show the ability of simulating the time history for each point. Figure 10.14 shows that the pressures at $z/D = 0.25$ and $z/D = 0.50$ will increase to the effective normal stress while the pressure at $z/D = 0.75$ cannot reach the effective normal stress. At $z/D = 0.75$, the generation of pore pressure is initially greater than the dissipation, however it becomes smaller than the dissipation after a certain time. Thus, the liquefaction front will not reach that point.

10.5 Summary

From the numerical analysis in this paper, the following conclusions can be made regarding the wave-induced progressive liquefaction in the seabed:

1. It is found there is a significant problem in the inviscid two-layered wave model used by Sassa *et al.* [6]. The assumption that P_0 is constant will lead to the increase of wave height, which conflicts with the energy conservation law;
2. The authors proposed to set m_0 as constant as a simplified approximation. Comparing to the results obtained by the proposed method, Sassa's method will significantly overestimate the final liquefaction depth;
3. A two-layered wave model based on the Navier-Stokes equations is developed to consider the viscous effects of liquefied soil. It is found that the viscosity of the liquefied layer will decrease with the final liquefaction depth;
4. The finite solution for shear stress is adopted in this paper. It is found that the infinite solution will underestimate the liquefaction depth for a shallow seabed.

It is necessary to point out that the source term is the key point in this problem. More effort should be put into the research of the pore pressure build-up pattern to simulate the progressive liquefaction nature.

References

1. Dalrymple, R., Liu, P.L.F.: Waves over soft muds: A two-layer fluid model. J. Phys. Oceanogr. **8**, 1121–1131 (1978)
2. Jeng, D.S., Hsu, J.R.C.: Wave-induced soil response in a nearly saturated seabed of finite thickness. Géotechnique **46**(3), 427–440 (1996)
3. Kawamura, J., Hattori, K., Akagi, H.: Experimental evaluation of viscid properties of liquefied sand. In: Proceedings of the Eighth US-Japan Workshop on Earthquake Resistant Design of Lifeline Facilities and Countermeasures Against Liquefaction, pp. 405–412 (2002)
4. Lamb, H. (ed.): Hydrodynamics. Cambridge University Press, Cambridge (1932)
5. Sassa, S., Sekiguchi, H.: Wave-induced liquefaction of beds of sand in a centrifuge. Géotechnique **49**(5), 621–638 (1999)
6. Sassa, S., Sekiguchi, H., Miyamamot, J.: Analysis of progressive liquefaction as moving-boundary problem. Géotechnique **51**(10), 847–857 (2001)
7. Towhata, I., Vargas-Monge, W., Orense, R.P., Yao, M.: Shaking table tests on subgrade reaction of pipe embedded in sandy liquefied subsoil. Soil Dyn. Earthq. Eng. **18**, 347–361 (1999)
8. Yamamoto, T., Koning, H., Sellmeijer, H., Hijum, E.V.: On the response of a poro-elastic bed to water waves. J. Fluid Mech. **87**(1), 193–206 (1978)

Index

D.-S. Jeng, *Porous Models for Wave-seabed Interactions*,
DOI 10.1007/978-3-642-33593-8,

Printed by Printforce, the Netherlands